Springer Series in Reliability Engineering

Volume 9

Series Editor

Hoang Pham

For further volumes:
http://www.springer.com/series/6917

Tadashi Dohi · Toshio Nakagawa
Editors

Stochastic Reliability and Maintenance Modeling

Essays in Honor of Professor Shunji Osaki on his 70th Birthday

Springer

Editors
Tadashi Dohi
Department of Information Engineering
Hiroshima University
Higashi-Hiroshima
Japan

Toshio Nakagawa
Department of Business Administration
Aichi Institute of Technology
Toyota
Japan

ISSN 1614-7839
ISBN 978-1-4471-4970-5 ISBN 978-1-4471-4971-2 (eBook)
DOI 10.1007/978-1-4471-4971-2
Springer London Heidelberg New York Dordrecht

Library of Congress Control Number: 2013934841

Printed on acid-free paper

Springer is part of Springer Science+Business Media (www.springer.com)

Preface

In Honor and Celebration of Professor Osaki's 70th Birthday and Retirement

Dr. Shunji Osaki is an internationally recognized researcher in the field of system reliability engineering spanning over four decades. In 2001–2011, he worked in Nanzan University, Nagoya, as Professor in the Department of Mathematical and Information Engineering. Prior to 2001, Dr. Osaki was Professor (1986–2001) and Associate Professor (1970–1986) in the Department of Industrial and Systems Engineering, Hiroshima University from 1970 to 2001. Dr. Osaki earned a Doctor of Engineering, a.k.a, a Ph.D. from Kyoto University in 1970. We have provided his academic accomplishments over the last four decades, following this preface.

Thanks to Dr. Osaki's endless contributions, system reliability engineering has made significant advancements in Japan. Active in both research and academic, he has built a wealth of technical networks through publications and technical discussions with international colleagues, scholars, friends, and students. He published over 20 books and over 400 papers in the area of system reliability engineering since joining the graduate program in Kyoto University in mid-1960. He also helped and supported 8 Ph.D. students, 72 M.S. students, and 223 B.S. students during his tenure at Nanzan University and Hiroshima University.

In honor and celebration of Dr. Osaki's 70th birthday and retirement, this book represents a collection of recent research topics by numerous distinguished researchers, all of whom have been inspired by Dr. Osaki's great contributions. The topics include system reliability and maintenance modeling, dependable computing, system performance analysis, and software engineering.

This book, containing both his legacy and the state-of-the-art contributions in the technical field, will help readers explore new ideas and topics on stochastic reliability and maintenance modeling. It will also serve as a useful guide for researchers to apply reliability and maintenance theory to computer and communication systems.

Our special thanks to all the authors and reviewers of the respective articles involved in this book. We would also like to thank Professor Hoang Pham, Rutgers, The State University of New Jersey, USA, the Editor of Springer Series in Reliability Engineering, Ms. Claire Protherough and Ms. Grace Quinn, Springer Senior Editorial Assistants (Engineering), and Dr. Kazu Okumoto, Alcatel-Lucent Technologies, USA. We would not have completed this book without their kind supports and encouragements.

Once again, Happy 70th Birthday to Professor Shunji Osaki and Congratulations and Best Wishes for a very successful retirement.

Professor Shinji Osaki's Academic Achievements

- Earned a Ph.D. from Kyoto University in 1970. His Ph.D. thesis is "Studies on system analysis and synthesis by Markov renewal process."
- Joined Hiroshima University in 1970 as Lecturer and became Associate Professor that same year in the Department of Industrial Engineering. Full Professor in 1986, where he was responsible for research and teaching in applied probability, quality control and reliability engineering.
- Took a two-year leave of absence in 1970 to join the University of Southern California as a Post-Doctoral Research Fellow to pursue a joint research on applied stochastic modeling with Professor Richard Bellman.
- Returned from USC and established a laboratory in the Department of Industrial Engineering, Hiroshima University to help and support graduate and undergraduate students.
- Visited the Manchester University, UK in 1976–1977 as a Simon Visiting Research Fellow.
- Joined Nanzan University, Nagoya in 2002 as Professor in the Faculty of Mathematical and Information Engineering, in anticipation of the retirement age at Hiroshima University.

In addition, Dr. Osaki exhibited a wide range of research interests and continued to publish books and technical papers in the field. The following is a summary of his research contributions.

- As a Ph.D. candidate, co-authored with Professor Hisashi Mine a well-referenced book, "Morkovian Decision Processes," American Elsevier Publishing Company Inc. Also published three key papers on Markov decision processes in Journal of Mathematical Analysis and Applications.
- Based on his Ph.D. thesis in 1970, published 11 papers in Japanese technical journals and 10 papers in international journals, including IEEE Transactions on Reliability, Management Science, and Journal of Applied Probability.
- One of his key contributions was a reliability analysis of two-unit standby redundant systems based on Markov renewal processes. A series of work in this technically exciting area was pursued with Dr. Toshio Nakagawa and his students in Hiroshima University.
- Was the winner of the fourth Ohnishi Memorial Best Paper Award in 1971, which is the most prestigious research award in the Operations Research Society of Japan.
- These research accomplishments constitute as the fundamental theory and were then applied to fault-tolerant computer systems. They resulted in:

  - A heavily cited monograph, "Reliability Evaluation of Some Fault-Tolerant Computer Architectures", Lecture Notes in Computer Science, Springer-Verlag, co-authored with his M.S. student, Toshihiko Nishio. It became a technical base for probabilistic analyses in modern dependable computing.

- Authored two well-known textbooks:

  - "Stochastic System Reliability Modeling," World Scientific
  - "Applied Stochastic System Modeling," Springer-Verlag.

Dr. Osaki also worked on a large number of stochastic maintenance models such as replacement models, inspection models, order-replacement models, and shock models. In late 1970, he worked in these research areas with support from Dr. Toshio Nakagawa and several students such as Dr. Kazuhira Okumoto and Dr. Naoto Kaio. In early 1980, he started a research project on software reliability modeling with his student, Dr. Shigeru Yamada, and developed several well-known software reliability growth models such as the delayed S-shaped model, which is often called the Yamada, Ohba, and Osaki Model. Over the last 20 years, he was a co-author of numerous papers in this well-established research area. His research interest in the 1990s moved on to other research topics such as financial engineering, production/inventory analysis, neuro computing, quality control, dynamic power management, etc. He was very sensitive and instrumental to new research trends in applied stochastic modeling, and generated many ideas and hints that are helpful to other researchers and students.

He founded several international workshops during the last three decades.

- Organized the Reliability Symposium on Stochastic Models in Reliability Theory, on April 1984, in Nagoya, Japan, with the co-chair, Dr. Yukio Hatoyama, who was the Japanese Prime Minister during September 2009 to June 2010.
- Served as the General Co-Chair with Professor Jinhua Cao for The China–Japan Reliability Symposium in China in September 1987 and realized the so-called "Moving-Events" by bringing many Chinese and Japanese colleagues in three different places, Shanghai, Xi'an and Beijing.
- In 1993 and 1996, initiated Australia–Japan Workshop on Stochastic Model in Engineering, Technology and Management, with Professor D. N. Pra Murthy, and continued until the third event in 1999.
- Worked with Professors A. H. Christer, L. C. Thomas, N. Balakrishnan, Nikolaos Limnios and Katsushige Sawaki to found the following workshops. Effectively bridged Japanese researchers in applied stochastic modeling and European/American researchers.

  - UK-Japanese Research Workshop on Stochastic Modelling in Innovative Manufacturing in 1995
  - Euro-Japanese Workshops on Stochastic Risk Modelling for Finance, Insurance, Production and Reliability in 1998 and 2002
  - International Workshops on Recent Advances in Stochastic Operations Research in 2005 and 2007.

As demonstrated above, Dr. Osaki enjoys organizing relatively small workshops to stimulate pure academic discussions with his colleagues/friends, apart from politics in big academic societies.

Dr. Osaki has been a key member of several academic societies such as The Operations Research Society of Japan, The Institute of Electronics, Information and Communication Engineers, The Institute of Systems, Control and Information Engineers, Japan Industrial Management Association, Information Processing Society of Japan, Reliability Engineering Association of Japan, and IEEE Reliability Society. He served as an Associate Editor or an Editorial Board Member of several international journals such as Journal of Mathematical Analysis and Applications, International Journal of Policy and Information, Applied Stochastic Models and Data Analysis, Computers & Mathematics with Applications, Revue Francaise d'Automatique, d'Informatique et Rechereche Operationelle Recherche Operationnelle, International Journal of Reliability, Quality and Safety Engineering, IIE Transactions on Quality and Reliability Engineering, Communications in Applied Analysis, Applied Stochastic Models in Business and Industry, among others.

Japan, August 2012

Tadashi Dohi
Toshio Nakagawa

References

Books/Editions

H. Mine and S. Osaki, Markovian Decision Processes, American Elsevier Publishing Company, Inc., 1970.
H. Yoda, S. Osaki and T. Nakagawa, Applied Probability (in Japanese). Asakura Publishing Co. Ltd., 1977.
S. Osaki and T. Nishio, Reliability Evaluation of Some Fault-Tolerant Computer Architectures, Lecture Notes in Computer Science, vol. 97, Springer-Verlag, 1980.
S. Osaki and Y. Hatoyama (eds.), Stochastic Models in Reliability Theory, Lecture Notes in Economics and Mathematical Systems, vol. 235, Springer-Verlag, 1984.
S. Osaki, Stochastic System Reliability Modeling, World Scientific, 1985.
S. Osaki and J. Cao (eds.), Reliability Theory and Applications, World Scientific, 1987.
S. Osaki, J. Hishitani and H. Kawakami, Introduction of Japanese Micro TeX (in Japanese), Keigaku Publishing Co. Inc., 1989.
S. Osaki, N. Kaio and T. Ichimori, Management Systems Science with Operations Research (in Japanese), Asakura Publishing Co. Ltd., 1989.
S. Osaki, Applied Stochastic System Modeling, Springer-Verlag, 1992.
S. Osaki and D. N. P. Murthy (eds.), Stochastic Models in Engineering, Technology and Management, World Scientific, 1993.
S. Osaki, Statistical Handbook in Quality Control and Reliability Engineering (in Japanese), Japanese Standards Association, 1994.
S. Osaki, Introduction of Stochastic Models (in Japanese), Asakura Publishing Co. Ltd., 1996.
R. J. Wilson, D. N. P. Murthy and S. Osaki (eds.), Stochastic Models in Engineering, Technology and Management, Technology Management Centre, The University of Queensland, 1996.
A. H. Christer, S. Osaki and L. C. Thomas (eds.), Stochastic Models in Innovative Manufacturing, Lecture Notes in Economics and Mathematical Systems, vol. 445, Springer-Verlag, 1997.
R. J. Wilson, S. Osaki and M. J. Faddy (eds.), Stochastic Models in Engineering, Technology and Management,Technology Management Centre, The University of Queensland, 1999.
S. Osaki (ed.), Proceedings of International Conference on Applied Stochastic System Modeling, Hiroshima University, 2000.
T. Dohi, N. Limnios and S. Osaki (eds.), Proceedings of The Second Euro-Japanese Workshop on Stochastic Risk Modelling for Finance, Insurance, Production and Reliability, Hiroshima University, Japan, 2002.
T. Dohi, S. Osaki and K. Sawaki (eds.), Proceedings of 2005 International Workshop on Recent Advances in Stochastic Operations Research, Hiroshima University.
T. Dohi, S. Osaki and K. Sawaki (eds.), Recent Advances in Stochastic Operations Research, World Scientific, Singapore, 2006.
T. Dohi, S. Osaki and K. Sawaki (eds.), Proceedings of 2007 International Workshop on Recent Advances in Stochastic Operations Research, Hiroshima University, Japan, 2007.
T. Dohi, S. Osaki and K. Sawaki (eds.), Recent Advances in Stochastic Operations Research II, World Scientific, Singapore, 2009.

English Journal Papers

S. Osaki and H. Mine, Linear Programming Algorithms for Semi-Markovian Decision Processes, Journal of Mathematical Analysis and Applications, vol. 22, pp. 356–381, 1968.
S. Osaki and H. Mine, Some Remarks on a Markovian Decision Problem with an Absorbing State, Journal of Mathematical Analysis and Applications, vol. 23, pp. 327–333, 1968.
H. Mine, S. Osaki and T. Asakura, Some Considerations for Multiple-Unit Redundant System with Generalized, IEEE Transactions on Reliability, vol. R-17, pp. 170–174, 1968.

H. Mine and S. Osaki, On Failure-Time Distributions for Systems of Dissimilar Units, IEEE Transactions on Reliability, vol. R-18, pp. 165–168, 1969.

S. Osaki and H. Mine, Linear Programming Considerations on Markovian Decision Processes with No Discounting, Journal of Mathematical Analysis and Applications, vol. 26, pp. 222–232, 1969.

S. Osaki, A Note on a Two-Unit Standby Redundant System, Journal of the Operations Research Society of Japan, vol. 12, pp. 43–51, 1970.

S. Osaki, Reliability Analysis of a Two-Unit Redundant System with Priority, Canadian Operational Research Journal, vol. 8, pp. 60–62, 1970.

S. Osaki, Reliability Analysis of a Two-Unit Standby Redundant System with Standby Failure, Opsearch, vol. 7, pp. 13–22, 1970.

S. Osaki, System Reliability Analysis by Markov Renewal Processes, Journal of the Operations Research Society of Japan, vol. 12, pp. 127–188, 1970.

H. Mine, K. Yamada and S. Osaki, On Terminating Stochastic Games, Management Science, vol. 16, pp. 560–571, 1970.

S. Osaki, A Note on a Probability Problem Arising in Reliability and Traffic Studies, Journal of the Operations Research Society of Japan, vol. 13, pp. 17–22, 1970.

S. Osaki, Renewal Theoretic Aspects of Two-Unit Redundant Systems, IEEE Transactions on Reliability, vol. R-19, pp. 105–110, 1970.

S. Osaki and T. Asakura, A Two-Unit Standby Redundant System with Repair and Preventive Maintenance, Journal of Applied Probability, vol. 7, pp. 641–648, 1970.

S. Osaki, A Note on a Two-Unit Standby-Redundant System with Imperfect Switchover, Revue Francaise d'Informatique et de Recherche Operationnelle, vol. 2, pp. 103–109, 1971.

T. Nakagawa and S. Osaki, On a Two-Unit Standby Redundant System with Standby Failure, Operations Research, vol. 19, pp. 510–523, 1971.

S. Osaki, Notes on Renewal Processes and Neuronal Spike Trains, Mathematical Biosciences, vol. 12, pp. 33–39, 1971.

D. L. Jaquette and S. Osaki, Initial Provisioning of a Standby System with Deteriorating and Repairable Spares, IEEE Transactions on Reliability, vol. R-21, pp. 245–247, 1972.

S. Osaki and R. Vasudevan, On a Model of Neuronal Spike Trains, Mathematical Biosciences, vol. 14, pp. 337–341, 1972.

S. Osaki, On a Two-Unit Standby-Redundant System with Imperfect Switchover, IEEE Transactions on Reliability, vol. R-21, pp. 20–24, 1972.

S. Osaki, Reliability Analysis of a Two-Unit Standby-Redundant System with Preventive Maintenance, IEEE Transactions on Reliability, vol. R-21, pp. 24–29, 1972.

S. Osaki, An Intermittently Used System with Preventive Maintenance, Journal of the Operations Research Society of Japan, vol. 15, pp. 102–111, 1972.

S. Osaki, Note on a Simple Inspection System, Logistics and Transportation Review, vol. 8, pp. 77–82, 1974.

T. Nakagawa and S. Osaki, The Optimum Repair Limit Replacement Policies, Operational Research Quarterly, vol. 25, pp. 311–317, 1974.

T. Nakagawa and S. Osaki, A Note on Delays Induced by Random Events, Transportation Science, vol. 8, pp. 190–192, 1974.

T. Nakagawa and S. Osaki, Optimum Preventive Maintenance Policies for a 2-Unit Redundant System, IEEE Transactions on Reliability, vol. R-23, pp. 86–91, 1974.

T. Nakagawa and S. Osaki, Optimum Preventive Maintenance Policies Maximizing the Mean Time to the First System Failure for a Two-Unit Standby Redundant System, Journal of Optimization Theory and Applications, vol. 14, pp. 115–129, 1974.

T. Nakagawa and S. Osaki, Combining Drift and Catastrophic Failure Modes, IEEE Transactions on Reliability, vol. R-23, pp. 253–257, 1974.

T. Nakagawa and S. Osaki, Some Aspects of Damage Models, Microelectronics and Reliability, vol. 13, pp. 253–257, 1974.

S. Osaki, Signal-Flow Graphs in Reliability Theory, Microelectronics and Reliability, vol. 13, pp. 539–541, 1974.

T. Nakagawa and S. Osaki, Stochastic Behaviour of a Two-Unit Stand-by Redundant System, INFOR (Canadian Journal of Operational Research and Information Processing), vol. 12, pp. 66–70, 1974.

T. Nakagawa and S. Osaki, Stochastic Behaviour of a Two-Dissimilar-Unit Standby Redundant System with Repair Maintenance, Microelectronics and Reliability, vol. 13, pp. 143–148, 1974.

T. Nakagawa and S. Osaki, A Model for Interaction of Two Renewal Processes with Threshold Level, Information and Control, vol. 24, pp. 1–10, 1974.

T. Nakagawa and S. Osaki, Off Time Distributions in an Alternating Renewal Process with Reliability Applications, Microelectronics and Reliability, vol. 13, pp. 181–184, 1974.

T. Nakagawa and S. Osaki, Optimum Replacement Policies with Delay, Journal of Applied Probability, vol. 11, pp. 102–110, 1974.

T. Nakagawa and S. Osaki, Applications of the Sojourn-Time Problem to Reliability, IEEE Transactions on Reliability, vol. R-24, pp. 301–302, 1975.

T. Nakagawa and S. Osaki, The Discrete Weibull Distribution, IEEE Transactions on Reliability, vol. R-24, pp. 300–301, 1975.

T. Nakagawa, A. l. Goel and S. Osaki, Stochastic Behavior of an Intermittently Used System, Fevue Francaise d'Automatique et Recherche Operationnelle, vol. 12, pp. 101–112, 1975.

T. Nakagawa and S. Osaki, Optimal Dental Scheduling, Mathematical Biosciences, vol. 25, pp. 91–104, 1975.

T. Nakagawa and S. Osaki, Stochastic Behavior of a Two-Unit Priority Standby Redundant System with Repair, Microelectronics and Reliability, vol. 14, pp. 309–313, 1975.

T. Nakagawa and S. Osaki, Stochastic Behavior of 2-Unit Redundant Systems with Imperfect Switchover, IEEE Transactions on Reliability, vol. R-24, pp. 143–146, 1975.

T. Nakagawa and S. Osaki, A Note on Age Replacement, IEEE Transactions on Reliability, vol. R-24, pp. 92–94, 1975.

T. Nakagawa and S. Osaki, On a Terminating Renewal Process with Reliability Applications, IEEE Transactions on Reliability, vol. R-24, pp. 88–90, 1975.

T. Nakagawa and S. Osaki, The Busy Period of a Repairman for Redundant Repairable Systems, Revue Francaise d'Automatique et Recherche Operationnelle, vol. 3, pp. 69–73, 1975.

T. Nakagawa and S. Osaki, Stochastic Behavior of Two-Unit Paralleled Redundant Systems with Repair Maintenance, Microelectronics and Reliability, vol. 14, pp. 457–461, 1975.

T. Nakagawa and S. Osaki, Stochastic Behavior of a 2-Unit Parallel Fuel Charging System, IEEE Transactions on Reliability, vol. R-24, pp. 302–304, 1975.

T. Nakagawa and S. Osaki, Analysis of a Repairable System which Operates at Discrete Times, IEEE Transactions on Reliability, vol. R-25, pp. 110–112, 1976.

T. Nakagawa and S. Osaki, Joint Distributions of Uptime and Downtime for Some Repairable Systems, Journal of the Operations Research Society of Japan, vol. 19, pp. 209–216, 1976.

T. Nakagawa and S. Osaki, A Summary of Optimum Preventive Maintenance Policies for a Two-Unit Standby Redundant System, Zeitschrift fur Operations Research, vol. 20, pp. 171–187, 1976.

S. Osaki and T. Nakagawa, Bibliography for Reliability and Availability of Stochastic Systems, IEEE Transactions on Reliability, vol. R-25, pp. 284–287, 1976.

T. Nakagawa and S. Osaki, Markov Renewal Processes with Some Non-Regeneration Points and their Applications to Reliability Theory, Microelectronics and Reliability, vol. 15, pp. 633–636, 1976.

S. Osaki and S. Yamada, Age Replacement with Lead Time, IEEE Transactions on Reliability, vol. R-25, pp. 344–345, 1976.

A. Tsurui and S. Osaki, On a First-Passage Problem for a Cumulative Process with Exponential Decay, Stochastic Processes and Their Applications, vol. 4, pp. 79–88, 1976.

F. Sugimoto and S. Osaki, Optimum Repair Limit Replacement Policies for a System Subject to Intermittent Blows, Cahiers du Centre D'Etudes de Recherche Operationnelle, vol. 18, pp. 485–491, 1976.

T. Nakagawa and S. Osaki, Reliability Analysis of a One-Unit System with Unrepairable Spare Units and Its Optimization Applications, Operational Research Quarterly, vol. 27, pp. 101–110, 1976.

K. Okumoto and S. Osaki, Repair Limit Replacement Policies with Lead Time, Zeitschrift fur Operations Research, vol. 20, pp. 133–142, 1976.

S. Osaki, An Ordering Policy with Lead time, International Journal of Systems Science, vol. 8, pp. 1091–1095, 1977.

S. Yamada and S. Osaki, Optimum Number of Checks in Checking Policy, Microelectronics and Reliability, vol. 16, pp. 589–591, 1977.

T. Nakagawa and S. Osaki, Discrete Time Age Replacement Policies, Operational Research Quarterly, vol. 28, pp. 881–885, 1977.

S. Osaki and K. Okumoto, Repair Limit Suspection Policies for a Two-Unit Standby Redundant System with Two Phase Repairs, Microelectronics and Reliability, vol. 16, pp. 41–45, 1977.

K. Okumoto and S. Osaki, Optimum Policies for a Standby System with Preventive Maintenance, Operational Research Quarterly, vol. 28, pp. 415–423, 1977.

N. Kaio and S. Osaki, Ordering Policies with Two Types of Lead Times, Microelectronics and Reliability, vol. 16, pp. 225–229, 1977.

L. C. Thomas and S. Osaki, A Note on Ordering Policy, IEEE Transactions on Reliability, vol. R-27, pp. 380–381, 1978.

L. C. Thomas and S. Osaki, An Optimal Ordering Policy for a Spare Unit with Lead Time, European Journal of Operational Research, vol. 2, pp. 409–419, 1978.

T. Nakagawa and S. Osaki, Optimum Ordering Policies with Lead Time for an Operation Unit, Revue Francaise d'Automatique et Recherche Operationnelle, vol. 12, pp. 383–393, 1978.

S. Yamada and S. Osaki, Optimum Checking Request Policies, International Journal of Systems Science, vol. 9, pp. 579–593, 1978.

N. Kaio and S. Osaki, Optimum Planned Maintenance with Salvage Costs, International Journal of Production Research, vol. 16, pp. 249–257, 1978.

N. Kaio and S. Osaki, Optimum Ordering Policies with Two Kinds of Lead Times and Non-Liner Ordering Costs, International Journal of Systems Science, vol. 9, pp. 265–272, 1978.

N. Kaio and S. Osaki, Optimum Ordering Polices When Order Costs depend on Time, Revue Francaise d'Automatique et Recherche Operationnelle, vol. 12, pp. 93–99, 1978.

N. Kaio and S. Osaki, Optimum Inspection-Ordering Policies with Salvage Cost, Microelectronics and Reliability, vol. 18, pp. 253–257, 1978.

N. Kaio and S. Osaki, Optimum Ordering Policies with Lead Time for an Operating Unit in Preventive Maintenance, IEEE Transactions on Reliability, vol. R-27, pp. 270–271, 1978.

S. Kuchii, N. Kaio and S. Osaki, Simulation Comparison of Point Estimation Methods in the 2-Parameter Weibull Distribution, Microelectronics and Reliability, vol. 19, pp. 333–336, 1979.

N. Kaio and S. Osaki, Discrete-Time Ordering Policies, IEEE Transactions on Reliability, vol. R-28, pp. 405–406, 1979.

N. Kaio and S. Osaki, Optimum Age Replacement Policy with Two Failure Modes, Revue Francaise d'Informatique et de Recherche Operationnelle, vol. 13, pp. 205–209, 1979.

N. Kaio and S. Osaki, Optimum Ordering Policies with Discounting When Order Costs Depend on Time, International Journal of Systems Science, vol. 10, pp. 539–555, 1979.

M. Takeda and S. Osaki, A Two-Unit Paralleled Redundant System with Bivariate Exponential Failure Law and Allowed Down Time, Cahiers du Centre D'Etudes de Recherche Operationnelle, vol. 21, pp. 55–62, 1979.

S. Osaki and Y. Nishio, Availability Evaluation of Redundant Computer Systems, Computers and Operations Research, vol. 6, pp. 55–62, 1979.

N. Kaio and S. Osaki, Extended Optimum Ordering Policies with Discounting, International Journal of Systems Science, vol. 11, pp. 149–157, 1980.

S. Osaki, A Two-Unit Parallel Redundant System with Bivariate Exponential Lifetimes, Microelectronics and Reliability, vol. 20, pp. 521–523, 1980.

N. Kaio and S. Osaki, Optimum Planned Maintenance with Discounting, International Journal of Production Research, vol. 14, pp. 257–263, 1980.

N. Kaio and S. Osaki, Discrete Time Ordering Policies with Minimal Repair, Revue Francaise d'Automatique et de Recherche Operationnelle, vol. 14, pp. 257–263, 1980.

S. Yamada and S. Osaki, Reliability Evaluation of a Two-Unit Unrepairable System, Microelectronics and Reliability, vol. 20, pp. 589–597, 1980.

S. Yamada and S. Osaki, Checking Request Policies for a One-Unit System and Their Comparisons, Microelectronics and Reliability, vol. 20, pp. 859–874, 1980.

N. Kaio and S. Osaki, Comparisons of Point Estimation Methods in the 2-Parameter Weibull Distribution, IEEE Transactions on Reliability, vol. R-29, p. 21, 1980.

S. Osaki, N. Kaio and H. Arita, The Weibull Probability Papers by Microcomputer, International Journal of Policy and Information, vol. 5, pp. 1–13, 1981.

N. Kaio and S. Osaki, Optimum Repair Limit Policies with a Cost Constraint, Microelectronics and Reliability, vol. 21, pp. 597–599, 1981.

N. Kaio and S. Osaki, Optimum Planned Maintenance Policies with Lead Time, IEEE Transactions on Reliability, vol. R-30, p. 79, 1981.

S. Yamada and S. Osaki, A Note on Two Inspection Policies, OMEGA (The International Journal of Management Science), vol. 9, pp. 99–101, 1981.

S. Yamada and S. Osaki, Optimum Replacement Policies for a System Composed of Components, IEEE Transactions on Reliability, vol. R-30, pp. 278–283, 1981.

S. Osaki, N. Kaio and S. Yamada, A Summary of Optimal Ordering Policies, IEEE Transactions on Reliability, vol. R-30, pp. 272–277, 1981.

N. Kaio and S. Osaki, A Discrete-Time Repair Limit Policy, Advances in Management Studies, vol. 1, pp. 157–160, 1982.

Y. Yonehara, M. Nakamura and S. Osaki, Reliability Analysis of a 2-out-of-n: F System with Repairable Primary and Degradation Units, Microelectronics and Reliability, vol. 22, pp. 1081–1096, 1982.

S. Osaki and M. Kinugasa, Performance-Related Reliability Evaluation of a Three-Unit Hybrid Redundant System, International Journal of Systems Science, vol. 13, pp. 1–19, 1982.

N. Kaio and S. Osaki, Optimum Repair Limit Policies with a Time Constraint, International Journal of Systems Science, vol. 13, pp. 1345–1350, 1982.

S. Osaki, Reliability Evaluation of a TMR Computer System with Multivariate Exponential Failures and a General Repair, Microelectronics and Reliability, vol. 22, pp. 781–787,1982.

S. Yamada and S. Osaki, Cumulative Process Models for a Software Failure Process and Their Comparisons, Transactions of the Institute of Electronics and Communication Engineers of Japan, vol. E65, pp. 457–463, 1982.

S. Yamada and S. Osaki, Reliability Growth Models for Hardware and Software Systems Based on Nonhomogeneous Poisson Processes: A Survey, Microelectronics and Reliability, vol. 23, pp. 91–112, 1983.

S. Yamada and S. Osaki, S-Shaped Software Reliability Growth Models with Four Types of Software Error Data, International Journal of Systems Science, vol. 14, pp. 683–692, 1983.

S. Yamada, M. Ohba and S. Osaki, S-Shaped Reliability Growth Modeling for Software Error Detection, IEEE Transactions on Reliability, vol. R-32, pp. 475–484, 1983.

S. Yamada, M. Ohba and S. Osaki, S-Shaped Software Reliability Growth Models and Their Applications, IEEE Transactions on Reliability, vol. R-33, pp. 289–292, 1984.

S. Osaki, Performance/Reliability Measures for Fault-Tolerant Computing Systems, IEEE Transactions on Reliability, vol. R-33, pp. 268–271, 1984.

S. Yamada, H. Narihisa and S. Osaki, Optimum Release Policies for a Software System with a Scheduled Software Delivery Time, International Journal of Systems Science, vol. 15, pp. 905–914, 1984.

M. Nakamura and S. Osaki, Performance/Reliability Evaluation for Multi-Processor Systems with Computational Demands, International Journal of Systems Science, vol. 15, pp. 95–105, 1984.

N. Kaio and S. Osaki, Extended Block Replacement Models, Revue Francaise d'Automatique et Recherche Operationnelle, vol. 18, pp. 59–70, 1984.

S. Yamada and S. Osaki, Software Reliability Growth Modeling with Number of Test Runs, Transactions of the Institute of Electronics and Communication Engineers of Japan, vol. E67, pp. 79–83,1984.

N. Kaio and S. Osaki, Some Remarks on Optimum Inspection Policies, IEEE Transactions on Reliability, vol. R-33, pp. 277–279, 1984.

N. Kaio and S. Osaki, The Computer-Aided Weibull Hazard Paper by Microcomputer, International Journal of Policy and Information, vol. 8, pp. 65–71, 1984.

S. Yamada and S. Osaki, Discrete Software Reliability Growth Models, Applied Stochastic Models and Data Analysis, vol. 1, pp. 65–77, 1985.

S. Yamada and S. Osaki, An Error Detection Rate Theory for Software Reliability Growth Models, Transactions of the Institute of Electronics and Communication Engineers of Japan, vol. E68, pp. 292–296, 1985.

S. Yamada, S. Osaki and H. Narihisa, A Software Reliability Growth Model with Two Types of Errors, Revue Francaise d'Automatique et Recherche Operationnelle, vol. 19, pp. 87–104, 1985.

S. Yamada and S. Osaki, Cost-Reliability Optimal Release Policies for Software Systems, IEEE Transactions on Reliability, vol. R-34, pp. 422–424, 1985.

S. Yamada and S. Osaki, Software Reliability Growth Modeling: Models and Applications, IEEE Transactions on Software Engineering, vol. SE-11, pp. 1431–1437, 1985.

N. Kaio and S. Osaki, A Note on Optimum Checkpointing Policies, Microelectronics and Reliability, vol. 25, pp. 451–453, 1985.

H. Ohshimo and S. Osaki, Reliability/Performance Evaluation for a Multisystem with Preventive Maintenance, Microelectronics and Reliability, vol. 25, pp. 841–846, 1985.

S. Yamada and S. Osaki, Optimal Software Release Policies for a Non-Homogeneous Software Error Detection Rate Model, Microelectronics and Reliability, vol. 26, pp. 691–702, 1986.

N. Kaio and S. Osaki, Optimal Inspection Policies: A Review and Comparison, Journal of Mathematical Analysis and Applications, vol. 119, pp. 3–20, 1986.

T. Kitaoka, S. Yamada and S. Osaki, A Discrete Non-homogeneous Error Detection Rate for Software Reliability, Transactions of the Institute of Electronics and Communication Engineers of Japan, vol. E69, pp. 859–865, 1986.

H. Ohshimo and S. Osaki, Stochastic Modeling of a Multisystem from the View-Point of Reliability and Performance, International Journal of Systems Science, vol. 17, pp. 619–627,1986.

N. Kaio and S. Osaki, Optimal Inspection Policy with Two Types of Imperfect Inspection Probabilities, Microelectronics and Reliability, vol. 26, pp. 935–942 1986.

S. Yamada and S. Osaki, Optimal Software Release Policies with Simultaneous Cost and Reliability Requirements, European Journal of Operational Research, vol. 31, pp. 46–51, 1987.

N. Kaio and S. Osaki, Review of Discrete and Continuous Distributions in Replacement Models, International Journal of Systems Science, vol. 19, pp. 171–177, 1988.

S. Osaki and X.-X. Li, Characterizations of Gamma and Negative Binomial Distributions, IEEE Transactions on Reliability, vol. 37, pp. 379–382, 1988.

N. Kaio and S. Osaki, Optimum Planned Policies with Minimal Repair, Microelectronics and Reliability, vol. 28, pp. 287–293, 1988.

N. Kaio and S. Osaki, Inspection Policies: Comparisons and Modifications, Revue Francaise d'Automatique et Recherche Operationnelle, vol. 22, pp. 387–400, 1988.

K. Yasui, T. Nakagawa and S. Osaki, A Summary of Optimum Replacement Policies for a Parallel Redundant System, Microelectronics and Reliability, vol. 28, pp. 635–641, 1988.

N. Kaio and S. Osaki, Comparison of Inspection Policies, Journal of the Operational Research Society, vol. 40, pp. 499–503, 1989.

S. Osaki, H. Ohshimo and S. Fukumoto, Effect of Software Maintenance Policies for a Hardware-Software System, International Journal of Systems Science, vol. 20, pp. 331–338, 1989.

S. Osaki, S. Yamada and J. Hishitani, Availability Theory for Two-unit Nonindependent Series Systems Subject to Shut-Off Rules, Reliability Engineering and System Safety, vol. 25, pp. 33–42, 1989.

N. Kaio and S. Osaki, Optimal Ordering Policies with Two Types of Randomized Lead Times, Computers & Mathematics with Applications, vol. 19, pp. 43–52, 1990.

H. Ohshimo, S. Fukumoto and S. Osaki, Reliability/Performance Evaluation for Multisystems from the Viewpoint of Job Assignments, Transactions of the Institute of Electronics and Communication Engineers of Japan, vol. E73, pp. 1257–1263, 1990.

N. Kaio and S. Osaki, Optimum Ordering Policies with Non-Linear Running and Salvage Costs, Microelectronics and Reliability, vol. 30, pp. 785–793, 1990.

N. Kaio and S. Osaki, Modified Age Replacement Policies, Microelectronics and Reliability, vol. 21, pp. 1733–1738, 1990.

S. Yamada, J. Hishitani and S. Osaki, Test-effort Dependent Software Reliability Measurement, International Journal of Systems Science, vol. 22, pp. 73–83, 1991.

J. Hishitani, S. Yamada and S. Osaki, Reliability Assessment Measures Based on Software Reliability Growth Model with Normalized Method, Journal of Information Processing, vol. 14, pp. 178–183, 1991.

S. Fukumoto, N. Kaio and S. Osaki, Evaluation for a Database Recovery Action with Periodical Checkpoint Generations, IEICE Transactions, vol. E74, pp. 2076–2082, 1991.

T. Dohi, N. Kaio and S. Osaki, A Note on Optimal Inventory Policies Taking Account of Time Value, Revue Francaise d'Automatique et Recherche Operationnelle, vol. 26, pp. 1–14, 1992.

S. Fukumoto, N. Kaio and S. Osaki, A Study of Checkpoint Generations for a Database Recovery Mechanism, Computers & Mathematics with Applications, vol. 24, pp. 63–70, 1992.

M. Kimura, S. Yamada and S. Osaki, Software Reliability Assessment for an Exponential-S-shaped Reliability Growth Phenomenon, Computers & Mathematics with Applications, vol. 24, pp. 71–78, 1992.

H. Tanaka, T. Dohi, H. Fujiwara and S. Osaki, Construction of a Decision Support System for a Combination of Options, Computers & Mathematics with Applications, vol. 24, pp. 135–140, 1992.

T. Dohi and S. Osaki, Optimal Trading of Stock Options under Alternative Strategy, Computers & Mathematics with Applications, vol. 24, pp. 127–134, 1992.

N. Kaio, T. Dohi and S. Osaki, Optimal Maintenance Policies with Lead Times and Repair, International Journal of Systems Science, vol. 23, pp. 1299–1308, 1992.

K. Tokunou, S. Yamada and S. Osaki, A Markovian Imperfect Debugging Model for Software Reliability Measurement, IEICE Transactions on Fundamentals, vol. E75-A, pp. 1590–1596, 1992.

S. Yamada, J. Hishitani and S. Osaki, Software Reliability Measurement and Assessment Based on Nonhomogeneous Poisson Process Models: A Survey, Microelectronics and Reliability, vol. 32, pp. 1763–1773, 1992.

S. Yamada, K. Tokunou and S. Osaki, Imperfect Debugging Models with Fault Introduction Rate for Software Reliability Assessment, International Journal of Systems Science, vol. 23, pp. 2241–2252, 1992.

S. Fukumoto, N. Kaio and S. Osaki, Optimal Checkpointing Policies Using the Checkpointing Density, Journal of Information Processing, vol. 15, pp. 87–92, 1992.

S. Yamada, K. Tokunou and S. Osaki, Software Reliability Measurement in Imperfect Debugging Environment and Its Application, Reliability Engineering and System Safety, vol. 40, pp. 139–147, 1993.

M. Kimura, S. Yamada, H. Tanaka and S. Osaki, Software Reliability Measurement with Prior-Information on Initial Fault Content, Transactions of Information Processing Society of Japan, vol. 34, pp. 1601–1609, 1993.

T. Dohi and S. Osaki, A Note on Portfolio Optimization with Path-dependent Utility, Annals of Operations Research, vol. 45, pp. 77–90, 1993.

T. Dohi and S. Osaki, A Comparative Study of Stochastic EOQ Models with Discounting, IMA Journal of Mathematics Applied in Business & Industry, vol. 5, pp. 171–174, 1993.

S. Yamada, J. Hishitani and S. Osaki, Software-Reliability Growth with a Weibull Test-Effort: A Model & Application, IEEE Transactions on Reliability, vol. 42, no. 1, pp. 100–106, 1993.

T. Dohi, N. Kaio and S. Osaki, Continuous Time Inventory Control for Wiener Process Demand, Computers & Mathematics with Applications, vol. 26, pp. 11–22, 1993.

S. Yamada, M. Kimura, H. Tanaka and S. Osaki, Software Reliability Measurement and Assessment with Stochastic Differential Equations, IEICE Transactions on Fundamentals, vol. E77-A, no. 1, pp. 109–116, 1994.

H. Koshimae, H. Tanaka and S. Osaki, Some Remarks on MTBF's for Non-homogeneous Poisson Processes, IEICE Transactions on Fundamentals, vol. E77-A, pp. 144–149, 1994.

N. Kaio, T. Dohi and S. Osaki, Inspection Policy with Failure Due to Inspection, Micro electronics and Reliability, vol. 34, pp. 599–602, 1994.

T. Dohi, A. Watanabe and S. Osaki, A Note on Risk Averse Newsboy Problem, Revue Francaise d'Automatique et Recherche Operationnelle, vol. 28, pp. 181–202, 1994.

T. Dohi, N. Kaio and S. Osaki, Optimal Order Limit Policy under Cost Effectiveness Criterion, Computers & Industrial Engineering, vol. 27, pp. 197–200, 1994.

M. Odagiri, N. Kaio and S. Osaki, A Note on Optimal Checkpoint Sequence Taking Account of Preventive Maintenance, IEICE Transactions on Fundamentals, vol. E77-A, pp. 244–246, 1994.

S. Yamada, M. Kimura, H. Tanaka and S. Osaki, Software Reliability Measurement and Assessment with Stochastic Differential Equations, IEICE Transactions on Fundamentals, vol. E77-A, pp. 109–116, 1994.

T. Dohi, E. Kitaoka and S. Osaki, Alternative Optimality Criteria of Portfolio Selection Based upon Threshold Stopping Rule, Applied Stochastic Models and Data Analysis, vol. 10, pp. 257–268, 1994.

T. Dohi, N. Kaio and S. Osaki, Optimizing Monitoring Time in a Continuous-Review Cyclic Inventory System, IMA Journal of Mathematics Applied in Business & Industry, vol. 6, pp. 223–237, 1995.

T. Dohi, H. Tanaka, N. Kaio and S. Osaki, Alternative Growth versus Security in Continuous Dynamic Trading, European Journal of Operational Research, vol. 84, pp. 430–443, 1995.

T. Dohi and S. Osaki, Optimal Inventory Policies Under Product Obsolescent Circumstance, Computers & Mathematics with Applications, vol. 29, pp. 23–30, 1995.

M. Kimura, S. Yamada, H. Tanaka, T. Nagaike and S. Osaki, Quality Assessment Models for Initial Production Control based on Stochastic Differential Equations, Microelectronics and Reliability, vol. 35, no. 4, pp. 657–668, 1995.

T. Dohi, N. Kaio and S. Osaki, Solution Procedure for a Repair Limit Problem Using TTT-transform, IMA Journal of Mathematics Applied in Business & Industry, vol. 6, pp. 101–111,1995.

T. Dohi, N. Kaio and S. Osaki, Optimal Control of a Finite Dam with a Sample Path Constraint, Mathematical and Computer Modelling, vol. 22, pp. 45–51, 1995.

K. Okuhara and S. Osaki, A Study on the Characteristics in a Symmetry Boltzmann Machine Composed of Two Boltzmann Machines, Mathematical and Computer Modelling, vol. 22, pp. 273–278, 1995.

M. Kimura, S. Yamada and S. Osaki, Statistical Software Reliability Prediction based on Mean Time between Failures, Mathematical and Computer Modelling, vol. 22, pp. 149–155, 1995.

T. Dohi, N. Kaio and S. Osaki, Continuous Review Cyclic Inventory Models with Emergency Order, Journal of the Operations Research Society of Japan, vol. 38, pp. 212–229, 1995.

T. Dohi, N. Kaio and S. Osaki, Optimal Production Planning under Diffusion Demand Pattern, Mathematical and Computer Modelling, vol. 21, no. 11, pp. 35–46, 1995.
T. Dohi, N. Matsushima, N. Kaio and S. Osaki, Nonparametric Repair-limit Replacement Policies with Imperfect Repair, European Journal of Operational Research, vol. 96, pp. 260–273, 1996.
H. Koshimae, T. Dohi, N. Kaio and S. Osaki, Graphical/Statistical Approach to Repair Limit Replacement Problem, Journal of the Operations Research Society of Japan, vol. 39, pp. 230–246, 1996.
T. Dohi, N. Kaio and S. Osaki, Optimal Ordering Policies with Time-dependent Delay Structure, Journal of Quality in Maintenance Engineering, vol. 2, pp. 50–62, 1996.
M. Odagiri, T. Dohi, N. Kaio and S. Osaki, An Economic Analysis for a Hybrid Data Backup System, IEICE Transactions on Fundamentals, vol. E79-A, pp. 118-125, 1996.
T. Dohi, N. Kaio and S. Osaki, Optimal Planned Maintenance with Salvage Cost for a two-unit standby redundant system, Microelectronics and Reliability, vol. 36, pp. 1581–1588, 1996.
T. Dohi, T. Aoki, N. Kaio and S. Osaki, Computational Aspects of Optimal Checkpoint Strategy in Fault-Tolerant Database Management, IEICE Transactions on Fundamentals, vol. E80-A, pp. 2006–2015, 1997.
S. Fukumoto, S. Nakagawa, N. Kaio and S. Osaki, Optimum Checkpoint Policies Attending with Unsuccessful Rollback Recovery, International Journal of Reliability, Quality and Safety Engineering, vol. 4, pp. 427–439 1997.
Y. Shinohara, T. Dohi and S. Osaki, Comparisons of Optimal Release Policies for Software Systems, Computers & Industrial Engineering, vol. 33, pp. 813–816, 1997.
T. Dohi, T. Shibuya and S. Osaki, Models for 1-out-of-Q Systems with Stochastic Lead Times and Expedited Ordering Options for Spares Inventory, European Journal of Operational Research, vol. 103, pp. 255–272, 1997.
T. Dohi, N. Kaio and S. Osaki, Optimal Software Release Policies with Debugging Time Lag, International Journal of Reliability, Quality and Safety Engineering, vol. 4, pp. 241–255, 1997.
T. Dohi, H. Koshimae, N. Kaio and S. Osaki, Geometrical Interpretations of Repair Cost Limit Replacement Policies, International Journal of Reliability, Quality and Safety Engineering, vol. 4, pp. 309–333, 1997.
T. Dohi, Y. Yamada, N. Kaio and S. Osaki, The Optimal Lot Sizing for Unreliable Economic Manufacturing Models, International Journal of Reliability, Quality and Safety Engineering, vol. 4, pp. 413–426, 1997.
Y. Shinohara, Y. Nishio, T. Dohi and S. Osaki, An Optimal Software Release Problem under Cost Rate Criterion: Artificial Neural Network Approach, Journal of Quality in Maintenance Engineering, vol. 4, pp. 236–247, 1998.
T. Dohi, N. Kaio and S. Osaki, Minimal Repair Policies for An Economic Manufacturing Process, Journal of Quality in Maintenance Engineering, vol. 4, pp. 248–262, 1998.
T. Dohi, N. Kaio and S. Osaki, On the Optimal Ordering Policies in Maintenance Theory-Survey and Applications, Applied Stochastic Models and Data Analysis, vol. 14, pp. 309–321, 1998.
T. Shibuya, T. Dohi and S. Osaki, Spare Part Inventory Models with Stochastic Lead Times, International Journal of Production Economics, vol. 55, pp. 257–271, 1998.
T. Dohi, T. Aoki, N. Kaio and S. Osaki, Nonparametric Preventive Maintenance Optimization Models under Earning Rate Criteria, IIE Transactions on Quality and Reliability Engineering, vol. 30, pp. 1099–1108, 1998.
K. Okuhara, M. Kijima and S. Osaki, Learning to Design Synergetic Computers with an Extended Symmetric Diffusion Network, Neural Computation, vol. 11, pp. 1475–1491, 1999.
H. Okamura, T. Dohi and S. Osaki, Optimal Order-Limit Policies for an (r, Q) Inventory System, IMA Journal of Mathematics Applied in Business & Industry, vol. 10, pp. 127–145, 1999.
K. Amasaka and S. Osaki, The Promotion of the New Statistical Quality Control Internal Education at Toyota Motor: A Proposal of 'Science Statistical Quality Control' for Improving the Principle of Total Quality Management, European Journal of Engineering Education, vol. 24, pp. 259–276, 1999.

T. Dohi, Y. Nishio and S. Osaki, Optimal Software Release Scheduling Based on Artificial Neural Networks, Annals of Software Engineering, vol. 8, pp. 167–185, 1999.
H. Okamura, T. Dohi and S. Osaki, Optimal Policies for a Controlled Queueing System with Removable Server under a Random Vacation Circumstance, Computers & Mathematics with Applications, vol. 39, pp. 215–227, 2000.
T. Dohi, Y. Yatsunami, Y. Nishio and S. Osaki, The Effective Smoothing Techniques to Estimate the Optimal Software Release Schedule based on Artificial Neural Network, IEICE Transactions on Fundamentals, vol. E83-A, pp. 796–803, 2000.
T. Dohi, K. Nomura, N. Kaio and S. Osaki, A Simulation Study to Analyze Unreliable File Systems with Checkpointing and Rollback Recovery, IEICE Transactions on Fundamentals, vol. E83-A, pp. 804–811, 2000.
T. Dohi, N. Kaio and S. Osaki, A Graphical Method to Repair-Cost Limit Replacement Policies with Imperfect Repair, Mathematical and Computer Modelling, vol. 31, pp. 99–106, 2000.
T. Dohi, Y. Teraoka and S. Osaki, Software Release Games, Journal of Optimization Theory and Applications, vol. 105, pp. 325–346, 2000.
T. Dohi, N. Kaio and S. Osaki, The Optimal Age-Dependent Checkpoint Strategy for a Stochastic System Subject to General Failure Mode, Journal of Mathematical Analysis and Applications, vol. 249, pp. 80–94, 2000.
T. Dohi, K. Takeita and S. Osaki, Graphical Methods for Determining/Estimating Optimal Repair-Limit Replacement Policies, International Journal of Reliability, Quality and Safety Engineering, vol. 7, pp. 43–60, 2000.
T. Dohi, N. Kaio and S. Osaki, A Graphical Method to Repair—Cost Limit Replacement Policies with Imperfect Repair, Mathematical and Computer Modelling, vol. 31, pp. 99–106 2000.
T. Dohi, H. Morishita and S. Osaki, A Statistical Estimation Method of Optimal Software Release Timing Applying Autoregressive Models, IEICE Transactions on Fundamentals, vol. E84-A, pp. 331–338, 2001.
T. Dohi, A. Ashioka, N. Kaio and S. Osaki, Optimizing the Repair-Time Limit Replacement Schedule with Discounting and Imperfect Repair, Journal of Quality in Maintenance Engineering, vol. 7, pp. 71–84, 2001.
T. Dohi, F. S. Othman, N. Kaio and S. Osaki, The Lorenz Transform Approach to the Optimal Repair-Cost Limit Replacement Policy with Imperfect Repair, Revue Francaise d'Automatique, Informatique et Recherche Operationnelle, vol. 35, pp. 21–36, 2001.
T. Dohi, N. Kaio and S. Osaki, Determination of Optimal Repair-cost Limit on the Lorenz Curve, Journal of the Operations Research Society of Japan, vol. 44, pp. 207–219, 2001.
H. Okamura, S. Miyahara, T. Dohi and S. Osaki, Performance Evaluation of Workload-based Software Rejuvenation Scheme, IEICE Transactions on Information and Systems (D), vol. E84-D, pp. 1368–1375, 2001.
T. Dohi, N. Kaio and S. Osaki, Optimal Periodic Maintenance Strategy under an Intermittently Used Environment, IIE Transactions on Quality and Reliability Engineering, vol. 33, pp. 1037–1046, 2001.
T. Dohi, H. Okamura and S. Osaki, Optimal Control of Preventive Maintenance Schedule and Safety Stocks in an Unreliable Manufacturing Environment, International Journal of Production Economics, vol. 74, pp. 147–155, 2001.
T. Dohi, N. Kaio and S. Osaki, A New Graphical Method to Estimate the Optimal Repair-time Limit with Incomplete Repair and Discounting, Computers & Mathematics with Applications, vol. 46, pp. 999–1007, 2003.
H. Okamura, T. Dohi and S. Osaki, A Structural Approximation Method to Generate the Optimal Auto-sleep Schedule for a Computer System, Computers & Mathematics with Application, vol. 46, pp. 1103–1110, 2003.
T. Dohi, A. Ashioka, N. Kaio and S. Osaki, The Optimal Repair-time Limit Replacement Policy with Imperfect Repair: Lorenz Transform Approach, Mathematical and Computer Modelling, vol. 38, pp. 1169–1176, 2003.

T. Dohi, K. Iwamoto, N. Kaio and S. Osaki, A Generalized Discrete-time Order-replacement Model, IMA Journal of Management Mathematics, vol. 15, pp. 125–138, 2004.

T. Dohi, A. Ashioka, N. Kaio and S. Osaki, A Simulation Study on the Cost Distribution under Age Replacement Policy with Discounting, Industrial Engineering and Management System, vol. 3, pp. 143–148, 2004.

T. Danjou, T. Dohi, N. Kaio and S. Osaki, Analysis of Periodic Software Rejuvenation Policies based on Net Present Value Approach, International Journal of Reliability, Quality and Safety Engineering, vol.11, pp. 313–327, 2004.

T. Dohi, A. Ashioka, N. Kaio and S. Osaki, Statistical Estimation Algorithms for Some Repair–limit Replacement Scheduling Problems under Earning Rate Criteria, Computers & Mathematics with Applications, vol. 51, pp. 345–356, 2006.

T. Dohi, H. Suzuki and S. Osaki, Transient Cost Analysis of Non-Markovian Software Systems with Rejuvenation, International Journal of Performability Engineering, vol. 2, no. 3, pp. 233–243, 2006.

T. Dohi, N. Kaio and S. Osaki, Estimating Cost-effective Checking Request Policies, Quality Technology and Quantitative Management, vol. 4, pp. 1–13, 2007.

T. Dohi, N. Kaio and S. Osaki, Optimal (T, S)-policies in a Discrete-time Opportunity-based Age Replacement: An Empirical Study, International Journal of Industrial Engineering, vol.14, pp. 340–347, 2007.

Contents

Generalized Logit-Based Proportional Hazards Models and Their Applications in Survival and Reliability Analyses

N. Balakrishnan, M. C. Pardo and M. L. Avendaño

Abstract We introduce a flexible family of generalized logit-based regression models for survival and reliability analyses. We present its parametric as well as its semiparametric versions. The method of maximum likelihood and the partial likelihood approach are applied to estimate the parameters of the parametric and semiparametric models, respectively. This new family of models is illustrated with male laryngeal cancer data and compared with Cox regression.

1 Introduction

Data arising from survival and reliability analyses often consist of a response variable that measures the duration of time until the occurrence of a specific event and a set of variables (covariates) thought to be associated with the event-time variable. These data arise in a number of applied fields, such as medicine, biology, public health, epidemiology, engineering, economics, and demography, and they have some features that pose difficulties to traditional statistical methods. The first is that the data are generally asymmetrically distributed, while the second feature is that lifetimes are frequently censored (the end-point of interest has not been observed for that individual). Regression models for survival and reliability data have traditionally been based on the proportional hazards model of Cox [3] which is defined through the hazard function $h\left(t \mid \mathbf{x}\right)$ of the form

$$h\left(t \mid \mathbf{x}\right) = h_0\left(t\right) \exp\left(\mathbf{x}'\beta\right),$$

N. Balakrishnan(✉)
Department of Mathematics and Statistics, McMaster University,
Hamilton, Ontario, Canada
e-mail: bala@univmail.cis.mcmaster.ca

M.C. Pardo · M.L. Avendaño
Department of Statistics and Operational Research I,
Faculty of Mathematics, Complutense University of Madrid, Madrid, Spain

T. Dohi and T. Nakagawa (eds.), *Stochastic Reliability and Maintenance Modeling*, Springer Series in Reliability Engineering 9, DOI: 10.1007/978-1-4471-4971-2_1,

where $h_0(t)$ is an arbitrary function of time called baseline hazard function, $\mathbf{x}' = (x_1, \ldots, x_p)$ is a vector of covariates for the individual at time t, and $\boldsymbol{\beta}' = (\beta_1, \ldots, \beta_p)$ is a vector of unknown parameters to be estimated. In the case when the baseline hazard function is treated nonparametrically, then this model becomes a semiparametric model. Instead, if we assume that the baseline hazard function is specified up to a few unknown parameters, which is usually accomplished with a specific parametric distribution such as Weibull distribution, we obtain a parametric proportional hazards model.

Some recent research has focused on developing extended regression models that include Cox model as a special case. In this line, we can find the model introduced by Etezadi-Amoli and Ciampi [4] of the form

$$h(t \mid \boldsymbol{\alpha}, \boldsymbol{\beta}, \mathbf{x}) = \exp(\mathbf{x}'\boldsymbol{\alpha})\, h_0\left(t \exp(\mathbf{x}'\boldsymbol{\beta})\right),$$

where $\boldsymbol{\alpha}$ and $\boldsymbol{\beta}$ are vectors of regression parameters. For $\boldsymbol{\beta} = \mathbf{0}$, we deduce the Cox model, while for $\boldsymbol{\alpha} = \boldsymbol{\beta}$ we obtain the accelerated failure time (AFT) model which is also a popular model in the analysis of survival and reliability data. These authors then show that a better fit is obtained with this new model than with the Cox and AFT models in two examples based on artificial and real data. Nevertheless, the main emphasis of their work is on a spline approximation for the baseline hazard function.

A different family of models with smooth background hazard or survival functions have been proposed by Younes and Lachin [10] and Royston and Parmar [9], which includes the proportional hazards and proportional odds models as special cases. The class of these models is based on transformation of the survival function by a link function $g(\cdot)$ of the form

$$g(S(t \mid \boldsymbol{\beta}, \mathbf{x})) = g(S_0(t)) + \mathbf{x}'\boldsymbol{\beta},$$

where $S_0(t) = S(t \mid \mathbf{0}, \mathbf{x})$ is the baseline survival function. The former tackled the estimation problem by using B-splines to estimate the baseline hazard function while the latter utilized natural cubic splines to model $g(S_0(t))$. Here, again the main focus of the work was to check the advantage of a smooth modeling of the background hazard or survival functions, respectively.

An alternative model to the Cox model is based on the hazard function

$$h(t \mid \boldsymbol{\beta}, \mathbf{x}) = h_0(t) \frac{\exp(\mathbf{x}'\boldsymbol{\beta})}{1 + \exp(\mathbf{x}'\boldsymbol{\beta})}, \tag{1}$$

where the covariate effects are modeled on the logarithmic scale rather than on the log odds scale. In spite of the simplicity of this model, it has not been studied much in the literature. Recently, MacKenzie [7, 8] has considered this logit link-based model with a constant baseline survival function and nonproportional hazards and displayed its applicability, which is given by

$$h\left(t \mid \alpha, \boldsymbol{\beta}, \mathbf{x}\right) = h\frac{\exp\left(t\alpha + \mathbf{x}'\boldsymbol{\beta}\right)}{1+\exp\left(t\alpha + \mathbf{x}'\boldsymbol{\beta}\right)}.$$

The flexibility shown by MacKenzie's model gives us an impetus to extend the Cox model in a similar manner. First of all, to assume that the baseline hazard model is a constant is to limit the flexibility the model. In fact, the aim of the previous papers was to estimate in a proper way the baseline hazard function. Secondly, to measure the influence of the unknown parameters on a generalized log-odds scale instead of a log-odds scale. Therefore, this model is a particular case of our models introduced in Sect. 2 without time-dependence. The reason for not considering time-dependence is to start with a very general family of models but then focus on its simplest form. We hope to consider in our future study time-dependence and also to estimate the background hazard with splines.

In this chapter, we not only study the logit link-based model in (1), but also generalize it to a flexible parametric family of proportional hazards model based on a generalization of the logistic distribution (see Balakrishnan [1]) called Type-I generalized logistic model. The formulation of the model and estimation methods for parametric and semiparametric models are then discussed in Sect. 2. Measures of fitting this model are discussed in Sect. 3. Next, an illustrative example is presented in Sect. 4. Finally, some concluding remarks are made in Sect. 5.

2 The Generalized Logit Link Proportional Hazards Model

The logit link-based model in (1) can be generalized by replacing the logistic distribution function in (1) by a generalization of the logistic distribution called Type-I generalized logistic which is given by

$$F\left(y\right) = \frac{1}{\left(1+e^{-y}\right)^{a}}, \quad -\infty < y < \infty,\ a > 0;$$

see Balakrishnan [1].

By utilizing this form, we propose a proportional hazards model defined through the hazard function

$$h\left(t \mid \boldsymbol{\beta}, a, \mathbf{x}\right) = h_0\left(t\right) K(\boldsymbol{\beta}, a, \mathbf{x}), \tag{2}$$

with

$$K(\boldsymbol{\beta}, a, \mathbf{x}) = \frac{1}{\{1+\exp\left(-\mathbf{x}'\boldsymbol{\beta}\right)\}^{a}}, \quad a > 0.$$

For two covariate profiles $\mathbf{x}_i$ and $\mathbf{x}_j$, the hazards are proportional and the relative risk does not depend on t as

$$\rho\left(t \mid \boldsymbol{\beta}, a, \mathbf{x}_i, \mathbf{x}_j\right) = \frac{h\left(t \mid \boldsymbol{\beta}, a, \mathbf{x}_i\right)}{h\left(t \mid \boldsymbol{\beta}, a, \mathbf{x}_j\right)} = \left(\frac{1+\exp\left(-\mathbf{x}_j'\boldsymbol{\beta}\right)}{1+\exp\left(-\mathbf{x}_i'\boldsymbol{\beta}\right)}\right)^a.$$

Note that in the special case when $a = 1$, we deduce the proportional hazards model with a logit link function in (1).

The survival function corresponding to the hazard model in (2) is

$$S\left(t \mid \boldsymbol{\beta}, a, \mathbf{x}\right) = \exp\left\{-K\left(\boldsymbol{\beta}, a, \mathbf{x}\right) H_0\left(t\right)\right\}, \tag{3}$$

where $H_0(t) = \int_0^t h_0(u)du$ is the baseline cumulative hazard function.

Equation (2) characterizes the generalized logit link proportional hazards model with density given by

$$f\left(t \mid \boldsymbol{\beta}, a, \mathbf{x}\right) = \exp\left\{-H\left(t \mid \boldsymbol{\beta}, a, \mathbf{x}\right)\right\} h\left(t \mid \boldsymbol{\beta}, a, \mathbf{x}\right),$$

where $H(t|\boldsymbol{\beta}, a, \mathbf{x})$ is the cumulative hazard function.

At this point, we have not made any assumption about the baseline hazard function $h_0(t)$, so that the model is parametric only for the covariate effect, and consequently the model is semiparametric. Instead, if we assume a parametric form for the function $h_0(t)$, the model becomes parametric. Now, we will describe the statistical inferential methods for both these cases.

2.1 Parametric Model

We may assume that the baseline hazard function is specified up to a few unknown parameters. This is usually accomplished with a specific parametric distribution such as the Weibull distribution. In this case, we get a Weibull generalized logit link proportional hazards model WGLPH with hazard function

$$h\left(t \mid \lambda, \gamma, \boldsymbol{\beta}, a, \mathbf{x}\right) = \lambda\gamma t^{\gamma-1} K\left(\boldsymbol{\beta}, a, \mathbf{x}\right) \tag{4}$$

which is fully parametric in form. This model contains as a special case the generalized logit link exponential proportional hazards model for the case when $\gamma = 1$.

Then, the cumulative hazard is given by

$$H(t \mid \lambda, \gamma, \boldsymbol{\beta}, a, \mathbf{x}) = \lambda K(\boldsymbol{\beta}, a, \mathbf{x})\, t^{\gamma}$$

and Eq. (4) characterizes the WGLPH with density given by

$$f(t \mid \lambda, \gamma, \boldsymbol{\beta}, a, \mathbf{x}) = \lambda K(\boldsymbol{\beta}, a, \mathbf{x})\, \gamma t^{\gamma-1} \exp\left\{-\lambda K(\boldsymbol{\beta}, a, \mathbf{x})\, t^{\gamma}\right\}.$$

Note that this is a Weibull density function with parameters γ and $\lambda K(\boldsymbol{\beta}, a, \mathbf{x})$.

When we assume such a fully parametric form for the distribution of survival times, the estimation of the unknown parameters of the model is by full maximum likelihood method. Consider a sample of n independent individuals with data $(t_i, \mathbf{x}_i, \delta_i)$, where $\delta_i = 1$ for an event and 0 otherwise, for $i = 1, \ldots, n$. Accordingly, under the assumption that the censoring mechanism is non-informative, the full likelihood for a random sample of n individuals is given by

$$L(\boldsymbol{\beta}, a) = \prod_{i=1}^{n} \{h(t_i \mid \boldsymbol{\beta}, a, \mathbf{x}_i)\}^{\delta_i} \{S(t_i \mid \boldsymbol{\beta}, a, \mathbf{x}_i)\}.$$

For the Weibull baseline hazard, the log-likelihood function simply becomes

$$\begin{aligned} l(\lambda, \gamma, \boldsymbol{\beta}, a) &= \ln(L(\lambda, \gamma, \boldsymbol{\beta}, a)) \\ &= \sum_{i=1}^{n} \left[\delta_i \ln\left\{\lambda \gamma t_i^{\gamma-1} K(\boldsymbol{\beta}, a, \mathbf{x}_i)\right\} - \lambda K(\boldsymbol{\beta}, a, \mathbf{x}_i)\, t_i^{\gamma}\right]. \end{aligned} \quad (5)$$

To obtain the maximum likelihood estimates, the log-likelihood function in (5) must be maximized numerically by using a procedure for constrained optimization. In order to maximize (5), we obtain its first derivatives with respect to all the parameters which are presented in Appendix A.

Observe that in this case, the corresponding survival function can then be estimated as

$$S(t|\hat{\lambda}, \hat{\gamma}, \hat{\boldsymbol{\beta}}, \hat{a}, \mathbf{x}) = \exp\left\{-K\left(\hat{\boldsymbol{\beta}}, \hat{a}, \mathbf{x}\right) \hat{\lambda} t^{\hat{\gamma}}\right\}.$$

2.2 Semiparametric Model

On the other hand, when we assume an unknown functional form for the baseline survival function, the estimation of the unknown parameters of the model is done by maximum partial likelihood method. Consider a sample of n independent individuals with data $(t_i, \mathbf{x}_i, \delta_i)$ as before and when the censoring mechanism is non-informative. In this case, the partial likelihood for a random sample of n individuals can be written as

$$L(\boldsymbol{\beta}, a) = \prod_{i=1}^{n} \left[\frac{K(\boldsymbol{\beta}, a, \mathbf{x}_i)}{\sum\limits_{l \in R(t_i)} K(\boldsymbol{\beta}, a, \mathbf{x}_l)} \right]^{\delta_i},$$

where $R(t_i)$ is the risk set at time t_i.

Then, the partial log-likelihood function is given by

$$\begin{aligned} l(\boldsymbol{\beta}, a) &= \ln(L(\boldsymbol{\beta}, a)) \\ &= \sum_{i=1}^{n} \delta_i \left[\ln(K(\boldsymbol{\beta}, a, \mathbf{x}_i)) - \ln\left(\sum_{l \in R(t_i)} K(\boldsymbol{\beta}, a, \mathbf{x}_l) \right) \right]. \end{aligned} \tag{6}$$

For the purpose of maximizing (6) and obtaining the partial maximum likelihood estimates, we use numerical methods for carrying out the required constrained optimization. Its first derivatives with respect to all the parameters are presented in Appendix B.

Once we have fitted a generalized likelihood proportional hazards model, it may be of interest to estimate the survival probability. The estimator of the survival function is based on Breslow's estimator of the baseline cumulative hazard rate, which proceeds as follows:

Let the full likelihood function be

$$\begin{aligned} L(\boldsymbol{\beta}, a, h_0(t)) &= \prod_{i=1}^{n} \{h_0(t_i) K(\boldsymbol{\beta}, a, \mathbf{x}_i)\}^{\delta_i} \{\exp[-K(\boldsymbol{\beta}, a, \mathbf{x}_i) H_0(t_i)]\} \\ &= \prod_{j=1}^{D} \{h_0(t_j) K(\boldsymbol{\beta}, a, \mathbf{x}_j)\} \prod_{i=1}^{n} \{\exp[-K(\boldsymbol{\beta}, a, \mathbf{x}_i) H_0(t_i)]\}, \end{aligned}$$

where $j = 1, \ldots, D$ correspond to the times without censoring. We then obtain

$$L(\boldsymbol{\beta}, a, h_0(t)) = \prod_{j=1}^{D} h_0(t_j) K(\boldsymbol{\beta}, a, \mathbf{x}_j) \exp\left\{ -\sum_{i=1}^{n} K(\boldsymbol{\beta}, a, \mathbf{x}_i) H_0(t_i) \right\}.$$

Defining $H_0(t) = \sum\limits_{t^* < t} h_0(t^*)$, and supposing that $\boldsymbol{\beta}$ and a are fixed, we have

$$L(h_0(t)) = \prod_{j=1}^{D} h_0(t_j) K(\boldsymbol{\beta}, a, \mathbf{x}_j) \exp\left\{ -\sum_{i=1}^{n} K(\boldsymbol{\beta}, a, \mathbf{x}_i) \left(\sum_{t_i^* < t_i} h_0(t_i^*) \right) \right\}.$$

Taking $h_0(t) = 0$ when the event is censored, then we get

$$
\begin{aligned}
L\left(h_0(t)\right) &= \left[\prod_{j=1}^{D} h_0\left(t_j\right) K(\beta, a, \mathbf{x}_j)\right] \exp\left\{-\sum_{j=1}^{D} h_0(t_j) \sum_{i \in R(t_j)} K\left(\beta, a, \mathbf{x}_i\right)\right\} \\
&= \prod_{j=1}^{D} h_0\left(t_j\right) K(\beta, a, \mathbf{x}_j) \exp\left\{-h_0(t_j) \sum_{i \in R(t_j)} K\left(\beta, a, \mathbf{x}_i\right)\right\},
\end{aligned}
$$

so that

$$
L\left(h_0(t_1), \cdots, h_0(t_D)\right) \propto \prod_{j=1}^{D} h_0\left(t_j\right) \exp\left\{-h_0(t_j) \sum_{i \in R(t_j)} K\left(\beta, a, \mathbf{x}_i\right)\right\}.
$$

For the determination of the maximum likelihood estimate, we take the derivative with respect to $h_0(t_j)$ which is given by

$$
\begin{aligned}
\frac{\partial L}{\partial h_0(t_j)} &= \exp\left\{-h_0(t_j) \sum_{i \in R(t_j)} K\left(\beta, a, \mathbf{x}_i\right)\right\} \\
&\quad + h_0(t_j) \exp\left\{-h_0(t_j) \sum_{i \in R(t_j)} K\left(\beta, a, \mathbf{x}_i\right)\right\} \left\{-\sum_{i \in R(t_j)} K\left(\beta, a, \mathbf{x}_i\right)\right\}.
\end{aligned}
$$

Upon equating this to zero, we obtain the maximum likelihood estimate to be

$$
\hat{h}_0(t_j) = \frac{1}{\sum_{i \in R(t_j)} K\left(\beta, a, \mathbf{x}_i\right)},
$$

and consequently

$$
\hat{H}_0(t) = \sum_{t^* < t} \frac{1}{w(t^*)},
$$

where $w(t^*) = \sum_{l \in R(t^*)} K(\beta, a, \mathbf{x}_l)$.

Since

$$
\hat{S}_0(t) = \exp\left[-\hat{H}_0(t)\right]
$$

is the estimator of the survival function of an individual with covariate vector $\mathbf{x} = \mathbf{0}$, for estimating the survival function of an individual with covariate vector $\mathbf{x} = \mathbf{x}^*$, we use the estimator

$$
\hat{S}(t|\hat{\beta}, \hat{a}, \mathbf{x} = \mathbf{x}^*) = \left\{\hat{S}_0(t)\right\}^{K\left(\hat{\beta}, \hat{a}, \mathbf{x} = \mathbf{x}^*\right)}. \tag{7}
$$

In both cases of parametric and semiparametric setting, the variance of the estimated parameters $\hat{\boldsymbol{\theta}} = (\hat{\lambda}, \hat{\gamma}, \hat{\boldsymbol{\beta}}, \hat{a})$ and $\hat{\boldsymbol{\theta}} = (\hat{\boldsymbol{\beta}}, \hat{a})$ obtained by maximizing Eqs. (5) and (6), respectively, can be estimated as

$$var\left(\hat{\theta}\right) = diag\left(I^{-1}\left(\hat{\theta}\right)\right),$$

where I is the observed information matrix.

3 Measures of Fit

After fitting several possible models for a given data, we will need to compare the fit of each model for selecting the best one. When we fit several non-nested models, we may use the Akaike information criterion (AIC) to choose the best one among them. The AIC is defined as

$$-2l\left(\hat{\theta}\right) + 2(\text{number of model parameters}).$$

Essentially, we compare the AIC scores for different models and then select the one with the smallest AIC score.

Another popular criterion for model selection among parametric models is the Bayesian information criterion (BIC). The BIC is given by

$$-2l\left(\hat{\theta}\right) + (\text{number of model parameters})\log(\text{sample size}),$$

and in the same way as with AIC scores, we select the one with the smallest BIC value.

On the other hand, to describe how well a model fits the observed data, we can do tests of goodness-of-fit for the estimated survival function. Such tests summarize the discrepancy between observed values and the expected values for the survival function under the model. We will use two well-known statistics for this purpose, the first one is the lack of fit sum of squares (SS) given by

$$\sum_i (\text{observed value}_i - \text{fitted value}_i)^2,$$

and the second is the Kolmogorov–Smirnov statistic (KS) defined as

$$\max_i |\text{observed value}_i - \text{fitted value}_i|.$$

4 Numerical Illustration

We illustrate the use of the family of proposed models by analyzing *death times of male laryngeal cancer patients*. Kardaun [5] reported data on 90 males diagnosed with cancer of the larynx during the period 1970–1978 at a Dutch hospital. Times recorded were the intervals (in years) between the first treatment and either death or the end of the study. Also recorded were the patient's age at the time of diagnosis and the stage of the patient's cancer, wherein the stage is a factor of four levels. The larynx data have been used by Klein and Moeschberger [6] to illustrate some techniques in survival analysis. The larynx data can be obtained from the `MKsurv` Package of the `R` software package.

4.1 Fit of a Fully Parametric Model

First, we fit the fully parametric proportional hazards model by means of three specific models, namely, the Weibull proportional hazards model (WPH), the Weibull logit link proportional hazards model (WLPH), and the WGLPH. To get a good fit of these models, we do a grid $\{0.5, 1.0, \ldots, 4.5, 5.0\} \times \{0.5, 1.0, \ldots, 4.5, 5.0\}$ for the initial values of the parameters λ and γ for the required maximization of the three specified models. Moreover, $a = 1$ is used as the initial value for the parameter a of the WGLPH model.

In the fitting of the WGLPH model, we found that the numerical methods, used to determining the maximum, in fact, find local maxima that can be far away from the global maximum. We therefore adopt a profile full log-likelihood method to solve this problem.

Let the profile log-likelihood function be

$$p\,l\,(a) = \sup_{\lambda,\gamma,\boldsymbol{\beta}} l\,(\lambda, \gamma, \boldsymbol{\beta}, a)\,.$$

We then calculate this function for $a \in \{5.0, 6.0, 7.0, 8.0, 9.0\}$ and the same grid for initial points of λ and γ as mentioned earlier. Finally, we found the value of $a = 8$ to be the choice of a that maximized the profile full log-likelihood (WGLPHaF). The parameter estimates (Est) and their standard errors (SE) so obtained for the four fitted models are presented in Table 1.

Now, we compare the empirical survival function with the estimated survival function for each of the four fitted models. The empirical survival function was calculated by Kaplan–Meier estimator, while the survival function of the WLPH, WGLPHaF, and WGLPH models were estimated with the corresponding function (3) of the parametric model. In Fig. 1, we have shown a plot of these estimated survival functions. To assess the goodness-of-fit of these models, we calculated the SS and KS statistics to compare the empirical survival function with the corresponding estimated survival function for all four models (WPH, WLPH, WGLPHaF, and WGLPH).

Table 1 Fit of the four fully parametric models

Variable	WPH Est	WPH SE	WLPH Est	WLPH SE	WGLPHaF ($a = 8$ fixed) Est	WGLPHaF ($a = 8$ fixed) SE	WGLPH Est	WGLPH SE
λ	0.01855	0.01893	0.32468	0.11235	3.16696	4.94242	2.56534	6.59244
γ	1.13020	0.13843	1.09364	0.13413	1.13038	0.13832	1.13034	0.13847
Age	0.01973	0.01420	-0.01638	0.00810	0.00740	0.00751	0.00785	0.00755
Stage II	0.16633	0.46111	-0.04115	0.60885	0.04320	0.15510	0.04406	0.16261
Stage III	0.66255	0.35545	0.75535	0.56191	0.23229	0.16294	0.24550	0.21569
Stage IV	1.74530	0.41471	11.26756	96.80031	0.73511	0.41099	0.79024	0.75908
a	—	—	—	—	—	—	7.74186	3.94820

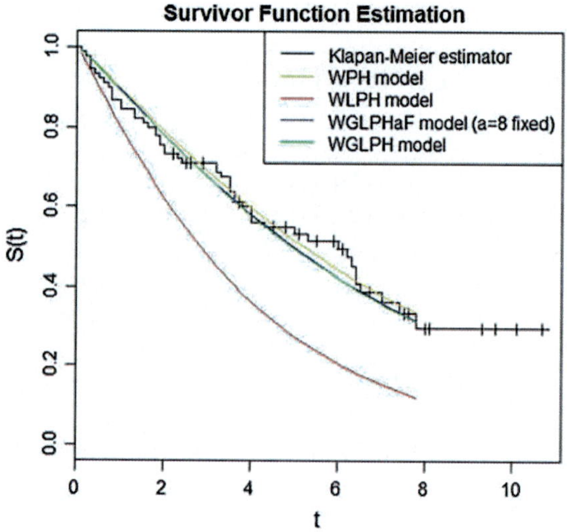

Fig. 1 Survival functions of the four fitted parametric models and the Kaplan–Meier estimator

In Table 2, the values of the SS and KS statistics obtained for these four fitted models are presented.

Also, in Table 3, two comparisons of these four models are made based on the (AIC) and the BIC.

All these results show that the WGLPH, with the parameter a determined by the profile likelihood method, provides overall the best fitting model for the considered data.

Table 2 Goodness-of-fit statistics for the four fitted fully parametric models

	WPH	WLPH	WGLPHaF	WGLPH
SS	0.035	1.034	0.034	0.034
KS	0.057	0.289	0.071	0.073

Table 3 AIC and BIC values for the four fitted parametric models

	WPH	WLPH	WGLPHaF	WGLPH
AIC	294.8468	300.4143	294.7641	296.7652
BIC	309.8456	315.4131	309.7630	314.2639

4.2 Fit of a Semiparametric Model

Now, we fit the proportional hazards model (PH), the logit link proportional hazards model (LPH), and the generalized logit link proportional hazards model (GLPH) in a semiparametric framework. In order to get the best fit for the GLPH model, we do a grid for the initial value of the parameter a in $\{0.5, 1.0, 1.5, 2.0\}$ and we look for the maximum value of the partial log-likelihood function in (6).

As in the parametric case, we consider a profile partial log-likelihood method maximizing the function

$$p\,l\,(a) = \sup_{\beta} l\,(\beta, a)\,.$$

We calculate this function for $a \in \{1, 2, \ldots, 8, 9, 10\}$ in GLPH with a fixed and we get $a = 8$ that maximizes the profile partial log-likelihood function (GLPHaF). In Table 4, we present the parameter estimates (Est) and their standard errors (SE) for the four fitted models. Note that we obtain similar estimation for GLPHaF ($a = 8$ fixed) and GLPH models, but we reduce substantially the SE in GLPHaF, as the parameter a is fixed in this case. Furthermore, we compare the empirical survival function and the estimated survival function for each of the four models. The empirical survival function was calculated by Kaplan–Meier estimator, the Breslow's estimator was used to estimate the survival function in Cox PH model, while the survival function of LPH, GLPHaF, and GLPH models were estimated with the corresponding function $S(t)$ of the semiparametric model in (7). In Fig. 2, we have presented a plot of these estimated survival functions. In Table 5 we present the values of the SS and KS statistics to compare the empirical survival function and the estimated survival function for each of the four fitted models (PH, LPH, GLPHaF and GLPH).

Table 4 Fit of the four semiparametric models

	PH		LPH		GLPHaF ($a = 8$ fixed)		GLPH	
Variable	Est	SE	Est	SE	Est	SE	Est	SE
Age	0.01890	0.01425	-0.01568	0.00828	0.00570	0.00565	0.00574	0.02170
Stage II	0.13856	0.46231	-0.07464	0.61227	0.03047	0.12829	0.03060	0.14502
Stage III	0.63835	0.35608	0.75403	0.57134	0.18337	0.12140	0.18453	0.67040
Stage IV	1.69306	0.42221	7.93424	19.34429	0.55128	0.23583	0.55537	2.35490
a	—	—	—	—	—	—	8.95955	22.95138

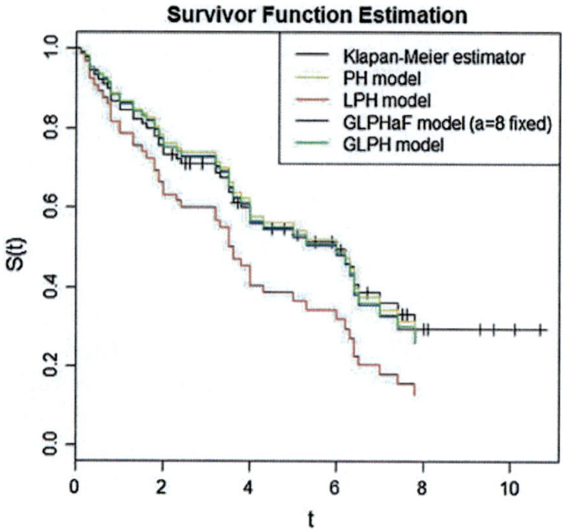

Fig. 2 Survival functions of the four fitted semiparametric models and the Kaplan–Meier estimator

Table 5 Goodness-of-fit statistics for the four fitted semiparametric models

	PH	LPH	GLPHaF	GLPH
SS	0.014	0.517	0.008	0.011
KS	0.029	0.182	0.033	0.033

Table 6 AIC and BIC values for the four fitted semiparametric models

	PH	LPH	GLPHaF	GLPH
AIC	384.3589	389.2912	384.3173	386.3173
BIC	394.3581	399.2905	394.3165	398.8163

In Table 6, the comparisons of the four fitted models are made based on the AIC and the BIC. From all these results, we draw the general conclusion that the generalized logit-link proportional hazards models are good competitors for the Cox model.

5 Final Remarks

The family of proposed models is quite flexible and seems to provide a good competitor for the Cox model. We derive the likelihood function and the partial likelihood function for the parametric and semiparametric models, respectively, for obtaining the parameter estimates and their standard errors. The estimated survival function of a member of the proposed family of models fits the empirical survival function better than the Cox model. Furthermore, this generalized logit-based proportional hazards model is the one with minimum AIC and BIC.

There are still some unresolved issues in this regard. First of all, the asymptotic properties of the parameter estimates have to be established, and associated statistical inferential issues need to be studied in detail. Another problem of interest is to introduce time-dependence in the models. In this case, the proposed models will provide an extension of [7, 8] models in two ways, with one coming from the non-constant hazard function and the other arising from the generalized logit-link function.

Appendix A: Derivatives of the Log-Likelihood Function in a Fully Parametric Model

To maximize (5), we obtain its first derivatives with respect to all the parameters as

$$\frac{\partial l\left(\lambda,\gamma,\boldsymbol{\beta},a\right)}{\partial\lambda} = \sum_{i=1}^{n}\left[\frac{\delta_i}{\lambda} - K\left(\boldsymbol{\beta},a,\mathbf{x}_i\right)t_i^{\gamma}\right],$$

$$\frac{\partial l\left(\lambda,\gamma,\boldsymbol{\beta},a\right)}{\partial\gamma} = \sum_{i=1}^{n}\left[\delta_i\left(\frac{1}{\gamma}+\ln(t_i)\right) - \lambda K\left(\boldsymbol{\beta},a,\mathbf{x}_i\right)t_i^{\gamma}\ln(t_i)\right],$$

$$\frac{\partial l\left(\lambda,\gamma,\boldsymbol{\beta},a\right)}{\partial\beta_j} = \sum_{i=1}^{n}\left[\frac{a x_{ij}\left(\delta_i - \lambda K\left(\boldsymbol{\beta},a,\mathbf{x}_i\right)t_i^{\gamma}\right)}{1+\exp(\mathbf{x}_i'\boldsymbol{\beta})}\right],$$

for $j = 1,\ldots,p$, and

$$\frac{\partial l\left(\lambda,\gamma,\boldsymbol{\beta},a\right)}{\partial a} = \sum_{i=1}^{n}\left[\ln\left(1+\exp\left(-\mathbf{x}_i'\boldsymbol{\beta}\right)\right)\left(\lambda K\left(\boldsymbol{\beta},a,\mathbf{x}_i\right)t_i^{\gamma} - \delta_i\right)\right].$$

To obtain the corresponding information matrix $I(\lambda,\gamma,\boldsymbol{\beta},a)$, we need the Hessian matrix $H(\lambda,\gamma,\boldsymbol{\beta},a)$ which is the matrix of second derivatives of the log-likelihood function in (5) with respect to its parameters.

We obtain them readily as follows:

$$\frac{\partial^2 l\left(\lambda,\gamma,\boldsymbol{\beta},a\right)}{\partial\lambda^2} = -\frac{1}{\lambda^2}\left[\sum_{i=1}^{n}\delta_i\right],$$

$$\frac{\partial^2 l\left(\lambda,\gamma,\boldsymbol{\beta},a\right)}{\partial\lambda\partial\gamma} = -\sum_{i=1}^{n}K\left(\boldsymbol{\beta},a,\mathbf{x}_i\right)t_i^{\gamma}\ln(t_i),$$

$$\frac{\partial^2 l(\lambda,\gamma,\boldsymbol{\beta},a)}{\partial\lambda\partial\beta_j} = -a\sum_{i=1}^{n}\frac{K(\boldsymbol{\beta},a,\mathbf{x}_i)\,t_i^{\gamma}x_{ij}}{1+\exp(\mathbf{x}_i'\boldsymbol{\beta})}, \qquad \text{with } j=1,\cdots,p,$$

$$\frac{\partial^2 l(\lambda,\gamma,\boldsymbol{\beta},a)}{\partial\lambda\partial a} = \sum_{i=1}^{n}K(\boldsymbol{\beta},a,\mathbf{x}_i)\,t_i^{\gamma}\ln\left(1+\exp(-\mathbf{x}_i'\boldsymbol{\beta})\right),$$

$$\frac{\partial^2 l(\lambda,\gamma,\boldsymbol{\beta},a)}{\partial\gamma^2} = \sum_{i=1}^{n}\left[-\frac{\delta_i}{\gamma^2} - \lambda K(\boldsymbol{\beta},a,\mathbf{x}_i)\,t_i^{\gamma}(\ln(t_i))^2\right],$$

$$\frac{\partial^2 l(\lambda,\gamma,\boldsymbol{\beta},a)}{\partial\gamma\partial\beta_j} = -a\lambda\sum_{i=1}^{n}\frac{K(\boldsymbol{\beta},a,\mathbf{x}_i)\,t_i^{\gamma}\ln(t_i)x_{ij}}{1+\exp(\mathbf{x}_i'\boldsymbol{\beta})}, \qquad \text{with } j=1,\cdots,p,$$

$$\frac{\partial^2 l(\lambda,\gamma,\boldsymbol{\beta},a)}{\partial\gamma\partial a} = \lambda\sum_{i=1}^{n}K(\boldsymbol{\beta},a,\mathbf{x}_i)\,t_i^{\gamma}\ln(t_i)\ln\left(1+\exp(-\mathbf{x}_i'\boldsymbol{\beta})\right),$$

for $j=1,\ldots,p$ and $l=j,\cdots,p$,

$$\frac{\partial^2 l(\lambda,\gamma,\boldsymbol{\beta},a)}{\partial\beta_j\partial\beta_l} = -a\sum_{i=1}^{n} \times\left[\frac{\delta_i\exp(\mathbf{x}_i'\boldsymbol{\beta})x_{ij}x_{il} + \lambda K(\boldsymbol{\beta},a,\mathbf{x}_i)\,t_i^{\gamma}x_{ij}x_{il}\left(a-\exp(\mathbf{x}_i'\boldsymbol{\beta})\right)}{\left(1+\exp(\mathbf{x}_i'\boldsymbol{\beta})\right)^2}\right],$$

$$\frac{\partial^2 l(\lambda,\gamma,\boldsymbol{\beta},a)}{\partial\beta_j\partial a} = \sum_{i=1}^{n}\left[\frac{\delta_i x_{ij} - \lambda K(\boldsymbol{\beta},a,\mathbf{x}_i)\,t_i^{\gamma}x_{ij}\left\{1-a\ln\left(1+\exp(-\mathbf{x}_i'\boldsymbol{\beta})\right)\right\}}{1+\exp(\mathbf{x}_i'\boldsymbol{\beta})}\right],$$

$$\frac{\partial^2 l(\lambda,\gamma,\boldsymbol{\beta},a)}{\partial a^2} = -\lambda\sum_{i=1}^{n}K(\boldsymbol{\beta},a,\mathbf{x}_i)\,t_i^{\gamma}\left\{\ln\left(1+\exp(-\mathbf{x}_i'\boldsymbol{\beta})\right)\right\}^2.$$

Appendix B: Derivatives of the Partial Log-Likelihood Function in the Semiparametric Model

To maximize (6), we obtain its first derivatives with respect to all the parameters as

$$\frac{\partial l(\boldsymbol{\beta},a)}{\partial\beta_j} = a\sum_{i=1}^{n}\delta_i\left[\frac{x_{ij}}{1+\exp(\mathbf{x}_i'\boldsymbol{\beta})} - \frac{1}{\sum_{l\in R(t_i)}K(\boldsymbol{\beta},a,\mathbf{x}_l)}\sum_{l\in R(t_i)}\frac{x_{lj}K(\boldsymbol{\beta},a,\mathbf{x}_l)}{1+\exp(\mathbf{x}_l'\boldsymbol{\beta})}\right]$$

for $j=1,\ldots,p$, and

$$\frac{\partial l(\boldsymbol{\beta}, a)}{\partial a} = \sum_{i=1}^{n} \delta_i \left[-\ln\left(1+\exp\left(-\mathbf{x}_i'\boldsymbol{\beta}\right)\right) \right.$$

$$\left. + \frac{1}{\sum_{l\in R(t_i)} K\left(\boldsymbol{\beta}, a, \mathbf{x}_l\right)} \sum_{l\in R(t_i)} K\left(\boldsymbol{\beta}, a, \mathbf{x}_l\right) \ln\left(1+\exp\left(-\mathbf{x}_l'\boldsymbol{\beta}\right)\right) \right].$$

Now, to obtain the information matrix, we need the second derivatives of the partial log-likelihood function (6) for $j = 1, \cdots, p$ and $m = j, \cdots, p$, which are as follows:

$$\frac{\partial^2 l\left(\boldsymbol{\beta}, a\right)}{\partial \beta_j \partial \beta_m} = a \sum_{i=1}^{n} \delta_i \left[\frac{-\exp(\mathbf{x}_i'\boldsymbol{\beta}) x_{ij} x_{im}}{\left(1+\exp(\mathbf{x}_i'\boldsymbol{\beta})\right)^2} - \frac{a v_m(t_i) v_j(t_i) - w(t_i) y_{jm}(t_i)}{(w(t_i))^2} \right],$$

$$\frac{\partial^2 l\left(\boldsymbol{\beta}, a\right)}{\partial \beta_j \partial a} = \sum_{i=1}^{n} \delta_i \left[\frac{x_{ij}}{1+\exp(\mathbf{x}_i'\boldsymbol{\beta})} - \frac{v_j(t_i)}{w(t_i)} - a\frac{v_j(t_i)u(t_i)}{(w(t_i))^2} + a\frac{z_j(t_i)}{w(t_i)} \right],$$

$$\frac{\partial^2 l\left(\boldsymbol{\beta}, a\right)}{\partial a^2} = \sum_{i=1}^{n} \delta_i \left[\frac{-(u(t_i))^2}{(w(t_i))^2} + \frac{\sum_{l\in R(t_i)} K\left(\boldsymbol{\beta}, a, \mathbf{x}_l\right) \left\{\ln\left(1+\exp(-\mathbf{x}_l'\boldsymbol{\beta})\right)\right\}^2}{w(t_i)} \right],$$

where

$$w(t) = \sum_{l\in R(t)} K\left(\boldsymbol{\beta}, a, \mathbf{x}_l\right),$$

$$v_j(t) = \sum_{l\in R(t)} \frac{K\left(\boldsymbol{\beta}, a, \mathbf{x}_l\right) x_{lj}}{1+\exp(\mathbf{x}_l'\boldsymbol{\beta})},$$

$$y_{jm}(t) = \sum_{l\in R(t)} \frac{K\left(\boldsymbol{\beta}, a, \mathbf{x}_l\right) x_{lj} x_{lm} \left(a - \exp(\mathbf{x}_l'\boldsymbol{\beta})\right)}{\left(1+\exp(\mathbf{x}_l'\boldsymbol{\beta})\right)^2},$$

$$u(t) = \sum_{l\in R(t)} K\left(\boldsymbol{\beta}, a, \mathbf{x}_l\right) \ln\left(1+\exp(-\mathbf{x}_l'\boldsymbol{\beta})\right),$$

$$z_j(t) = \sum_{l\in R(t)} \frac{K\left(\boldsymbol{\beta}, a, \mathbf{x}_l\right) x_{lj} \left\{\ln\left(1+\exp(-\mathbf{x}_l'\boldsymbol{\beta})\right)\right\}}{1+\exp(\mathbf{x}_l'\boldsymbol{\beta})}.$$

Acknowledgments This work was supported by research grant MTM2009-06997 and MAEC-AECID.

References

1. Balakrishnan N (ed) (1992) Handbook of the Logistic Distribution. Marcel Dekker, New York
2. Cheng SC, Wei LJ, Ying Z (1995) Analysis of transformation models with censored data. Biometrika 82:835–845
3. Cox DR (1972) Regression models and life-tables. J Roy Stat Soc B 34(2):187–220
4. Etezadi-Amoli J, Ciampi A (1987) Extended hazard regression for censored survival data with covariates: A spline approximation for the background hazard function. Biometrics 43:181–192
5. Kardaun O (1983) Statistical analysis of male larynx-cancer patients: A case study. Stat Nederlandica 37:103–126
6. Klein JP, Moeschberger ML (1997) Survival analysis: techniques for censored and truncated data. Springer-Verlag, New York
7. MacKenzie G (1996) Regression models for survival data: The generalized time-dependent logistic family. Statistician 45:21–34
8. MacKenzie G (1997) On a non-proportional hazards regression model for repeated medical random counts. Stat Med 16:1831–1843
9. Royston P, Parmar MKB (2002) Flexible parametric proportional hazards and proportional odds models for censored survival analysis, with application to prognostic modelling and estimation of treatment effects. Stat Med 21:2175–2197
10. Younes N, Lachin J (1997) Link-based models for survival data with interval and continuous time censoring. Biometrics 53:1199–1211

Design of Reliability Test Plans: An Overview

E. A. Elsayed

Reliability prediction of new components, products, and systems is a difficult task due to the lack of well-designed test plans that yield "useful" information during the test and due to the stochastic nature of the normal operating conditions. The accuracy of the reliability prediction has a major effect on the warranty cost and repair and maintenance strategies. Therefore, it is important to design efficient test plans. In this chapter, we present an overview of reliability testing with emphasis on accelerated testing and address issues associated with the design of optimal test plans, stress application methods, and reliability prediction models. We further discuss the concept of equivalence of test plans and how it could be used for test time reduction. Finally, we present accelerated degradation modeling and the design of accelerated degradation test plans.

1 Introduction

The high rate of technological advances and innovations are spurring the continuous introduction of new products and services. Moreover, the intensity of the global competition for the development of new products in a short time has motivated the development of new methods such as robust design, just-in-time manufacturing, and design for manufacturing and assembly. More importantly, both producers and customers expect the product to perform the intended functions satisfactorily for extended periods of time. Hence, extended warranties and similar assurances of product reliability have become standard features of the product and serve as implied indicators of the product's reliability. Likewise, recalls of products and recent failures of systems, such as air traffic control systems and autos (sudden acceleration

---

E. A. Elsayed
Department of Industrial and Systems Engineering, Rutgers, The State University of New Jersey, 96 Frelinghuysen Road, Piscataway, NJ 08854-8018, USA
e-mail: elsayed@rci.rutgers.edu

T. Dohi and T. Nakagawa (eds.), *Stochastic Reliability and Maintenance Modeling*, Springer Series in Reliability Engineering 9, DOI: 10.1007/978-1-4471-4971-2_2,

and brake failures) and products have emphasized the importance of testing. For example, a recent recall of a popular car is attributed, by the manufacturer, to lack of thoroughness in testing new cars and car parts under varying weather conditions, as demonstrated by the recently recalled gas-pedal mechanism that tended to stick more as humidity increased [40].

Careful reliability testing of systems, products, and components at the design stage is crucial to achieving the desired reliability at the field operating conditions. During the design stage of many products, especially those used in military, the elimination of design weaknesses inherent to intermediate prototypes of complex systems is conducted via the test, analyze, fix, and test (TAFT) process. This process is generally referred to as "reliability growth." Specifically, reliability growth is the improvement in the true but unknown initial reliability of a developmental item as a result of failure mode discovery, analysis, and effective correction. Corrective actions generally assume the form of fixes, adjustments, or modifications to problems found in the hardware, software, or human error aspects of a system [20]. Likewise, field test results are used in improving product design and consequently its reliability.

The above examples and requirements have magnified the need for providing more accurate estimates of reliability by performing testing of materials, components, and systems at different stages of product development.

There is a wide variety of reliability testing methodologies and objectives. They include testing to determine the potential failure mechanisms, reliability demonstration testing, reliability acceptance testing, reliability prediction testing using accelerated life testing (ALT), and others. This chapter focuses on ALT, reliability prediction models and the design of the ALT plans.

Testing under normal operating conditions requires a very long time especially for components and products with long expected lives, and it requires extensive number of test units, so it is usually costly and impractical to perform reliability testing under normal conditions.

In many cases, ALT might be the only viable approach to assess whether the product meets the expected long-term reliability requirements. ALT experiments can be conducted using three different approaches. The first is conducted by accelerating the "use" of the unit at normal operating conditions such as in cases of products that are used only a fraction of a time in a typical day which includes home appliances and auto tires. The second is conducted by subjecting a sample of units to stresses severer-than-normal operating conditions in order to accelerate the failure. The third is conducted by subjecting units that exhibit some type of degradation such as stiffness of springs, corrosions of metals, and wear out of mechanical components to accelerated stresses. The last approach is referred to as accelerated degradation testing (ADT).

The reliability data obtained from the experiments are then utilized to construct a reliability model for predicting the reliability of the product under normal operating conditions through a statistical and/or physics-based inference procedure. The accuracy of the inference procedure has a profound effect on the reliability estimates and the subsequent decisions regarding system configuration, warranties, and preventive maintenance schedules. Specifically, the reliability estimate depends on two factors, the ALT model and the experimental design of the ALT test plans. A "good" model can provide an appropriate fit to testing data and results in achieving accurate estimates at the normal conditions. Likewise, an optimal design of the test plans, which determines the stress loadings (constant-stress, ramp-stress, cyclic-stress, ...), allocation of test units number stress level, optimum test duration, and other experimental variables, can indeed improve the accuracy of the reliability estimates. Indeed, without an optimum test plan, it is likely that a sequence of expensive and time-consuming tests results in inaccurate reliability estimates. This might also cause delays in product release, or the termination of the entire product as has been observed by the author.

We describe briefly the methods of stress application, types of stresses, and focus on the reliability prediction models that utilize the failure data at stress conditions to obtain reliability information at normal conditions. We begin by describing the three important methods including two of the most commonly used prediction models that relate the test results at stress conditions to failure rate at the normal operating conditions.

2 Reliability Prediction Models Using ALT Data

Many ALT models have been developed and successfully implemented in a variety of engineering applications. The important assumption for relating the accelerated failures to those at normal operating conditions is that the components/products operating at the normal conditions experience the same failure mechanisms as those at the accelerated conditions. Elsayed [14] classifies the existing ALT models into three categories: *statistics-based models*, *physics-statistics-based models*, and *physics-experimental-based models*, as shown in Fig. 1. In particular, the statistics-based models are generally used when the relationship between the applied stresses and the failure time of the product is difficult to determine based on physics or chemistry principles. In this case, accelerated failure times are used to determine the model parameters statistically after assuming either a linear or nonlinear life-stress relationship.

The statistics-based models can be further classified into parametric models and semiparametric/nonparametric models. The most commonly used failure time distributions in the parametric models are the exponential, Weibull, normal, lognormal, gamma, and extreme value distributions. The underlying assumption of these models is that the failure times of the products follow the same distributions at different stress levels. In reality, however, when the failure process involves complex and/or inconsistent failure time distributions, the parametric models may not interpret the data satisfactorily and the reliability prediction will be far from accurate. Consequently, semiparametric or nonparametric models appear to be attractive and more suitable for reliability estimation due to their "distribution-free" property. We briefly review the two most commonly used ALT models as they will be used in the design

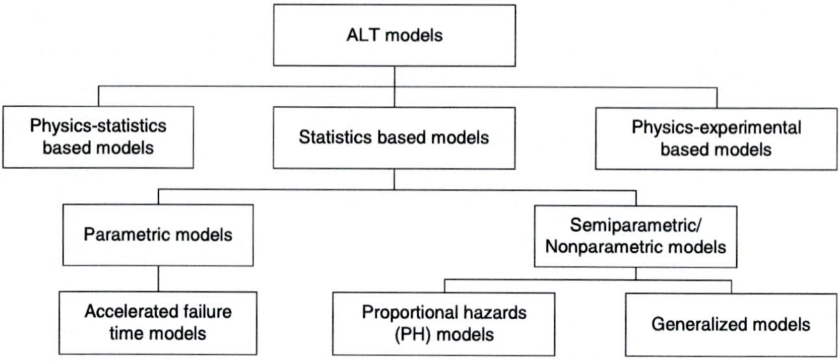

Fig. 1 Classification of ALT models [14]

of the ALT plans and describe a third model which relaxes the assumptions of the two models.

2.1 Proportional Hazards Model

Multiple regression models can be used to predict the time to failure (TTF) of a component under multiple covariates. A similar regression-based model that is widely used is the *proportional hazards* (PH) model introduced by Cox [8]. The PH model is generally expressed as:

$$\lambda(t; z) = \lambda_0(t)\exp(\boldsymbol{\beta} z)$$

where $z = (z_1, z_2, ...z_p)^T$ is a column vector of covariates (for ALT, it is the column vector of stresses and/or their interactions that components experience). $\boldsymbol{\beta} = (\beta_1, \beta_2, ...\beta_p)$ is a row vector of the unknown coefficients. $\lambda_0(t)$ is a baseline hazard rate function. Unlike standard regression models, the PH models assume that the applied stresses act multiplicatively, rather than additively, on the hazard rate—a much more realistic assumption in many cases [11, 16, 18]. The PH model is a class of models with the property that the hazard functions of two units at two different stress levels z_1 and z_2 are proportional to each other. In other words, the ratio of their hazard rates does not vary with time.

One of the advantages of the PH model is the ability to include time-dependent covariates. Let $z_i(t)$ be the covariate vector at time t for the ith individual unit under study, then the associated hazard rate function can be expressed as:

$$\lambda(t; z_i(t)) = \lambda_0(t)\exp(\boldsymbol{\beta} z_i(t))$$

where the hazard rate at time t depends only on the current stress level $z_i(t)$, and there is no effect caused by the previous stress history.

2.2 Accelerated Failure Time Models

Another widely used class of ALT models is *accelerated failure time* (AFT) models. For many products, there are well-established acceleration models that perform satisfactorily over the desired range of stresses. For instance, for temperature accelerated testing, the Arrhenius model has gained acceptance because of its many successful applications and general agreement of laboratory test results with long-term field performance. In an AFT model, it is assumed that for a unit under the applied stress vector z, the log-lifetime $Y = \log T$ has a distribution with a location parameter $\mu(z)$ depending on the stress vector z, and a constant scale parameter $\sigma > 0$ in the form of:

$$Y = \log T = \mu\,(z) + \sigma\varepsilon$$

where ε is a random variable whose distribution does not depend on z. The location parameter $\mu(z)$ follows some assumed life-stress relationship, e.g., $\mu(z_1, z_2) = \theta_0 + \theta_1 z_1 + \theta_2 z_2$, where z_1 and z_2 are some known functions of stresses. The popular Inverse Power law and Arrhenius model are special cases of this simple life-stress relationship. The AFT models assume that the covariates act multiplicatively on the failure time, or linearly on the log failure time, rather than multiplicatively on the hazard rate. The hazard function in the AFT model can be written in terms of the baseline hazard function $\lambda_0(\cdot)$ as:

$$\lambda(t; z) = \lambda_0(e^{\beta z} t) e^{\beta z}$$

The main assumption of the AFT models is that the TTFs are inversely proportional to the applied stresses, e.g., the TTF at high stress is shorter than the TTF at low stress. It also assumes that the failure time distributions are of the same type. In other words, if the failure time distribution at the higher stress is exponential then the distribution at the low stress is also exponential. Therefore, a general cumulative distribution function CDF for a two-parameters Weibull distribution under an applied stress vector z is

$$F(t; z) = 1 - \exp\left(-\left(\frac{t}{\theta\,(z)}\right)^{\beta}\right)$$

where β is the shape parameter and $\theta(z)$ is the scale parameter as a function of applied stresses which can be expressed as $\theta(z) = \theta_0 + \sum_{i=1}^{n} \theta_i z_i$, where θ_i is a coefficient of the covariate z_i. We illustrate the use of Weibull distribution for the true linear acceleration case in which the scale parameter at normal conditions θ_o is linearly related to the scale parameters at accelerated conditions θ_s using an acceleration factor A_F. The relationship between the failure time distributions at the accelerated and normal conditions can be derived as

$$F_s(t) = 1 - e^{-(\frac{t}{\theta_s})^{\beta_s}} \quad t \geq 0, \beta_s \geq 1, \theta_s > 0 \tag{1}$$

where β_s is the shape parameter of the Weibull distribution at stress conditions. The CDF at normal operating conditions is:

$$F_o(t) = F_s\left(\frac{t}{A_F}\right) = 1 - e^{-\left(\frac{t}{A_F\theta_s}\right)^{\beta_s}} = 1 - e^{-\left(\frac{t}{\theta_o}\right)^{\beta_o}} \tag{2}$$

As stated earlier, the underlying failure time distributions at both the accelerated stress and operating conditions have the same shape parameters, i.e., $\beta_s = \beta_o$, and $\theta_o = A_F\theta_s$. If the shape parameters at different stress levels are significantly different, then either the assumption of true linear acceleration is invalid or the Weibull distribution is inappropriate to use for analysis of such data.

Let $\beta_s = \beta_o = \beta \geq 1$. Then the probability density function at normal operating conditions is

$$f_o(t) = \frac{\beta}{A_F\theta_s}\left(\frac{t}{A_F\theta_s}\right)^{\beta-1} e^{-\left(\frac{t}{A_F\theta_s}\right)^{\beta}} \quad t \geq 0, \theta_s \geq 0 \tag{3}$$

The MTTF at normal operating conditions is

$$MTTF_o = \theta_o^{\frac{1}{\beta}}\Gamma\left(1 + \frac{1}{\beta}\right) \tag{4}$$

The failure rate at normal operating conditions is

$$\lambda_o(t) = \frac{\beta}{A_F\theta_s}\left(\frac{t}{A_F\theta_s}\right)^{\beta-1} = \frac{\lambda_s(t)}{A_F^{\beta}} \tag{5}$$

2.3 *Extended Linear Hazard Regression Model*

The PH and AFT models have very different assumptions (failure rate proportionality or failure time proportionality, respectively). The only model that satisfies both assumptions is the Weibull model. Assuming PH or AFT for a particular data set may lead to different results. Therefore, a simultaneous treatment of the two is of practical importance especially when the assumption regarding the PH or AFT is difficult to justify or does not hold. Ciampi and Etezadi-Amoli [6] propose the *extended hazard regression* (EHR) model which encompasses both the PH and AFT models as special cases. To further enhance the capability of modeling ALT, Elsayed et al. [18] propose a more generalized model - the *extended linear hazard regression* (ELHR) model by incorporating the time-varying coefficient effect into the EHR model. The ELHR model is expressed as:

$$\lambda(t; z) = \lambda_0(te^{(\beta_0+\beta_1 t)z})e^{(\alpha_0+\alpha_1 t)z} \tag{6}$$

The ELHR model encompasses all previous models—PH, AFT, and EHR as special cases. It incorporates the time-changing effects, proportional hazard effects, as well as time-varying coefficient effects into one model. The ELHR model outperforms the PH model and other extended models (e.g., Shyur et al. [35]) in that it can better interpret physical failure processes thus providing a better model fit to the corresponding failure time data. Furthermore, the ELHR model is essentially "distribution-free", and thus has a significant potential of dealing with complex failure processes. For example, by assuming the baseline hazard function $\lambda_0(\cdot)$ to be a quadratic function $\lambda_0(u) = \gamma_0 + \gamma_1 u + \gamma_2 u^2$, the model can be expressed as:

$$\lambda(t; z) = \gamma_0 e^{(\alpha_0+\alpha_1 t)z} + \gamma_1 t e^{(\theta_0+\theta_1 t)z} + \gamma_2 t^2 e^{(\omega_0+\omega_1 t)z} \tag{7}$$

where $\theta_0 = \alpha_0 + \beta_0$, $\theta_1 = \alpha_1 + \beta_1$, $\omega_0 = \alpha_0 + 2\beta_0$, $\omega_1 = \alpha_1 + 2\beta_1$. Then, the associated reliability is given by

$$\begin{aligned} R(t; z) &= \exp(-\Lambda(t; z)) \\ &= \exp\left(-\int_0^t \gamma_0 e^{(\alpha_0+\alpha_1 t)z} + \gamma_1 t e^{(\theta_0+\theta_1 t)z} + \gamma_2 t^2 e^{(\omega_0+\omega_1 t)z} du\right) \end{aligned}$$

where $\Lambda(t; z)$ is the cumulative hazard rate function. One of the drawbacks of the ELHR model is the number of parameters of the model. As the number increases it is likely that the accuracy of the estimated parameters decreases which might result in inaccurate reliability prediction at normal operating conditions. This drawback becomes more acute when the failure time data are small.

3 Accelerated Life Testing Plans

A detailed test plan is usually designed before conducting an accelerated life test. The plan requires determination of the type of stress, methods of applying stress, stress levels, the number of units to be tested at each stress level, and an applicable ALT model that relates the failure times at accelerated conditions to those at normal conditions. Of course, a clear objective of the test plan needs to be defined. We begin by the type of stresses followed by methods of stress loading.

3.1 Types of Stresses

In order to determine the type of stresses to be applied in ALT it is important to understand the potential failures of the components and the causes of such failures. This is usually based on engineering knowledge of the component's materials, function, and the stresses that induce such failures. A simplified design of experiments approach is usually conducted to study the effect of the type of stresses by using two levels of each stress (low and high). The high level of stress is the highest level that can be applied without causing a different failure mechanism other than that likely to occur at normal operating conditions. Therefore, a clear understanding of the physics of failure is necessary and testing such as highly accelerated life testing (HALT) is conducted to verify the failure mechanism and the magnitude of the highest stress. HALT subjects the test unit to vibration with random mode of frequency coupled with high temperature and shock in order to induce failures. The failure mechanism is investigated and the stress type and its maximum applied levels are determined accordingly.

In general, the type of applied stresses depends on the intended operating conditions of the product and the potential cause of failure.

We classify the types of stresses as:

1. Mechanical Stresses: *Fatigue* stress is the most commonly used accelerated test for mechanical components. Fatigue is the cause of failures of all rotating mechanical components. When the components are subject to elevated temperature, then *creep* testing (which combines both temperature and static or dynamic loads) should be applied. *Shock* and vibration testing is suitable for components or products subject to such conditions as in the case of bearings, shock absorbers, cell phones, tires, and circuit boards in airplanes and automobiles. Corrosion is another cause of failure of most ferrous material and is induced due to *humidity* and corrosive environment. Units that are subject to corrosion should then be tested using humidity and other corrosive environments as a stress. Wear out is another cause of moving mechanical parts. Depending on the actual use of the unit at normal operating conditions an accelerated test that mimics these conditions needs to be designed but with increased loads to cause significant wear out of the unit.
2. Electrical Stresses: These include power cycling, electric field, current density, and electromigration. Electric field is one of the most common electrical stresses as it induces failures in relatively short times as well as its effect is significantly higher than other types of stresses. Thermal fatigue which is induced by temperature cycling is another major cause of failure of electronic components.
3. Environmental Stresses: Temperature and thermal cycling are commonly used for most products. As stated earlier, it is important to use appropriate stress levels that do not induce different failure mechanisms than those at normal conditions. Humidity is as critical as temperature but its application usually requires a very long time before its effect is noticed. Other environmental stresses include ultraviolet light which affects the strength of elastomers, sulfur dioxide which causes corrosion in circuit boards, salt and fine particles and alpha rays which cause the failure of the read access memory (RAM) and similar components. Likewise, high levels of ionizing can cause electrons in outer orbits to be free which results in electronic noise and signal spikes in digital circuits. Therefore, radiation is an environmental stress that should be applied to the units subject to deployment in space and other similar environments.

3.2 Stress Loadings

Traditionally, ALT is conducted under constant stresses during the entire test duration. The test results are used to extrapolate the product life at normal conditions. In practice, constant-stress tests are easier to carry out but need more test units and a long time at low stress levels to yield sufficient degradation or failure data. However, in many cases the available number of test units and test duration are extremely limited. This has prompted the industry to consider different types of stress loading. Figure 2 shows examples of various stress loadings as well as their adjustable parameters. Some of these stress loadings have been widely utilized in ALT experiments. For instance, static-fatigue tests and cyclic-fatigue tests [23] have been frequently performed on optical fibers to study their reliability; dielectric-breakdown of thermal oxides [18] have been studied under elevated constant electrical fields and temperatures; the lifetime of ceramic components subject to slow crack growth due to stress corrosion have been investigated under cyclic stress by NASA [7]. These stress loadings are selected because of the ease and convenience of statistical analyses and familiarity of the existing analytical tools and industrial routines without following a systematic refinement procedure. Due to tight budgets and time constraints, there is an increasing need to determine the best stress loading in order to shorten the test duration and reduce the total cost while achieving an accurate reliability estimate. In the literature, most research has been focused on the design of optimum test plans when the stress loadings are given. However, until recently, fundamental research on the equivalency of these tests has not yet been investigated in reliability engineering literature. Without the understanding of such equivalency, it is difficult, if not impossible, for a test engineer to determine the best experimental settings before conducting actual ALT.

Furthermore, as is often the case, products are usually exposed to multiple stresses in actual use such as temperature, humidity, electric current, electric field, and various types of shocks and vibration. A typical example is automotive electronics located under the hood, where significant temperature fluctuation, vibration, corrosive gases, and dust contribute to various types of degradation leading to failures, such as cracks in solder joints, loss of connection of connectors, and sensor degradation. It is of interest to know with high confidence what the mileage of normal driving conditions is equivalent to each hour on test under accelerated conditions. Likewise, cellular phones are subject to different environmental conditions, shocks, and vibration. To study the reliability of such products, it is required to subject test units to multiple stresses simultaneously in ALT experiments. For constant-stress tests, it might not be difficult to extend the statistical methods for the design of optimum test plans for single stress to multiple stress scenarios. However, many practical and theoretical issues have to be dealt with when time-varying stresses such as step-stresses are considered. In a multi-stress multi-step test, when and in what order the levels of the stresses should be changed become challenging and unsolved problems. Figure 3 illustrates two example experimental settings out of thousands of choices as one can imagine in conducting a multi-stress multi-step ALT. In general, an arbitrary selection

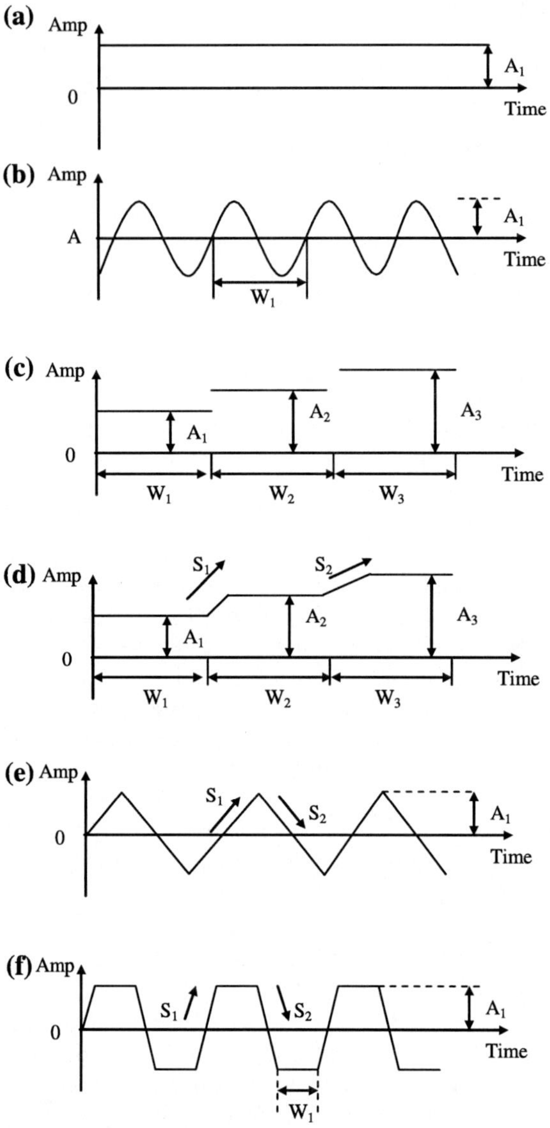

Fig. 2 Various loadings of a single type of stress; the vertical axis shows the amplitude of the applied stress. **a** Constant-stress. **b** Sinusoidal-cyclic-stress. **c** Step-stress. **d** Ramp-step-stress. **e** Triangular-cyclic-stress. **f** Ramp-soak-cyclic-stress

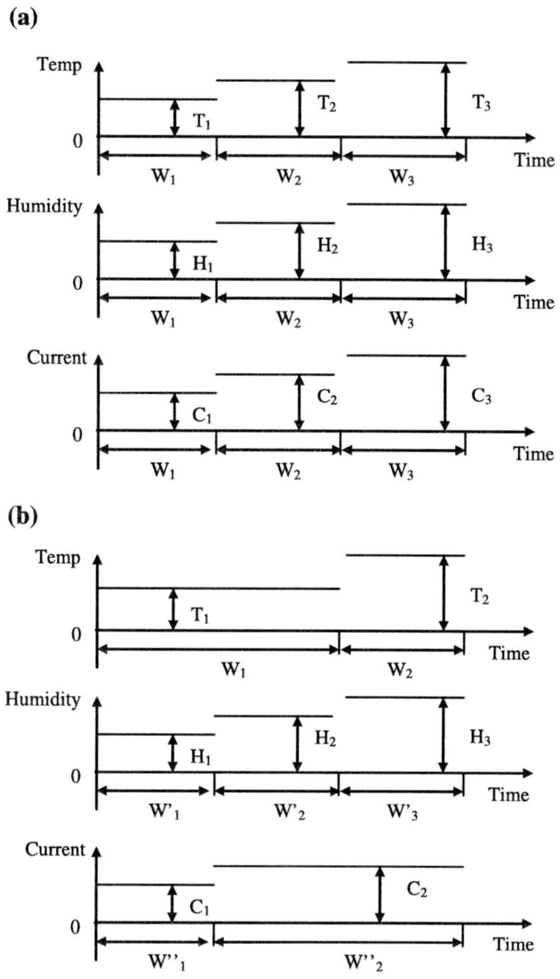

Fig. 3 Two example settings of an ALT involving temperature, humidity, and electric current. **a** Setting 1. **b** Setting 2

from combinations of multiple stress profiles may not result in accurate reliability estimates, especially when the effects of the stresses on the reliability of the product are highly correlated. Therefore, methods for tuning the high-dimensional decision variables under the constraints in time and cost need to be carefully researched and investigated.

3.3 Design of ALT Plans

An ALT plan requires the determination of the type of stress, method of applying stress, stress levels, the number of units to be tested at each stress level, and an applicable ALT model that relates the failure times at accelerated conditions to those at normal conditions.

When designing an *ALT*, we need to address the following issues: (a) Select the stress types to use in the experiment; (b) Determine the stress levels for each stress type selected; and (c) Determine the proportion of devices to be allocated to each stress level Elsayed and Jiao [17] and Elsayed [15]. In this chapter, we present an approach for designing test plans. We refer the reader to Meeker and Escobar [25], Escobar and Meeker [19] and Nelson [28–30] for other approaches for the design of ALT plans.

We consider the selection of the stress level z_i and the proportion of units p_i to allocate for each z_i such that the most accurate reliability estimate at use conditions z_D can be obtained. We consider two types of censoring: Type I censoring involves running each test unit for a prespecified time. The censoring times are fixed and the number of failures is random. Type II censoring involves simultaneously testing units until a prespecified number fails. The censoring time is random while the number of failures is fixed. We define the following notations:

ln	natural logarithm
ML	maximum likelihood
n	total number of test units
z_H, z_M, z_L	high, medium, low stress levels, respectively
z_D	specified design stress
p_1, p_2, p_3	proportion of test units allocated to z_L, z_M and z_L, respectively
T	prespecified period of time over which the reliability estimate is of interest at normal operating conditions
$R(t; z)$	reliability at time t, for given z
$f(t; z)$	*PDF* at time t, for given z
$F(t; z)$	*CDF* at time t, for given z
$\Lambda(t; z)$	cumulative hazard function at time t, for given z
$\lambda_0(t)$	unspecified baseline hazard function at time t

We assume the baseline hazard function $\lambda_0(t)$ to be linear with time:

$$\lambda_0(t) = \gamma_0 + \gamma_1 t$$

Substituting $\lambda_0(t)$ into the PH model described above, we obtain:

$$\lambda(t; z) = (\gamma_0 + \gamma_1 t)\exp(\boldsymbol{\beta} z)$$

We obtain the corresponding cumulative hazard function $\Lambda(t; z)$, and the variance of the hazard function as

$$\Lambda(t; z) = \left(\gamma_0 t + \frac{\gamma_1 t^2}{2}\right) e^{\beta z}$$

$$Var[(\hat{\gamma}_0 + \hat{\gamma}_1 t)e^{\hat{\beta} Z_D}] = (Var[\hat{\gamma}_0] + Var[\hat{\gamma}_1]t^2)e^{2(\beta z + Var[\hat{\beta}]z^2)}$$
$$+ e^{2\beta z + Var[\hat{\beta}]z^2}(e^{Var[\hat{\beta}]z^2} - 1)(\gamma_0 + \gamma_1 t)^2$$

3.3.1 Formulation of the Test Plan

Under the constraints of available test units, test time, and specification of the minimum number of failures at each stress level, the objective of the problem is to optimally allocate stress levels and test units so that the asymptotic variance of the hazard rate estimate at normal conditions is minimized over a prespecified period of time T. If we consider three stress levels, then the optimal decision variables $(z_L^*, z_M^*, p_1^*, p_2^*, p_3^*)$ are obtained by solving the following optimization problem with a nonlinear objective function and both linear and nonlinear constraints [15].

$$\text{Min} \int_0^T Var[(\hat{\gamma}_0 + \hat{\gamma}_1 t)e^{\hat{\beta} z_D}]dt$$

Subject to

$$\underset{\sim}{\Sigma} = F^{-1}$$

$$0 < p_i < 1, \quad i = 1, 2, 3$$

$$\sum_{i=1}^{3} p_i = 1$$

$$z_D < z_L < z_M < z_H$$

$$np_i \Pr[t \le \tau | z_i] \ge MNF, \quad i = 1, 2, 3$$

where, *MNF* is the minimum number of failures and $\underset{\sim}{\Sigma}$ is the inverse of the Fisher's information matrix.

Other objective functions can be formulated which result in different designs of the test plans. These functions include the D-Optimal design that provides efficient estimates of the parameters of the distribution. It allows relatively efficient determination of all quantiles of the population, but the estimates are distribution dependent.

3.3.2 Numerical Example

An accelerated life test is to be conducted at three temperature levels for MOS capacitors in order to estimate its life distribution at a design temperature of 50 °C. The test needs to be completed in 300 h. The total number of items to be placed under test is 200 units. To avoid the introduction of failure mechanisms other than those expected at the design temperature, it has been decided, through engineering judgment, that the testing temperature should not exceed 250 °C. The minimum number of failures for each of the three temperatures is specified as 25. Furthermore, the experiment should provide the most accurate reliability estimate over a 10-year period of time [15].

Consider three stress levels, then the formulation of the objective function and the test constraints follow the same formulation given in the above section. The plan derived that optimizes the objective function and meets the constraints is shown as follows:

$$z_L = 160\,^\circ\mathrm{C}, z_M = 190\,^\circ\mathrm{C}, z_H = 250\,^\circ\mathrm{C}$$

The corresponding allocations of units to each temperature level are:

$$p_1 = 0.5, \ p_2 = 0.4, \ p_3 = 0.1$$

3.3.3 Equivalent Accelerated Life Testing Plans

In design of ALT plans, estimate of one or more reliability characteristics, such as the model parameters, hazard rate, and the mean TTF at certain conditions are common. Accordingly, different optimization criteria might be considered. For instance, if the estimate of the model parameters is the main concern, D-optimality which maximizes the determinant of the Fisher information matrix is considered an appropriate criterion. When estimate of the time to quantile failure is of interest then the variance optimality that minimizes the asymptotic variance of time to quantile failure at normal operating conditions is commonly used. Meanwhile, different methods, e.g., maximum likelihood estimate (MLE) or Bayesian estimator can be used for estimation of the model parameters. However, each method has its inherent statistical properties and efficiencies. In light of this, we discuss equivalent test plans with respect to the same reliability characteristics and optimization criterion then determine equivalent test plans using the same inference procedure. In this chapter, we propose two possible definitions of equivalency as follows:

Definition 1 Two test plans are equivalent if the absolute difference of the objectives for reliability prediction is less than under the same set of constraints on the number of test units, expected number of failures, or total test time.

Definition 2 Two test plans are equivalent if they achieve the same objective for reliability prediction under the same constraints on the number of test units, expected number of failures, or total test time within a margin.

According to the above definitions, the equivalent test plans are not unique. Therefore, we recommend the following procedures for constructing equivalent plans [46].

The first step of the approach is to obtain an optimal baseline test plan. Since constant-stress test is the most commonly conducted ALT in industry and its statistical inference has been extensively investigated, we propose to use an optimal constant-stress plan as a baseline [45].

Suppose an optimal baseline test plan can be determined from the following general formulation:

$$\begin{aligned}&\text{Min} \quad f_B(x)\\&\text{s.t.} \quad Lb \le x \le Ub\\&C(x) \le 0, \;\; Ceq(x) = 0\end{aligned} \tag{8}$$

where $f_B(x)$ is the objective function (e.g., the asymptotic variance of mean TTF) and x is its decision variable which can be expressed as either a vector or a scalar, Lb and Ub are the corresponding lower and upper bounds of x. $C(x) \le 0$ and $Ceq(x) = 0$ are the possible inequality and equality constraints, respectively.

The second step is to determine the equivalent test plan based on Definitions 1 or 2 using formulations (8) or (9), respectively. Formulation (9) is given as follows:

$$\begin{aligned}&\text{Min} \quad \Pi_i(y)\\&\text{s.t.} \, |f_B(x) - f_E(y)| \le \delta\\&\Pi_j(x) - \Pi_j(y) = 0\\&Lb' \le y \le Ub'\\&C'(y) \le 0, \;\; Ceq'(y) = 0\end{aligned} \tag{9}$$

where $f_B(x)$ and $f_E(y)$ are the base and equivalent objective functions on reliability prediction, respectively, and y is the decision variable of the equivalent test plan. $\Pi(\cdot)$ represents the constraint of the total number of test units, expected number of failures, or the test time. When $\Pi_j(y)$ is the total number of test units, $\Pi_i(y)$ can be the censoring time under Type-I censoring or expected number of failures under Type-II censoring and vice versa. The idea is to set the allowed difference between objective values as a constraint as well as seek other merits.

Similarly, based on Definition 2, the optimal equivalent test plan can be determined as,

$$\begin{aligned}
&\text{Min } \Pi_i(y)\\
&\text{s.t. } f_B(x) - f_E(y) = 0\\
&\left|\Pi_j(x) - \Pi_j(y)\right| \le \delta\\
&Lb' \le y \le Ub'\\
&C'(y) \le 0,\\
&Ceq'(y) = 0
\end{aligned} \tag{10}$$

An example that demonstrates these methods and develops equivalent step-stress and ramp-stress test plans and the baseline constant-stress test plan is given in Zhu and Elsayed [46].

4 Accelerated Degradation Testing (ADT)

In this section, we present the concept of degradation, degradation modeling, and the design of accelerated degradation test plans.

4.1 Degradation Models

There are many instances where few or no failures are observed even under accelerated conditions making reliability inference via failure-time analysis significantly inaccurate, if not impossible. However, if a product's performance indices related to failure mechanism experience degradation over time, degradation analysis may be a viable alternative to traditional failure-time analysis and ALT. Indeed, degradation data may provide more reliability information than would be available from traditional censored failure-time data.

In general, degradation testing can be conducted by observing the degradation of the units at normal operating conditions and use appropriate models to predict the reliability of such units. Alternatively, if the degradation rate is "small" then an ADT is conducted instead. Again, an appropriate prediction model is needed to relate degradation data at stress conditions to reliability estimate of the units at normal operating conditions.

Moreover, to save time and cost, ADT experiments are commonly conducted to provide immediate degradation data for predicting the reliability under normal operating conditions. However, in ADT analysis, an inaccurate prediction will result unless an appropriate degradation model and a carefully designed test plan are used.

An appropriate ADT model is the one that accurately accounts for the influences of the stresses (covariates) on the degradation process based on the product's physical properties and the associated probability distributions. Nelson [27] briefly surveys the degradation behavior of various products and materials subject to degradation,

ADT models, and inference procedures. He also presents basic accelerated degradation models under constant stress. Meeker and Escobar [25] provide a review of degradation and describe the applications of ADT models. They propose mathematical models to analyze ADT data and suggest methods for estimating failure time distributions, distribution quantiles, and their confidence intervals. A part of the following is based on Liao [21].

Elsayed [14] provides a review of the degradation models and classifies ADT models into two types: physics-statistics-based models and statistics-based model. Furthermore, he classifies statistics-based model into two categories: parametric models and nonparametric/semiparametric models. This classification is summarized as follows.

1. Physics-statistics-based models
   Nelson [26] analyzes the degradation of an insulation material at different stress levels. He assumes that the temperature is the only acceleration factor that determines the degradation profile over time and presents a relationship among the absolute temperature, the median breakdown voltage, and time. He then estimates the lifetime distribution based on the performance degradation model. Based on Carey and Tortorella [4], Carey and Koenig [3] utilize ADT at higher temperature levels to infer the reliability of an integrated logic family, a component of a generation of submarine cables, at normal operating condition. They assume that the maximum propagation time delay (maximum degradation) and the absolute temperature are related by the Arrhenius law. The maximum likelihood estimator is then utilized to estimate the parameters of the Arrhenius relation, which is used for predicting the maximum degradation at normal operating conditions. Whitmore and Schenkelberg [41] model accelerated degradation process by a Brownian motion with a timescale transformation. The model incorporates the Arrhenius law for high stress testing. Inference methods for the model parameters based on ADT data are presented. Meeker et al. [24] use the Arrhenius law to describe the impact of temperature on the rate of a simple first-order chemical reaction and obtain a scale accelerated failure time model (SAFT). Approximate maximum likelihood estimation [33] is used to estimate model parameters. Confidence intervals for time-to-failure distribution are obtained by simulation-based methods. Chang [5] presents a generalized Eyring model to describe the dependence of performance aging on accelerated stresses in a power supply. The tests considered involve multiple measurements in a two-way design. The mean TTF of the power supply at the normal operating condition is estimated. Sometimes, the degradation indices (or rates) can be measured directly or by using surrogate indicators or by conducting destructive testing on the units.
2. Statistics-based models
   Statistics-based models consist of parametric models and nonparametric models. The parametric models assume that the degradation path of a unit follows a specific functional form with random parameters, or the degradation measure follows an assumed distribution with time-dependent parameters. Moreover, these

models assume that there is only a scaling transformation of the degradation paths or the degradation measure distributions at different stress levels but their forms remain unchanged. The nonparametric models relax the assumption about the form of the degradation paths or distribution of degradation and establish them in a nonparametric way. The models have greater flexibility in contrast to the parametric regression models, but they may not have explicit physical meaning.

a. Parametric models

   Based on the degradation paths, Crk [9] extends the methodology of the general degradation path approach to the development of the multivariate, multiple regression analysis of function parameters with respect to applied stresses.

   Tang and Chang [36] model nondestructive accelerated degradation data as a collection of stochastic processes for which the parameters depend on the stress levels. The model adopts the independent increment concept by assuming the incremental degradation within a time interval Δt is *i.i.d* random variable with mean $\mu_i \Delta t$ and variance $\sigma_i^2 \Delta t$. The constants μ_i and σ_i^2 are the parameters under the ith stress level, which are linked with applied stresses by a linear regression approach. The actual degradation path is the summation of these increments, whose first passage time to a threshold level D follows Birnbaum-Saunders distribution when $D \gg \mu_i \Delta t$. If the independent increment is s-normally distributed, then an inverse Gaussian distribution is used as it is a statistically more accurate model as discussed by Bhattacharyya and Fries [1] and Desmond [10].

   Among the approaches of degradation modeling by Brownian motion, Doksum and Hoyland [12] discuss ADT models for the variable-stress case and introduce a flexible class of models based on the concept of accumulated decay. The variable-stresses considered are simple-step-stress, multiple-step-stress, and progressive stress. The proposed model is a time-transformed Brownian motion with drift model, which assumes that certain deterministic stress level imposes the same scaling effect on drift and Brownian motion terms. Pieper et al. [32] propose a different model for the first passage time distribution under simple-step-stress condition. They also discuss an interesting extension that the time change point is random variable. However, the expression for the first passage probability density in this case cannot be obtained in an explicit form.

b. Nonparametric models

   Shiau and Lin [34] present a Nonparametric Regression Accelerated Life-stress (NPRALS) model for some groups of accelerated degradation curves (paths). They assume that various stress levels only influence the degradation rate but not the shape of the degradation curve. An algorithm is proposed to estimate the components of NPRALS such as the acceleration factor. By investigating the relationship between the acceleration factors and the stress levels, the mean TTF estimate of the product under the normal condition is obtained.

The nonparametric regression models bear the degradation-path-free property in contrast to the parametric models. They relax the specification of the form of the degradation path and perform much better than parametric models, if the assumed path function is far from true in the parametric modeling. However, nonparametric models require more data to obtain the same accuracy as that of the parametric models assuming that the parametric models are correct. In other words, the efficiency of nonparametric models is relatively low. Moreover, the time scaling assumption is important since it is required for predicting the form of degradation curve under normal operating conditions, but this assumption is rather weak. Moreover, to utilize the nonparametric regression model, the span of degradation curve under normal condition has to be covered by that of the accelerated degradation data after time scaling, and ADT must be conducted until test units fail.
Another nonparametric/semiparametric approach is to utilize the degradation hazard function. Eghbali [13] proposes an ADT model called proportional degradation hazards model (PDHM) assuming the logarithm of the degradation hazard is a linear function of the stress covariates $\underline{z}$, that is,

$$s(x; t, \underline{z}) = s_0(x; t) \exp(\underline{\beta}' \underline{z})$$

where $s_0(x; t) = g_0(x) q_0(t)$ can be expressed as two positive separable functions $g_0(x)$ and $q_0(t)$ of the degradation measure and the time, respectively. MLEs are utilized to obtain the model parameters. The model is applied to the ADT data of light emitting diode (LED) subject to accelerated temperature and current to predict reliability at normal operating conditions.

4.2 Design of ADT Plans

Design of ADT plans is similar to the design of ALT plans as both require the determination of the stress type, stress level, and allocation of test units to stresses. However, ADT plans require the identification of the degradation indicators, the frequency of measurements (sometimes the degradation can only be assessed via destructive testing). Of course, both ADT and ALT plans require the identification of the decision variables, constraints, and an optimization criterion such as the asymptotic variance of time to failure (TTF) estimate, variance of the reliability estimate, or variance of the estimated 100pth percentile of the lifetime distribution, etc. Although the optimization problem may be feasible, the obtained optimum test plan cannot correct the bias of a degradation model, therefore, a test plan is inappropriate if the degradation model is not accurate. We briefly discuss the common test plans.

4.2.1 Constant–Stress Degradation Test Plans

Boulanger and Escobar [2] present a method to determine the stress levels, sample size at each level, and observation times. However, their method is discussed under a predetermined termination time. Tseng and Yu [39] propose an intuitively appealing method for choosing the time to terminate a degradation test by analyzing the asymptotic convergence property of MTTF estimate but the termination rule is approximate since no constraint has been considered. Park and Yum [31] develop an optimal ADT plan under the assumptions of destructive testing and the simple constant rate relationship between the stress and the product performance. By solving a constrained nonlinear programming problem, the stress levels, the proportion of test units allocated to each stress level, and the inspection times are determined such that the asymptotic variance of the MLE of the MTTF at the normal operating conditions is minimized. Yu and Tseng [44] design an optimal degradation experiment under the constraint of the total experimental cost. They assume the degradation path can be transformed to a simple form. The optimal decision variables, sample size, inspection frequency, and termination time are determined by minimizing the variance of the estimated 100pth percentile of the lifetime distribution. As an application, Yu and Chiao [43] design an optimal degradation experiment for improving LED reliability. Wu and Chang [42] investigate the Nonlinear Mixed-effect Model and propose a step-by-step enumeration algorithm to determine the optimal sample size, inspection frequency, and termination time under the cost constraint. The variance of the estimator of percentile of the failure time distribution is minimized. They also study the sensitivity of the optimal plan to the changes of model parameters and cost. It shows that the optimal solution is slightly sensitive to the changes in the values of model parameters. Recently, Liao and Elsayed [22] propose the *Geometric Brownian Motion Degradation Rate* (GBMDR) model and inference procedure to estimate field reliability for a population and a specific individual unit.

4.2.2 Variable–Stress Degradation Test Plans

Since conducting a constant-stress ADT is costly due to the test duration, it may not be applicable for assessing the lifetime of a newly developed product because typically only a few test units are available. To overcome this difficulty, a variable-stress such as step-stress ADT experiment can be carried out. Tseng and Wen [38] provide an illustration of a statistical inference procedure for a step-stress ADT using a case study of LEDs. However, in the literature, variable-stress degradation test plans are rare. Tang et al. [37] investigates planning of an optimum step-stress ADT experiment where the test stress is increased in steps from a lower stress to a higher stress during the test. Based on the maximum likelihood theory, the asymptotic variance of TTF estimate at the normal operating conditions is then derived and used as a constraint instead of an objective function. The optimum testing plan which minimizes the testing cost gives the optimal sample size, number of inspections at each stress level, and number of total inspections. It is important to note that in such step-stress testing the sequence of load application has a significant impact on the reliability prediction at normal operating conditions, a fact that is rarely considered by researchers.

5 Summary

Reliability prediction of new components, products, and systems is a difficult task due to the lack of well-designed test plans that yield "useful" information during the test and due to the stochastic nature of normal operating conditions. The accuracy of the reliability prediction has a major effect on the warranty cost and repair and maintenance strategies. Therefore, it is important to design efficient test plans. In this chapter, we present an overview of reliability testing with emphasis on accelerated testing and address issues associated with the design of optimal test plans, stress application methods, and reliability prediction models. We further discuss the concept of equivalence of test plans and how it could be used for test time reduction. Finally, we present accelerated degradation modeling and the design of accelerated degradation test plans.

Dedication

This chapter is dedicated to my colleague and friend Dr. Shunji Osaki on his 70th Birthday for his contributions and leadership in the field of Reliability Engineering.

References

1. Bhattacharyya GK, Fries A (1982) Fatigue-failure models—Birnbaum-Saunders vs Inverse Gaussian. IEEE Trans Reliab 31:439–440
2. Boulanger M, Escobar LA (1994) Experimental design for a class of accelerated degradation tests. Technometrics 36:260–272
3. Carey MB, Koenig RH (1991) Reliability assessment based on accelerated degradation: a case study. IEEE Trans Reliab 40:499–506
4. Carey MB, Tortorella M (1988) Analysis of degradation data applied to MOS devices. In: Paper presented at the 6th international conference on reliability and maintainability. Strasbourg, France
5. Chang DS (1993) Analysis of accelerated degradation data in a two-way design. Reliab Eng Sys Saf 39:65–69
6. Ciampi A, Etezadi-Amoli J (1985) A general model for testing the proportional hazards and the accelerated failure time hypotheses in the analysis of censored survival data with covariates. Commun Statist Theor Meth 14:651–667
7. Choi SR, Salem JA (1997) Error in flexure testing of advanced ceramics under cyclic loading. Ceram Eng Sci Proc 18:495–502
8. Cox DR (1972) Regression models and life tables (with discussion). J Roy Stat Soc Ser B 34:187–220

9. Crk V (2000) Reliability assessment from degradation data. In: Proceedings annual reliability and maintainability, symposium, pp 155–161
10. Desmond AF (1986) On the relationship between two fatigue-life models. IEEE Tran Reliab 35:167–169
11. Dale CJ (1985) Application of the proportional hazards model in the reliability field. Reliab Eng 10:1–14
12. Doksum KA, Hóyland A (1992) Models for variable-stress accelerated life testing experiments based on Wiener Process and the Inverse Gaussian distribution. Technometrics 34:74–82
13. Eghbali G (2000) Reliability estimate using accelerated degradation data. Dissertation, Ph.D, Department of Industrial and Systems Engineering, Rutgers University
14. Elsayed EA (1996) (1996) Reliability engineering. Addison-Wesley Longman, Inc., New York
15. Elsayed EA (2008) Reliability prediction and accelerated testing. In: Kobbacy KAH, Murthy DNP (eds) Complex System Maintenance Handbook, Springer-Verlag, pp 155–178
16. Elsayed EA, Chan CK (1990) Estimation of thin-oxide reliability using proportional hazards models. IEEE Trans Reliab 39:329–335
17. Elsayed EA, Jiao L (2002) Optimal design of proportional hazards based accelerated life testing plans. Int J Mater Prod Technol 17(5/6):411–424
18. Elsayed EA, Liao HT, Wang XD (2006) An extended linear hazard regression model with application to time-dependent-dielectric-breakdown of thermal oxides. IIE Trans Qual Reliab Eng 38:1–12
19. Escobar LA, Meeker WQ (2006) A review of accelerated test models. Stat Sci 21:552–577
20. Hall JB, Ellner PM, Mosleh A (2010) Reliability growth management metrics and statistical methods for discrete-use systems. Technometrics 52:379–408
21. Liao H (2005) Degradation models and design of accelerated degradation testing plans. Rutgers The State University of New Jersey, New Brunswick
22. Liao H, Elsayed E (2010) Equivalent accelerated life testing plans for log-location-scale distributions. Naval Res Logistics 57:472–488
23. Matthewson MJ, Yuce HH (1994) Kinetics of degradation during fatigue and aging of fused silica optical fiber. Proc SPIE 2290:204–210
24. Meeker WQ, Escobar LA (1998) Statistical methods for reliability data. Wiley, New York
25. Meeker WQ, Escobar LA, Lu JC (1998) Accelerated degradation tests: modeling and analysis. Technometrics 40:89–99
26. Nelson WB (1981) Analysis of performance-degradation data from accelerated tests. IEEE Trans Reliab 30:149–155
27. Nelson WB (1990) Accelerated testing: statistical methods, test plans, and data analysis. Wiley, New York
28. Nelson WB (2004) Accelerated testing: statistical models, test plans, and data analyses. Wiley, New York
29. Nelson WB (2005a) A Bibliography of accelerated test plans. IEEE Trans Reliab 54:194–197
30. Nelson WB (2005b) A bibliography of accelerated test plans part II–references. IEEE Trans Reliab 54:370–373
31. Park JI, Yum BJ (1997) Optimal design of accelerated degradation tests for estimating mean lifetime at the use condition. Eng Optim 28:199–230
32. Pieper V, Domine M, Kurth P (1997) Level crossing problem and drift reliability. Math Methods Oper Res 45:347–354
33. Pinheiro JC, Bates DM (1995) Approximation to the loglikelihood function in the nonlinear mixed effects model. J Comput Graph Stat 4:12–35
34. Shiau JJH, Lin HH (1999) Analyzing accelerated degradation data by nonparametric regression. IEEE Trans Reliab 48:149–158
35. Shyur Huan-Jyh, Elsayed EA, Luxhoj JT (1999) A General model for accelerated life testing with time-dependent covariates. Naval Res Logistics 46:303–321
36. Tang LC, Chang DS (1995) Reliability prediction using nondestructive accelerated-degradation data: case study on power supplies. IEEE Trans Reliab 44:562–566

37. Tang LC, Yang GY, Xie M (2004) Planning of step-stress accelerated degradation test. In: Proceedings annual reliability and maintainability symposium
38. Tseng ST, Wen ZC (2000) Step-stress accelerated degradation analysis for highly reliable products. J Qual Technol 32:209–216
39. Tseng ST, Yu HF (1997) A termination rule for degradation experiment. IEEE Trans Reliab 46:130–133
40. Vlasic B, Bunkley N (2009) Toyota will fix or replace 4 million gas pedals. http://www.nytimes.com/2009/11/26/business/26toyota.html?_r=0
41. Whitmore GA, Schenkelberg F (1997) Modelling accelerated degradation data using wiener diffusion with a scale transformation. Lifetime Data Anal 3:27–45
42. Wu SJ, Chang CT (2002) Optimal design of degradation tests in presence of cost constraint. Reliab Eng Sys Saf 76:109–115
43. Yu HF, Chiao CH (2002) An optimal designed degradation experiment for reliability improvement. IEEE Trans Reliab 51:427–433
44. Yu HF, Tseng ST (1999) Designing a degradation experiment. Naval Res Logistics 46:689–706
45. Zhu Y (2010) Optimal design and equivalency of accelerated life testing plans. Ph.D. Dissertation, Department of Industrial and Systems Engineering, Rutgers University
46. Zhu Y, Elsayed EA (2011) Design of equivalent accelerated life testing plans under different stress applications. Quality Technol Quantitative Management 8(4):463–478

Maintenance Outsourcing: Issues and Challenges

D. N. P. Murthy, N. Jack and U. Kumar

Abstract All products and systems are unreliable in the sense that they degrade and fail. Corrective maintenance (CM) restores a failed item to an operational state and effective preventive maintenance (PM) reduces the likelihood of failure. These maintenance actions can be done either in-house or can be outsourced to an external agent. We focus on the maintenance being outsourced and look at the issues involved from the perspectives of the owner of the asset and the agent providing the maintenance service.

1 Introduction

Every business (mining, processing, manufacturing, and service-oriented businesses such as transport, health, utilities, communication) needs a variety of equipment to deliver its outputs. Equipment is an asset that is critical for business success in the fiercely competitive global economy. Equipment degrades with age and usage and ultimately becomes non-operational. Rapid changes in technology have resulted in equipment becoming larger, more complex, and expensive. Businesses incur heavy

---

D. N. P. Murthy (✉)
School of Mechanical and Mining Engineering, The University of Queensland, Queensland, QLDQ 4072, Australia
e-mail: p.murthy@uq.edu.au

N. Jack
Dundee Business School, University of Abertay Dundee, Dundee DD1 1HG, UK
e-mail: n.jack@abertay.ac.uk

U. Kumar
Division of Operation and Maintenance Engineering, Luleå University of Technology, SE-971 87 Luleå, Sweden
e-mail: uday.kumar@ltu.se

T. Dohi and T. Nakagawa (eds.), *Stochastic Reliability and Maintenance Modeling*, Springer Series in Reliability Engineering 9, DOI: 10.1007/978-1-4471-4971-2_3,

losses when their equipment is not in full operational mode—delays in delivery of goods lead to higher customer dissatisfaction and loss of goodwill.

Maintenance activities are actions to reduce the likelihood of equipment becoming non-operational and to restore non-operational units to operational state. For most businesses, it is no longer economical to carry out the maintenance in-house. There are a variety of reasons for this including the need for a specialist workforce and diagnostic tools that often require constant upgrading. In these situations, it is more economical to outsource the maintenance (in part or total) to an external agent through a service contract. Campbell [7] gives details of a survey where it was reported that 35 % of North American companies had considered outsourcing some of their maintenance.

Governments (local, state, or national) and private businesses own infrastructure (roads, rail, and communication networks, public buildings, dams, etc) that were traditionally maintained by in-house maintenance departments. Here also, there is a growing trend toward outsourcing these maintenance activities to external agents so that the owners can focus on their core activities.

In maintenance outsourcing the maintenance of an asset (equipment or infrastructure) owned by the first party (the owner or customer) is carried out by the second party (the service agent who is also referred to as the "contractor" in many technical papers) under a service contract. In this chapter, we look at maintenance outsourcing from both the owner and service agent perspectives and discuss the issues involved, review the literature, and discuss some of the challenges for future research.

The outline of the chapter is as follows. We start with a brief discussion of maintenance and of outsourcing in Sects. 2 and 3, respectively. Section 4 reviews the current status of maintenance outsourcing and gives a brief literature review. In Sect. 5 we propose a framework to study maintenance outsourcing and discuss several relevant issues. Section 6 deals with the game theoretic approach to maintenance outsourcing and Sect. 7 looks at Agency Theory and its relevance to maintenance outsourcing. We deal with the criteria for the selection of service agents to carry out maintenance in Sects. 8 and 9 deals with some topics for future research. We conclude with some comments in Sect. 10.

2 Maintenance

Maintenance actions can be broadly divided into two categories.

Corrective Maintenance (CM): These are maintenance actions performed when the asset has a failure (in the case of equipment) or has degraded sufficiently (in the case of infrastructure). The most common form of CM is "minimal repair" where the state of the asset after repair is nearly the same as that just before failure. The other extreme is "as good as new" repair and this is seldom possible unless one replaces the failed asset with a new one. Any repair action that restores the asset state to better than that before failure and not as good as that of a new asset is referred to as "imperfect repair".

Preventive Maintenance (PM): These are actions carried out to fix minor problems in case of infrastructure (e.g., small potholes in a section of a road) or components that have degraded in the case of equipment due to age and/or usage. The policy used for initiating such actions can be age, usage, and/or condition. As a result, there are several different kinds of PM policies and in the context of equipment some of the well-known ones are the following:

- Age-based maintenance
- Clock-based maintenance
- Opportunistic maintenance
- Condition-based maintenance

The more investment made in PM actions the more likely CM costs are reduced. But, for any asset there is an optimal level of PM effort that will achieve a proper balance between these costs. Most books on maintenance [4, 26, 28] include models to obtain the optimal PM effort.

Maintenance of an asset involves carrying out several activities as indicated in Fig. 1 (adapted from Dunn [9]).

The three key issues are:

- (D-1): **What** (components) need to be maintained?
- (D-2): **When** should the maintenance be carried out?
- (D-3): **How** should the maintenance be carried out?

3 Outsourcing

Businesses (producing products and/or services) need to come up with new solutions and strategies to develop and increase their competitive advantage. Outsourcing is one of these strategies that can lead to greater competitiveness [11]. It can be defined as a managed process of transferring activities performed in-house to some external agent. The conceptual basis for outsourcing (see, Campbell [7]) is as follows:

1. Domestic (in-house) resources should be used mainly for the core competencies of the company.

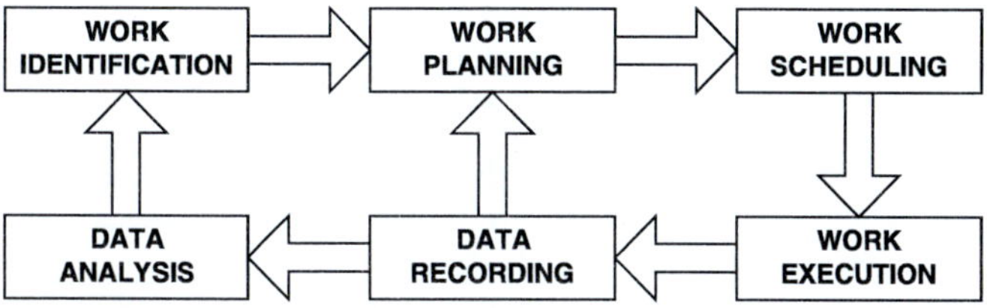

Fig. 1 Activities in asset maintenance

2. All other (support) activities that are not considered strategic necessities and/or whenever the company does not possess the adequate competences and skills should be outsourced (provided there is an external agent who can carry out these activities in a more efficient manner).

However, there are some disadvantages of outsourcing the maintenance and these are indicated below.

- Dependency on the external party carrying out the activities.
- Cost of outsourcing.
- Loss of maintenance knowledge (and personnel).
- Becoming locked into a single external party when the cost of switching is high.

4 Maintenance Outsourcing: Current Status and Literature Review

Outsourcing of maintenance involves some or all of the maintenance actions (PM and/or CM) being carried out by an external service agent under a service contract. The contract specifies the terms of maintenance and the cost issues. It can be simple or complex and can involve penalty and incentive terms. We look at the issues in outsourcing from both the owner and service agent perspectives.

4.1 Owner Perspective

4.1.1 Outsourcing Equipment Maintenance

The advantages of outsourcing maintenance are as follows:

- Better maintenance due to the expertise of the service agent.
- Access to high-level specialists on an "as and when needed" basis.
- Fixed cost service contract removes the risk of high costs.
- Service providers respond to changing customer needs.
- Access to latest maintenance technology.
- Less capital investment for the customer.
- Managers can devote more resources to other facets of the business by reducing the time and effort involved in maintenance management.

For very specialized (and custom built) products, the knowledge to carry out the maintenance and the spares needed for replacement must be obtained from the original equipment manufacturer (OEM). In this case, the customer is forced into having a maintenance service contract with the OEM and this may result in a non-competitive market. In the USA, Section II of the Sherman Act [16] deals with this problem by making it illegal for OEMs to act in this manner.

When the maintenance service is provided by an agent other than the OEM often the cost of switching prevents customers from changing their service agent. In other words, customers get "locked in" and are unable to do anything about it without a major financial consequence.

4.1.2 Outsourcing of Infrastructure Maintenance

As mentioned above, it used to be the case that infrastructures were owned and operated by governments. Recently, there has been a growing trend toward selling these assets to private businesses that either lease them back to the government or to the operator of the asset. The maintenance of the asset is often outsourced as it is again viewed as not being the core activity of the business that owns the asset. A complicating factor is the additional parties involved and these are shown in Fig. 2.

For example, in the case of a rail network, the operators are the different rail companies that use the track and the maintenance is outsourced to specialist contractors. The government plays a critical role in terms of providing loans to and/or acting as a guarantor for the owner and the regulators are independent authorities responsible for ensuring public safety. The role of maintenance now becomes important in the context of safety and risk. For further discussion, see Vickerman [38].

4.1.3 Decision Problems

There are three different outsourcing scenarios that depend on which of the three activities in maintenance (D-1, D-2, and D-3) are being outsourced. These are indicated in Table 1.

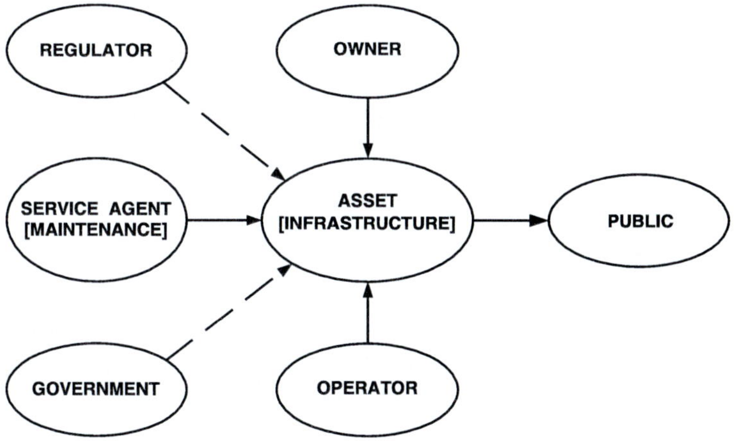

Fig. 2 Different parties involved in the maintenance of infrastructure

Table 1 Different contract scenarios

Scenarios	Decisions	
	Customer	Service agent
S-1	D-1, D-2, D-3	-
S-2	D-1	D-2, D-3
S-3	-	D-1, D-2, D-3

In scenario S-1, the service agent is only providing the resources (workforce and material) to execute the work. This corresponds to the minimalist approach to outsourcing. In scenario S-2, the service agent decides on **how** and **when** and **what** is to be done is decided by the customer. Finally, in scenario S-3 the service agent makes all three decisions.

For the owners of both equipment and infrastructure the decision problems are (i) whether to outsource or not, (ii) what maintenance activities to outsource, and (iii) how to implement and manage the process.

4.2 Service Agent Perspective

The service agent who provides the maintenance needs to operate as a service business. This implies that issues such as return on investment (ROI), number of customers to service (market share), location of operations, and range of service contracts to offer are some of the variables that are important in the context of strategic management of the business. The type of contract depends on the needs of the customers and this can be either standard or customized. At the operational level, the service agent needs to deal with issues such as scheduling of maintenance tasks, spare part inventory control, etc.

The pricing of the different service contracts offered is critical for business profitability. If the price is too low, the service agent might end up making a loss instead of a profit. On the other hand, if it is too high there might be no customers for the service. The price of a contract must cover the maintenance costs and estimating the cost is a challenge due to information uncertainties.

4.3 Literature Review

The literature on maintenance outsourcing deals mainly with the owner perspective and is focussed on management issues. More specifically, attempts are made to address one or more of the following questions in a qualitative manner.

- Does outsourcing make sense?
- Are the objectives achievable?
- Is the organization ready?

- What are the outsourcing alternatives?
- What maintenance activities should be outsourced?
- How should the best service agent be selected?
- What are the negotiating tactics for contract formation?

Some of the relevant papers are Campbell [7], Judenberg [15], Martin[23], Levery [20] and Sunny [32]. Stremersch et al. [31] look at the industrial maintenance market.

Unfortunately, cost has been the sole basis used by businesses for making maintenance outsourcing decisions. Sunny [32] looks at what activities are to be outsourced by looking at the long strategic dimension (core competencies) as well as the short-term cost issues.

Bertolini et al. [5] took a quantitative approach and used the analytic hierarchy process (AHP) to make decisions regarding the outsourcing of maintenance. On the application side, Armstrong and Cook [2] look at clustering of highway sections for awarding maintenance contracts to minimize the cost and use a fixed-charge goal programming model to determine the optimal strategy. Bevilacqua and Braglia [6] illustrate their AHP model in the context of an Italian brick manufacturing business having to make decisions regarding maintenance outsourcing.

Tarakci et al. [33] investigated the coordination issues between an equipment owner and a service agent in a long-term maintenance outsourcing contract scenario. The equipment has an increasing failure rate and the agent performs both CM and PM. Incentive contracts that induce the agent to choose the maintenance policy that optimizes the expected total profit for both parties are studied. It is shown that a contract based on a combination of a target uptime level and a bonus produces the desired win–win situation. Tarakci et al. [34] extend the analysis to the situation where the owner has multiple pieces of equipment and uses multiple service agents to perform the maintenance.

Tarakci et al. [35] study the effects of learning when the contract between an owner and an agent consists of a fixed fee plus a cost subsidy for each maintenance action (CM and PM) performed. Learning occurs on the part of the agent which leads to cost and time reductions for PM actions. They demonstrate that a well-designed payment scheme can induce the agent to use a maintenance strategy which maximizes the owner's expected total profit.

Tseng et al. [36] look at a maintenance outsourcing problem where, in the contract terms, one or more time points are specified at which the owner can replace the equipment with new technology if it becomes available. If a replacement occurs then the agent has the flexibility to change the maintenance schedule for the remaining part of the contract period. The value of these switch points is analyzed for different types of contract payment methods.

In Almeida [1], the owner has more than one objective to optimize and is faced with choosing a contract from a set of alternatives. Each contract alternative specifies different values for response time, service quality, dependability, and cost. The best alternative is selected using the ELECTRE I method for multi-criteria decision making combined with utility functions. Lisnianski et al. [21] consider aging equipment with an increasing failure rate. With a piecewise constant approximation for the failure rate, a Markov process is used to model the operating times and repair times. The service agent offers a number of options involving different repair rates and costs to the owner and the optimal choice is made by comparing expected costs over a specified contract period.

A few game-theoretic models have also been proposed and these are discussed in a later section.

5 Framework for Maintenance Outsourcing Study

A proper framework to study maintenance outsourcing needs to include both owner and service agent perspectives and involves several interlinked elements. This is indicated in Fig. 3 for the case of single owner and service agent. Section 4 looked at two of the elements— namely, the owner (customer for the maintenance service) and the service agent (provider of maintenance service).

The number of owners and service agents can be one, few, or many and these lead to different markets for maintenance outsourcing (see Sect. 5.3). In Sects. 5.1 and 5.2 we look at the remaining elements and related issues. Also, the owner population can be homogeneous or heterogeneous in relation to factors such as usage profiles, attitude to risk, etc. Similarly, the service agents can be either homogeneous or heterogeneous in relation to factors such as size, competency, quality of service, reputation, risk profile, etc.

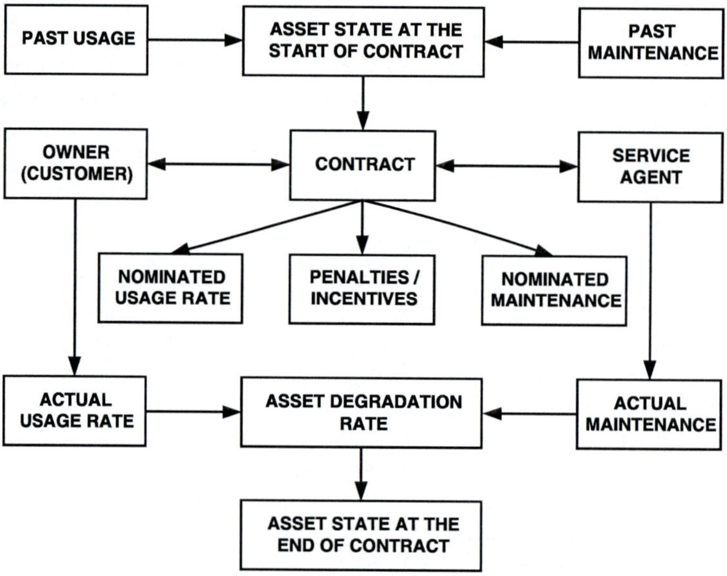

Fig. 3 Framework for study of maintenance outsourcing

5.1 Asset and Asset State

In the case of a new asset, the initial state is determined by the decisions made during its design and construction (or manufacture). The asset reliability characterizes the probability of no failure and decreases with age. The field reliability also depends on the operating stress (load) on the asset and the operating environment. The stress can be thermal, mechanical, electrical, etc., and the reliability decreases as the stress increases and/or the environment gets harsher.

The asset state at any given time (subsequent to it being put into operation) is a function of its inherent reliability and past history of usage and maintenance. This information is important in the context of maintenance service contracts for used assets. The information that the service agent (and the customer) has can vary from very little to a lot (if detailed records of past usage and maintenance have been kept).

Finally, for some assets, the delivery of maintenance requires the service agent to visit the site where the asset is located (for example, lifts in buildings and roads) and for others (mainly industrial equipment) the failed asset can be brought to a service center to carry out the maintenance actions.

5.2 Contract

The contract is a legal document that is binding on both parties (customer and service agent) and it needs to deal with technical, economic, and other issues.

5.2.1 Technical Issues

There is a growing trend toward functional guarantee contracts. Here, the contract specifies a level for the output generated from equipment, for example, the amount of electricity produced by a power plant, or the total length of flights and number of landings and takeoffs per year for aircraft. The service agent has the freedom to decide on the maintenance needed (subject to operational constraints) with incentives and/or penalties whether the target levels are exceeded or not. For more on this, see Kumar and Kumar [18].

In the context of infrastructures, there is a trend toward giving the service agent the responsibility for ongoing upgrades or the responsibility for the initial design resulting in a Build, Own, Operate, and Maintain (BOOM) contract.

5.2.2 Economic Issues

There are a number of alternative contract payment structures. The following list is from Dunn [9]:

- Fixed or Firm price
- Variable Price
- Price ceiling incentive
- Cost plus incentive fee
- Cost plus award fee
- Cost plus fixed fee
- Cost plus Margin

Each of these price structures represents different levels of risk sharing between the customer and the service agent. According to Vickerman [38], an increasing issue in privatized infrastructure is the appropriate incentives needed to ensure adequate maintenance of the infrastructure as a public resource.

5.2.3 Other Issues

Some other issues are as follows:

Requirements: Both parties might need to meet some stated requirement. For example, the customer needs to ensure that the usage intensity and operating loads of the asset do not exceed the levels specified in the contract. These can lead to greater degradation (due to higher stresses on the components) and higher servicing costs to the service agent. Similarly, the service agent needs to ensure proper data recording.

Contract Duration: This is usually fixed with options for renewal at the end of the contract.

Cheating: In maintenance outsourcing cheating by both owner and service agent are issues that need to be addressed. Cheating by the owner occurs when the nominated usage is higher than the actual usage and the service agent is not able to observe this. Similarly, cheating by the service agent occurs when the actual maintenance is below the nominated maintenance and the owner cannot observe this. Information, monitoring, and penalties/incentives can reduce and eliminate the potential for cheating.

Dispute Resolution: This specifies the avenues to follow when there is a dispute. The dispute can involve going to a third party (legal courts).

Unless the contract is written properly and relevant data (relating to equipment and collected by the service agent) are analyzed properly by the customer the long-term costs and risks will escalate.

5.3 Maintenance Outsourcing Market

Whether the maintenance outsourcing market is competitive or not depends on the number of customers and service agents. Table 2 indicates the different market scenarios. These have an impact on issues such as the types of service contracts available to customers and the pricing of the contracts.

Table 2 Maintenance outsourcing market scenarios

Number of customers	Number of service agents	
	One	Few
One	A-1	B-1
Few	A-2	B-2
Many	A-3	B-3

6 Game Theoretic Approach

Game theory is a set of ideas and principles that provide an effective guide to strategic business decision making. Any game must have at least two players (individuals or businesses) with the payoffs to the players being interdependent. The optimal decision by a particular player depends on what that player expects the other players involved to do. An important assumption of game theory is that players will always act rationally (choose their best action).

In a static game, the players have a single 'move' and do not know the actions taken by their rivals. This may be because the players move simultaneously. The players in a dynamic (sequential move) game make their decisions in a well-defined order and the game proceeds in a sequence of stages. In any type of game, an action is the decision that a player makes at a particular move. A strategy specifies what actions a player takes at each move in the game and so is a complete and exact plan, detailing what the player will do in any contingency that may arise.

In games with complete information, the payoffs are common knowledge among all the players. In games of incomplete information, some players do not know the payoffs of some of the other players. In a dynamic game with perfect information, all the players know the entire history of the game when it is their turn to move. Imperfect information implies that some players have only a partial idea of the history of the game. Games may be either cooperative or non-cooperative. In cooperative games players can communicate and, most importantly, make binding agreements. In non-cooperative games players may communicate, but binding agreements are not possible.

The most well-known and widely used solution concept in game theory is Nash equilibrium (NE). An NE is a set of strategies for all the players such that no player has an incentive to change their strategy unilaterally, given the strategies chosen by the other players. Dynamic games are solved using the technique of backward induction where optimal strategies are determined while proceeding from the final stage to the initial stage of the game.

Various applications of game theory can be found in Chatterjee and Samuelson [8], Osborne [27] and Watson [40]. The game theoretic approach allows maintenance outsourcing to be studied from both the customer and service agent perspectives. The information available to each player and their attitudes to uncertainty and risk also need to be taken into account.

6.1 One Customer and One Service Agent

First consider the case where there are only two players—one customer and one service agent. This is scenario A-1 from Table 2. When there is a dominant player then we have a leader–follower situation where the actions of the follower depend on the actions taken by the leader. This situation can be formulated as a two-stage dynamic or 'Stackelberg' game.

Let the service agent be the leader in this particular formulation. Given a set of options $\{A_1, A_2, ..., A_n\}$ offered by the agent (with the value of the decision variable for option i being θ_i), the customer chooses the option which optimizes his/her objective. This generates the customer's best response function $A^*(\theta_1, \theta_2, \ldots, \theta_n)$ as shown in Fig. 4. Using this response function, the service agent then optimally selects the values of the decision variables $\theta_1, \theta_2, ..., \theta_n$ to optimize his/her objective.

Murthy and Ashgarizadeh [24] use this type of formulation for the case where the equipment has a useful life L, failures occur according to a homogeneous Poisson process and repair times are exponentially distributed. The two options offered by the service agent are

- Repair all failures over the useful life L for a fixed fee P but also incur a penalty cost of α for each unit of equipment downtime that is incurred above the value τ
- Repair each failure over the useful life L at cost C_s for each repair

Murthy and Ashgarizadeh [24] give a complete characterization of the agent's optimal pricing strategy $\left(P^*, C_s^*\right)$ and also discuss the effect of varying the model parameters on the optimal strategy.

6.2 Multiple Customers and One Service Agent

Murthy and Ashgarizadeh [25] again use the Stackelberg game formulation with the same two pricing options offered by the agent. They extend their earlier model by assuming that the agent has also to decide the number of customers M to service. This is scenario A-2 from Table 2. In this case, a customer's failed equipment will have to wait for repair if one or more other pieces of equipment from other customers have already failed. M is now an extra decision variable for the agent and a complete characterization of the agent's optimal strategy $\left(P^*, C_s^*, M^*\right)$ is again given.

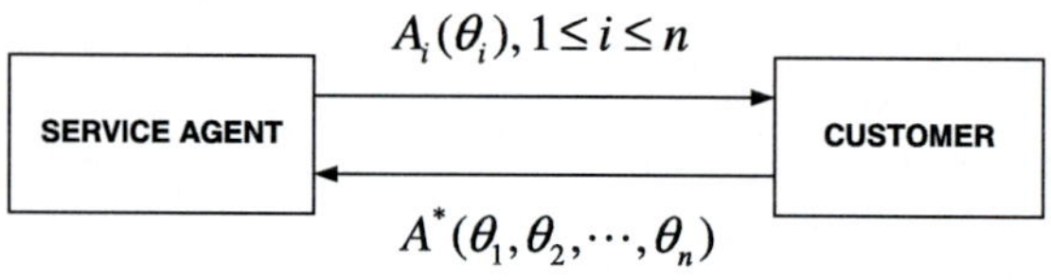

Fig. 4 Stackelberg game formulation

Ashgarizadeh and Murthy [3] extend the model further by assuming that the agent uses S repair facilities to service the M customers. This is scenario A-3 from Table 2. The agent's optimal strategy (P^*, C_s^*, M^*, S^*) with respect to pricing structure, number of customers to service, and number of repair facilities to use is specified.

6.3 Multiple Service Agents

So far, there are no game theory models in the literature which deal with the competition between service agents for the outsourcing of equipment maintenance. This is an open area for research.

6.4 Nash Formulation

If there is no dominant player and players choose their actions either in a cooperative or non-cooperative manner, then a static or 'Nash' game formulation is required.

Jackson and Pascual [14] consider a service contract for aging equipment with terms which specify the frequencies for PM actions and equipment replacement. In their model, the optimal price for the contract is determined by negotiation between the owner and the agent (a Nash bargaining solution) instead of by solving a Stackelberg game. Wang [39] looks at a maintenance contract problem for large and expensive equipment (aircraft, ships, power plant) where the OEM is the only possible service provider. The delay-time concept is used to model the failure behavior of the equipment. Three different contract options are considered, one where the agent is responsible solely for the maintenance and two which involve specified tasks being performed by the owner. Each option requires certain levels of reliability and availability to be satisfied and the optimal parameters for each are again found by negotiation. The cases where both parties have perfect information and where there is information asymmetry are also discussed.

7 Agency Theory

Agency Theory deals with the relationship that exists between two parties (a principal and an agent) where the principal delegates work to the agent which performs that work and a contract defines the relationship. Agency theory is concerned with resolving two problems that can occur in agency relationships. The first problem arises when the two parties have conflicting objectives and it is difficult or expensive for the principal to verify the actual actions of the agent and whether the agent has behaved properly or not. The second problem involves the risk sharing that takes place when the principal and agent have different attitudes to risk (due to various uncertainties).

According to Eisenhardt [10], the focus of the theory is on determining the optimal contract, behavior versus outcome, between the principal and the agent. Many different cases have been studied in-depth in the principal–agent literature and these deal with the range of issues indicated in Fig. 5. Agency theory has also been applied in many different disciplines. For an overview, see Acekere [37].

7.1 Issues in Agency Theory

Moral hazard: Moral hazard refers to the agent's lack of effort in carrying out the delegated tasks. The two parties in the relationship have different objectives and the principal cannot assess the effort level that the agent has actually used.

Adverse Selection: Adverse selection refers to the agent misrepresenting their skills to carry out the tasks and the principal being unable to completely verify this before deciding to hire them.

Information: To avoid adverse selection, the principal can try to obtain information about the agent's ability. One way of doing is contacting people for whom the agent has previously provided service.

Monitoring: The principal can counteract the moral hazard problem by closely monitoring the agent's actions.

Information Asymmetry: The overall outcome of the relationship is affected by several uncertainties. In general, the two parties will have different information to make an assessment of these uncertainties.

Risk: This results from the different uncertainties that affect the outcome of the relationship. For a variety of reasons, the risk attitude of the two parties will differ and a problem arises when they disagree over the allocation of the risk.

Costs: Both parties have various kinds of costs. Some of these depend on the outcome of the relationship (which is influenced by uncertainties), on acquiring information, monitoring, and on the administration of the contract.

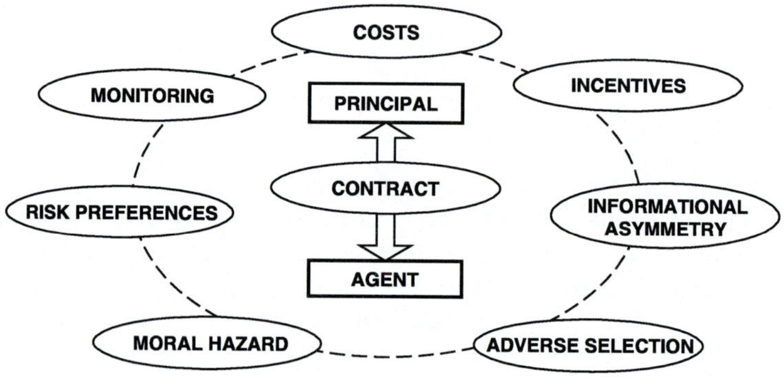

Fig. 5 Agency theory issues

The focus of principal–agent theory lies in the trade-off between (1) the cost of monitoring the agent's actions and (2) the cost of measuring the outcomes of the relationships and of transferring the risk to the agent.

7.2 Relevance to Maintenance Outsourcing

All of the above issues in Agency Theory are relevant to maintenance outsourcing problems. The customer is the principal and the maintenance service provider is the agent. The key factor is the contract which specifies what, when, and how maintenance is to be carried out. This contract needs to be designed taking account of all the relevant issues.

The customer and service agent both potentially face moral hazard. This can occur for the customer when the service agent does not do proper maintenance in order to reduce costs and it can occur for the agent when the customer uses the asset in a manner different to that stated in the contract. Adverse selection can also occur when the customer makes an inappropriate choice from an available pool of potential maintenance service providers (the B scenarios in Table 2). Both parties also possess different information about asset state, usage level, care and attention of the asset, and quality of maintenance used and this asymmetry affects the outcome of their relationship.

Scenario A-1 of Table 2 corresponds to the classical principal–agent model with a single principal (customer) and a single agent (maintenance provider). The interaction that takes place between the principal and the agent can be modeled as a multi-stage dynamic game with the principal as the dominant player. In stage1 of the game, the principal offers a contract to the agent. The agent decides whether to accept or reject this contract in stage 2. If the decision is accepted then, in stage 3, the agent chooses a 'work level' (e.g., service quality or capacity) for the contract period from a set of alternatives. The extra player 'Nature' is also involved during the contract period (the equipment is subject to random failure). What Nature does, together with the effort used by the agent, determines the outcome for the principal for the period (e.g., total equipment downtime and hence total profit).

If the principal cannot assess the agent's effort (moral hazard) then the contract offered must contain incentives for the agent to provide quality service. An example might be where the contract consists of a fixed fee plus penalties for excessive downtime. Kim et al. [17] discuss this type of principal–agent model involving performance-based contracting for equipment subject to infrequent Poisson failures. Plambeck and Zenios [29] use dynamic programming to solve a principal–agent problem where the equipment is used over a finite number of periods. In each period, the equipment can be in one of two states (working or failed) and the transitions between states are influenced by the actions taken by the agent. The agent performs both CM and PM and can exert high or low effort for each type of action. An optimal payment scheme is derived which induces the agent to maximize the principal's expected total discounted profit over the entire planning horizon. So far, Kim et al. [17] and

Plambeck and Zenios [29] are the only cases from the literature where appropriate stochastic formulations are used to model equipment failures.

In the remaining five scenarios of Table 2, there are multiple principals and/or multiple agents involved. In scenarios A-2 and A-3, the equipment under consideration could be a particular brand of lift installed in different buildings within a city. In this case, all the equipment is maintained either by the OEM or an agent of the OEM. There is an extensive literature dealing with the design of contracts for multiple principal/multiple agent problems (Macho-Stadler and Perez-Castrillo [22], and Laffont and Martimort [19] is a small sample of the papers from this literature) and all the Agency Theory issues are still relevant.

The results from the literature on multiple principal/multiple agent problems cannot be applied directly in the maintenance outsourcing context. Thus, new models which contain the required stochastic formulation for equipment failures need to be developed for this application area.

8 Criteria for Rating and Selection of Service Agents

A business is often faced with the strategic decision of whether to develop its own resources to perform maintenance or purchase the required skill and performance from external service agents. To make this decision the business needs to analyze whether maintenance forms a part of its core competencies or whether it only makes a minor contribution to the value chain. Once the business has decided to outsource it also needs to decide on the criteria to select the best service agents.

The selection criterion needs to be governed by the strategic intent of the business and the use of the outsourcing process to meet its goal. Therefore, the selection of the service agent is influenced by the reasons for outsourcing. These reasons can be one or more of the following:

- Concentrating on core activities
- Reducing the maintenance costs
- Spreading the business risk
- Downsizing the organization
- Supplementing the knowledge to achieve the business goals
- Bringing strategic knowledge to meet its requirement
- Facilitating the building up of competence outside the organization

In many contract situations with a large number of service agents participating, the selection of contractors is usually made in two phases (1) the pre-selection phase and (2) the final selection phase. We discuss these briefly and for more details, see, Straub and van Mossel [30].

8.1 Pre-selection Phase

In the pre-selection phase of a service contract process, the selection criteria are based on the following:

- Technical capabilities: The service agent must have the knowledge, the organizational structure, and the resource capabilities to meet the contractual agreements. That is, the service agent must have the correct organization (number of people and their competence) and equipment, etc., to carry out the maintenance as stated in the contract on time and correctly. Often, service agents enter a contract but lack the organizational capability to deliver the agreed performance and this creates bottleneck problems for the owner of the asset.
- Experience with similar equipment: Although the service agent might have the required manpower and competence, the agent may have had no experience in maintaining the asset under consideration. This can result in problems with the delivery and quality of service. Often it takes some time for the service agent to understand all the factors that can cause equipment downtime and this causes bottlenecks when the agent is dealing with a specific asset for the first time.
- Financial health of the service agent: Often owners are influenced by the reputation and capabilities of the service agent and fail to do a thorough analysis of the service agent's financial health. If the service agent is financially weak there is a risk that the agent might not be able to fulfill the contract or even go bankrupt due to cash flow problems.
- Innovative capability of service agent: In recent times, the innovative capability of the service agent has become a dominant factor in an agent being awarded the contract. If the agent has the reputation for being innovative, it provides assurance to the owners of the assets that new and innovative maintenance solutions will ensure better performance, higher quality, and/or reduced costs.
- Demonstrated good governance/moral integrity of the service agent: Good governance is reflected in factors such as transparency in action and moral integrity. Service agents who exhibit these characteristics are preferred to those who lack them.

8.2 Final Selection Phase

The final selection procedure involves a detailed and in-depth analysis of the criteria used in the pre-selection phase. Some of these are listed below.

- Business plan, vision for implementation of new and proven technology: The owners of assets should demand and examine the business plan of the service agents and assess these plans in terms of the implementation of new technologies, training of personnel, and other actions to facilitate innovations.
- Special focus should be given to evaluating the service agent's quality assurance process and its implementation.

- Past experience and performance of the service agent should be assessed by talking to previous customers of the service agent.
- Once short-listed, the owner of the asset must evaluate the team members that will be involved in carrying out the maintenance activities. This assessment is based on the qualification and experiences of each member with respect to the maintenance of similar assets.
- Proper data collection system for monitoring and reporting: The owner needs to pay special attention to this and use the information collected to improve the effectiveness of maintenance.

8.3 Selection of Service Agents: Practice at Swedish Rail Administration

In order to increase the effectiveness and efficiency of the maintenance process, the railway administration (Trafikverket), started to open up its maintenance contract for market competition [12, 13]. That is, anyone with the capability to deliver the contract could participate in the contract tendering process. Since railway maintenance is specialized and needs special tools and skills, there were only a few service agents in Sweden who could perform the service. This provided an opportunity for service agents from other European countries to bid for the contract. Today at least four service agents have been awarded contracts, based on their competence, capability, and price, for carrying out maintenance in different regions.

The selection of service agents at Trafikverket, in general, involves the following steps [12]:

1. Pre-qualification of contractors: This is performed at the Head office level and all the contractors or service agents planning to bid for a contract must register and be approved by the committee based on their capability, past performance, ethics, etc.
2. Announcement of contract: The contract is advertised in most of the listed major newspapers with a short description of the job and the contact details of the persons responsible for the contract.
3. Contract procurement process: During this step, potential contractors are informed about the type, scope, duration, and other relevant descriptions of the contract. Based on this information, interested contractors submit bids for the contract.
4. Pre-selection: Based on the details of the submitted bid and other relevant information about contractor, the client (infrastructure manager) selects 2–3 contractors to initiate the contract negotiation process.
5. Contract negotiations: During this step, the contract together with the scope of the work and the related price tags, etc., are discussed in detail with the selected potential service agents. This step also leads to the final selection of the service agent most suitable for the contract.

6. Study and analysis of contract: After selecting the service agent, the client and service agent both study and analyze the contract and enter into agreement whereby the contract is defined at a detailed level.
7. Signing of contract and its implementation as per the time and delivery plan.

9 Topics for New Research

As mentioned earlier, most of the literature on maintenance outsourcing is qualitative with only a small number of papers taking a quantitative approach. A proper study of maintenance outsourcing requires (1) an interdisciplinary approach involving science, engineering, technology, mathematics, and management and (2) a more quantitative approach to evaluate different maintenance contracts and identify the best contract taking into account the interests of the different parties involved.

Game theory and Agency Theory provide the foundations for building models to study maintenance outsourcing. However, most of the literature on game theory and Agency Theory consists of models that have very basic stochastic formulations. We suggest a multi-step approach to conduct new research of relevance to maintenance outsourcing.

Step 1: Develop a comprehensive framework that deals with the science, engineering, technology, and management issues in an integrated manner for a proper study of maintenance outsourcing.

Step 2: Identify the key elements, the variables to characterize these elements, and the interaction between the variables.

Step 3: Develop a simple model. This would imply a single stage (so that from a game theory perspective a static game formulation is used) and only two players—the owner of the asset and a single service agent. The objective functions for the two players would involve stochastic model formulations for failures and costs over a pre-specified contract period. The model formulation needs to look at contract specification (tasks to be carried out, incentives and penalties, monitoring schemes to detect cheating, etc.). Alternate scenarios can be considered which lead to different Stackelberg and Nash game formulations. The aim is to devise and evaluate contracts which ensure there are no incentives for cheating and that both parties reveal full information.

Step 4: Improve on the model of Step 3. This implies a multi-stage formulation and more than two players. This introduces new issues such as the owner having the option to change the service agent, competition between the agents, etc. These models need to incorporate learning effects and other factors such as customer satisfaction and loyalty (which lead to the renewal of contracts) and many other issues.

One further area where considerable research needed is a study of the role of data and information and their impact on the optimal strategies of the different players involved.

9.1 Maintenance Outsourcing in Railways

The rolling stock (engines, bogies, and wagons) interacts with the track and the degradation of the track and rolling stock is influenced by the interaction between them. It is affected by the condition of the rolling stock and of the infrastructure and by several other factors such as load, speed of travel, etc.

The owners of the infrastructure and the rolling stock can each outsource their maintenance so that there are several service agents involved. The different contracts between the owners and service agents are indicated in Fig. 6. This scenario implies several different players and the decision making needs to take into account the interaction between the different variables.

The need for an interdisciplinary approach to solve the maintenance outsourcing problem is highlighted through the following observations:

- Science: The degradation process due to the interaction between the track and the rolling stock.
- Engineering and Technology: The assessment of the condition of the track and the rolling stock (and for other variables such as axle load, etc.).
- Economic: The evaluation of the cost of maintenance; the consequence of failure resulting in the rolling stock and/or track being out of action, etc.
- Management: The drafting of the contract and the setting up of systems to collect relevant data and information.

The authors are currently looking at the structures of different contracts and models to evaluate each type of contract and to choose the best option.

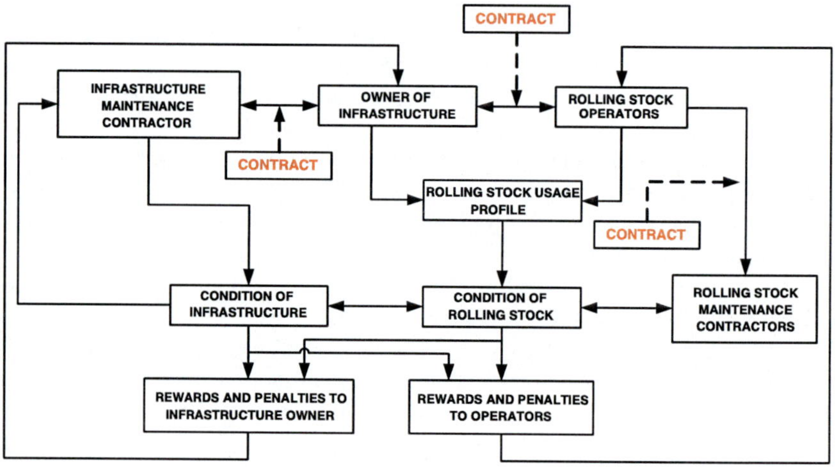

Fig. 6 Key elements and their interaction

10 Conclusions

We have looked at maintenance outsourcing and the issues that need to be addressed in the maintenance outsourcing context. A proper study and evaluation of maintenance outsourcing requires a quantitative approach. Game theory and Agency Theory provide good starting points to build new models which look at maintenance from both the owner and the service agent perspectives. This chapter gives a brief introduction to these two topics and defines some areas for possible future research.

Acknowledgments Dedicated to Professor Shunji Osaki on his seventieth birthday in recognition of his contributions and leadership of research in reliability and maintenance in Japan.

References

1. Almeida AT (2005) Multicrieria modelling of repair contract based on utility and ELECTRE I method with dependability and service quality criteria. Ann Oper Res 128:113–126
2. Armstrong RD, Cook WD (1981) The contract formation problem in preventive pavement maintenance: a fixed-charge goal-programming model. Comput Environ Urban Syst 6:147–155
3. Ashgarizadeh E, Murthy DNP (2000) Service contracts—a stochastic model. Math Comput Model 31:11–20
4. Ben-Daya M, Duffuaa S, Raouf A (2000) Maintenance, modeling, and optimization. Kluwer, Boston
5. Bertolini M, Bevilacqua M, Braglia M, Frosolini M (2004) An analytical method for maintenance outsourcing service selection. Int J Qual Reliab Manage 21:772–788
6. Bevilacqua M, Braglia M (2000) The analytic hierarchy process applied to maintenance strategy selection. Reliab Eng Syst Saf 70:71–83
7. Campbell JD (1995) Outsourcing in maintenance management: a valid alternative to self-provision. J Qual Maintenance Eng 1:18–24
8. Chatterjee K, Samuelson WF (2001) Game theory and business applications. Kluwer, Dordrecht
9. Dunn, S. (1999), Maintenance outsourcing—critical issues, available at: www.plant-maintenance.com/maintenance_articles_outsources.html
10. Eisenhardt KM (1989) Agency theory: an assessment and review. Acad Manag Rev 14:57–74
11. Embleton PR, Wright PC (1998) A practical guide to successful outsourcing. Empowerment Organ 6:94–106
12. Espling U (2007) Maintenance strategy for a railway infrastructure in a regulated environment. Ph D Thesis, Division of Operation and Maintenance Engineering, Luleå University of Technology, Luleå, Sweden 2007:54 (www.ltu.se/maintenance)
13. Espling U, Olsson U (2004) Partnering in railway infrastructure maintenance contract. J Qual Maintenance Eng 10:248–253
14. Jackson C, Pascual R (2008) Optimal service contract negotiation with aging equipment. Eur J Oper Res 189:387–398
15. Judenberg J (1994) Applications maintenance outsourcing. Inf Syst Manag 11:34–38
16. Blake HM (1984) In: Kintes EW (ed) The guide to American law. West Publishing Company, St Paul. Minn
17. Kim S-H, Cohen MA, Netessine S, Veeraraghavan S (2010) Contracting for infrequent restoration and recovery of mission-critical systems. Manage Sci 56:1551–1567
18. Kumar R, Kumar U (2004) Service delivery strategy: trends in mining industries. Int J Surf Min Reclam Environ 18:299–307

19. Laffont J, Martimort D (2002) The theory of incentives: the principal-agent model. Princeton University Press, Princeton
20. Levery M (1998) Outsourcing maintenance: a question of strategy. Eng Manage J, pp 34–40 (Feb)
21. Lisnianski A, Frenkel L, Khvatskin L, Ding Y (2008) Maintenance contract assessment for aging systems. Qual Reliab Eng Int 24:519–531
22. Macho-Stadler I, Perez-Castrillo D (1997) An introduction to the economics of information. Oxford University Press, Oxford
23. Martin HH (1997) Contracting out maintenance and a plan for future research. J Qual Maintenance Eng 3:81–90
24. Murthy DNP, Ashgarizadeh E (1998) A stochastic model for service contract. Int J Reliab Qual Saf Eng 5:29–45
25. Murthy DNP, Ashgarizadeh E (1999) Optimal decision making in a maintenance service operation. Eur J Oper Res 116:259–273
26. Osaki S (2002) Stochastic models in reliability and maintenance. Springer, Berlin
27. Osborne MJ (2004) An introduction to game theory. Oxford University Press, Oxford
28. Pham H (2003) Handbook of reliability engineering. Springer, Berlin
29. Plambeck E, Zenios SA (2000) Performance-based incentives in a dynamic principal-agent model. Manuf Serv Oper Manage 2:240–263
30. Straub A, Van Mossel HJ (2007) Contractor selection for performance based maintenance partnership. Int J Strateg Property Manage 11:65–76
31. Stremersch S, Wuyts S, Frambach RT (2001) The purchasing of full-service contracts: an exploratory study within the industrial maintenance market. Ind Mark Manage 30:1–12
32. Sunny I (1995) Outsourcing maintenance: making the right decisions for the right reasons. Plant Eng 49:156–157
33. Tarakci H, Tang K, Moskowitz H, Plante R (2006a) Incentive maintenance contracts for channel coordination. IIE Trans 38:671–684
34. Tarakci H, Tang K, Moskowitz H, Plante R (2006b) Maintenance outsourcing of a multi-process manufacturing system with multiple contractors. IIE Trans 38:67–78
35. Tarakci H, Tang K, Teyarachakul S (2009) Learning effects on maintenance outsourcing. Eur J Oper Res 192:138–150
36. Tseng F-S, Tang K, Moskowitz H, Plante R (2009) Maintenance outsourcing contracts for new technology adoptions. IEEE Trans Eng Manage 56:203–218
37. van Ackere A (1993) The principal-agent paradigm: its relevance to various functional fields. Eur J Oper Res 70:83–103
38. Vickerman R (2004) Maintenance incentives under different infrastructure regimes. Utilities Policy 12:315–322
39. Wang W (2010) A model for maintenance service contract design, negotiation and optimization. Eur J Oper Res 201:239–246
40. Watson J (2008) Strategy: an introduction to game theory. Norton, New York

Warranty/Maintenance: On Modeling Non-zero Rectification Times

Stefanka Chukova and Yu Hayakawa

Abstract This chapter revisits modelling of warranty/maintenance costs under the assumption that both, the warranty repairs and the maintenance actions, require non-negligible completion time. We provide an intuition on this topic by summarising our previous results, as well as the published work of other authors. We closely examine a case study that provides an excellent motivation for extending the research in this area. Also, again assuming non-negligible repair and maintenance times, we propose a simulation model for the expected warranty costs that integrates the concepts of reliability improvement and warranty. We conclude with a discussion on new directions for future research.

1 Introduction

A product warranty is an agreement offered by a producer to a consumer to repair or replace a faulty item, or to partially or fully reimburse the consumer in the event of a product failure. From the buyer's viewpoint, the product warranty assures free (partially or fully) of charge replacement or repair of a faulty product. It also provides information on the reliability and quality of the product. On the other hand, from the producer's viewpoint, the product warranty plays a protectional as well as a promotional role.

S. Chukova (✉)
School of Mathematics, Statistics and Operations Research, Victoria University of Wellington, Wellington PO Box 600, New Zealand
e-mail: Stefanka.Chukova@ecs.vuw.ac.nz

Y. Hayakawa
School of International Liberal Studies, Waseda University, 1-6-1 Nishi-Waseda, Shinjuku-ku, Tokyo, Japan
e-mail: yu.hayakawa@waseda.jp

T. Dohi and T. Nakagawa (eds.), *Stochastic Reliability and Maintenance Modeling*, Springer Series in Reliability Engineering 9, DOI: 10.1007/978-1-4471-4971-2_4,

Maintenance is an operation that involves fixing the product should it become faulty or out of order. It also includes performing routine actions which keep the product in working condition (a scheduled maintenance) or prevent operational problems from occurring (a preventive maintenance). Overall, maintenance consists of all actions that aim to retain or restore the product (or system) to a state in which it can perform the functions it is designed for.

In most published work on warranty and maintenance, the warranty repair times (and the maintenance times) are assumed to be negligible, i.e. the expected duration of the repair is small (negligible) compared to the expected lifetime of the product. And, yes, in many cases this is a reasonable assumption. But there are situations where the length of the repair (or the duration of the maintenance action) impacts significantly the operational cost. For example, if the maintenance is performed on an assembly line, which produces the main components of a system, the whole production process might be affected, e.g. put on hold, and it could lead to significant losses. If a taxi driver has to wait a couple of weeks until his car (taxi) undergoes a warranty repair, his loss of income could be quite high.

Why is it important to study models with non-negligible warranty repairs or maintenance times? First, the warranty period is a finite interval of time and the total repair time could be a significant portion of it. The total length of the repair time could be of importance in the warranty contract. Moreover, lengthy repairs/maintenance actions may lead to high penalty costs that have to be taken into account in the cost–benefit analysis. Therefore, taking into account the length and the type of the warranty repairs and the duration of the maintenance action is an important component in the warranty/maintenance cost modelling.

In this chapter, some results (see [2, 3]) regarding the evaluation of the expected warranty cost under non-renewing and renewing free replacement warranty policies over the warranty period and over the product life cycle are summarised. We allow for non-zero warranty repair time and assign costs, which are dependent on the length of the repair. Moreover, we review the advances in this area of modelling presented in [9] and [4]. We provide an insight into the importance of this type of modelling by summarising a case study presented in [6]. Lastly, again assuming non-negligible repair and maintenance times, we propose a simulation model for the expected warranty costs that integrates the concepts of reliability improvement and warranty. We conclude with a discussion on new directions for future research.

The outline of this chapter is as follows: In Sect. 2 we recall some basic warranty/maintenance terminology. In Sects. 3 and 4 we summarise the models for non-zero warranty repairs under non-renewing and renewing warranty policies. Sections 5 and 6 review two maintenance models with non-zero maintenance times. A case study is summarised in Sect. 7. In Sect. 8 we propose a new simulation model and Sect. 9 concludes this chapter.

2 Miscellaneous

This section provides the terminology used in warranty and maintenance analysis, that we need for our write-up.

2.1 Warranty Policy

The typical warranty coverage used in the industry can be classified as follows:

- Non-renewing warranty: The expenses associated with the failure of the product during the warranty period of length T are covered (fully or partially) by the warranter.
- Renewing warranty: The expenses associated with the failure of the product during the warranty period of length T are covered (fully or partially) by the warranter. In addition, after each warranty repair, the repaired item is warranted anew for a period T.

2.2 Maintenance Policy

The two classical models mostly studied in the maintenance literature are:

- Block-based model—In this model, a preventive maintenance action is performed periodically over a fixed time interval τ, i.e. at calendar times $\tau, 2\tau, 3\tau, \ldots$, a maintenance action is invoked. The block-based policy is proposed for a calendar-time-based maintenance model. At failure, the corrective maintenance is carried out.
- Age-based model—In this model, a preventive maintenance action is performed as soon as the product (system) reaches a pre-specified age κ. In addition, corrective maintenance is executed at failure.

2.3 Degree of Repair

In our presentation we consider different types of repairs. Pham and Wang [5] classified repairs according to the degree to which they restore the product. They propose the following classification:

- **Improved Repair:** A repair brings the product to a state *better* than when it was initially purchased. This is equivalent to the replacement of the faulty item by a new and improved item.

- **Perfect (or Complete) Repair:** A repair completely resets the performance of the product so that upon restart the product operates as a new one. This type of repair is equivalent to a replacement of the faulty item by a new one, identical to the original.
- **Imperfect Repair:** A repair contributes to some noticeable improvement of the product. It effectively sets back the clock for the repaired item. After the repair the performance and expected lifetime of the item are as they were at an earlier age.
- **Minimal Repair:** A repair has no impact on the performance of the item. The repair brings the product from a 'down' to an 'up' state without affecting its performance.
- **Worse Repair:** A repair contributes to some noticeable worsening of the product. It effectively sets forward the clock for the repaired item. After the repair the performance of the item is as it would have been at a later age.
- **Worst Repair:** A repair accidentally leads to the product's destruction.

In the following two sections, we summarise our results on modelling non-zero warranty repair times (as given in [2, 3]) based on the alternating renewal process.

3 Non-Renewing Warranty: Non-zero Repair Times

This section is concerned with the non-renewing warranty and incorporates non-zero warranty repair times.

3.1 The Model

We consider the following model: At the beginning the item is in operating ('on') condition for a time X_1. Then the repair ('off') condition starts and the item remains in it for a time Y_1. After the repair completion, the item is operative for a time X_2, which is followed by Y_2 long repair and so on.

The time between two consecutive returns of the virtual age, $V(t)$, of the item to 0 forms a renewal cycle, see Fig. 1. We suppose that both sequences of random

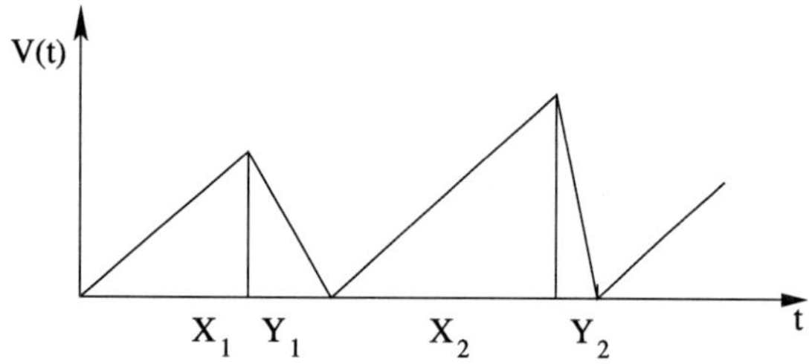

Fig. 1 The virtual age of the item

variables $\{X_i\}_1^\infty$ and $\{Y_i\}_1^\infty$ are independent and identically distributed. Moreover, we assume that X_i and Y_i are independent for $i = 1, 2, \ldots$. The above model is the well-known model of alternating renewal process [7] and we use it to model the warranty cost.

We assume that the cost of the ith repair is random and with form $C_i = A + \delta Y_i$, where A, and δ are known constants.

Moreover, we suppose that the cost is incurred at the end of the renewal cycle. Also, if the warranty coverage expires during a repair period, the corresponding repair is completed and its cost is fully incurred by the warranter. In this case we have a **complete renewal cycle**. If the warranty expires during an operating period, the cost of the following repair is not included in the total costs and we have an **incomplete renewal cycle**.

Life cycle of a product is defined as a time while the product is still usable and contemporary. It is assumed that during the life cycle, after the expiration of the warranty period for the initially purchased item, at the time of the first off warranty failure, the consumer purchases an identical item to the initial one with the same warranty coverage. We will assume that a life cycle can end only at off warranty time. The latter assumption is reasonable because the length of the life cycle is mainly determined by the consumer.

We aim to evaluate: (1) the warranty expenses under non-renewing free replacement warranty of duration T and (2) the expected total warranty costs over the life cycle L of the item. To achieve these goals, as a preliminary, we obtain some results regarding the alternating renewal process.

3.2 Alternating Renewal Process in Finite Horizon

Consider the length of a renewal cycle $X + Y$ with the cumulative distribution function (cdf) F_{X+Y}. Consider the alternating renewal process with "on" time distribution F_X and "off" time distribution F_Y. Denoting

$$S_n = \sum_{i=1}^{n} (X_i + Y_i) \quad \text{and} \quad S_0 = 0$$

it follows that S_n is the time of the completion of the nth repair and corresponding

$$N(t) = \max\{n : S_n \leq t\}$$

is the number of complete renewal cycles before time t (cf. [7]). Denote by $m_{X+Y}(y) = E(N(t))$ the corresponding renewal function. It is known (cf. [7]) that

$$P\,(\text{on at } t) = \bar{F}_X(t) + \int_0^t \bar{F}_X(t-y)\, dm_{X+Y}(y), \tag{1}$$

which is the probability of having operating item at time t. It is easy to see that $P\,(\text{off at } t) = 1 - P\,(\text{on at } t)$ is equivalent to

$$\begin{aligned} P\,(\text{off at } t) = &\int_0^t \bar{F}_Y(t-u)\, dF_X(u) \\ &+ \int_0^t \int_0^{t-u} \bar{F}_Y(t-u-v)\, dF_X(v)\, dm_{X+Y}(u). \end{aligned} \tag{2}$$

Theorem 3.1

$$P(S_{N(T)} \le t \mid \text{on at } T) = \frac{\bar{F}_X(T) + \int_0^t \bar{F}_X(T-u) dm_{X+Y}(u)}{\bar{F}_X(T) + \int_0^T \bar{F}_X(T-u) dm_{X+Y}(u)}, \quad 0 \le t \le T \tag{3}$$

Proof:

$$\begin{aligned} &P(S_{N(T)} \le t \mid \text{on at } T) P(\text{on at } \; T) \\ &\quad = P(\text{on at } \; T \mid S_{N(T)} = 0) P(S_{N(T)} = 0) \\ &\qquad + \int_0^t P(\text{on at } T \mid S_{N(T)} = u)\, dF_{S_{N(T)}}(u) \\ &\quad = \frac{P(X_1 + Y_1 > T, X_1 > T)}{P(X_1 + Y_1 > T)} P(X_1 + Y_1 > T) \\ &\qquad + \int_0^t P(\text{on at } T \mid X_n + Y_n > T - u)\, \bar{F}_{X+Y}(T-u)\, dm_{X+Y}(u) \\ &\quad = \bar{F}_X(T) + \int_0^t P(X_n > T - u \mid X_n + Y_n > T - u)\, \bar{F}_{X+Y}(T-u)\, dm_{X+Y}(u) \\ &\quad = \bar{F}_X(T) + \int_0^t \frac{\bar{F}_X(T-u)}{\bar{F}_{X+Y}(T-u)} \bar{F}_{X+Y}(T-u)\, dm_{X+Y}(u) \\ &\quad = \bar{F}_X(T) + \int_0^t \bar{F}_X(T-u)\, dm_{X+Y}(u) \end{aligned}$$

Therefore, using (1), the proof is completed. □

Corollary 3.2

$$P(S_{N(T)} = 0 \mid \text{on at } T) = \frac{\bar{F}_X(T)}{\bar{F}_X(T) + \int_0^T \bar{F}_X(T-u)\, dm_{X+Y}(u)} \tag{4}$$

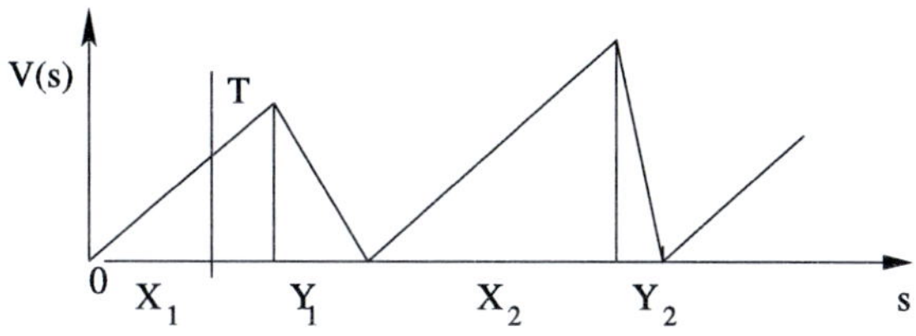

Fig. 2 $S_{N(T)} = 0$

Theorem 3.3 For $T \leq t$,

$$P(S_{N(T)} + X_{N(T)+1} \leq t \mid \text{on at } T)$$
$$= \frac{\bar{F}_X(T) - \bar{F}_X(t)}{\bar{F}_X(T) + \int_0^T \bar{F}_X(T-u)\, dm_{X+Y}(u)}$$
$$+ \frac{\int_0^T \left(\bar{F}_X(T-u) - \bar{F}_X(t-u)\right) dm_{X+Y}(u)}{\bar{F}_X(T) + \int_0^T \bar{F}_X(T-u)\, dm_{X+Y}(u)}$$

The proof is similar to that of Theorem 3.1, hence it is omitted.

We sketch another proof of Theorem 3.3 by utilising the multiplication rule and the total probability rule. Namely, by conditioning on $S_{N(T)}$, we consider the following two cases:

1. $S_{N(T)} = 0$, (**Fig.** 2).
   The following events are equivalent.

   $$\{S_{N(T)} + X_{N(T)+1} \leq t,\ S_{N(T)} = 0, \quad \text{on at } T\} \Longleftrightarrow \{T < X \leq t\}.$$

   The probability of the latter is equal to

   $$F_X(t) - F_X(T). \tag{5}$$

2. $S_{N(T)} = w \neq 0$, (**Fig.** 3).
   The following events are equivalent $\{S_{N(T)} + X_{N(T)+1} \leq t,\ S_{N(T)} = w \neq 0, \ \text{on at } T\} \Leftrightarrow$ {there is a renewal before T} say at time w with probability $dm_{X+Y}(w)$ and $\{T - w < X < t - w\}$, which will occur with probability $F_X(t-w) - F_X(T-w)$. The probability of the second event is

   $$\int_0^T (F_X(t-w) - F_X(T-w))\, dm_{X+Y}(w) \tag{6}$$

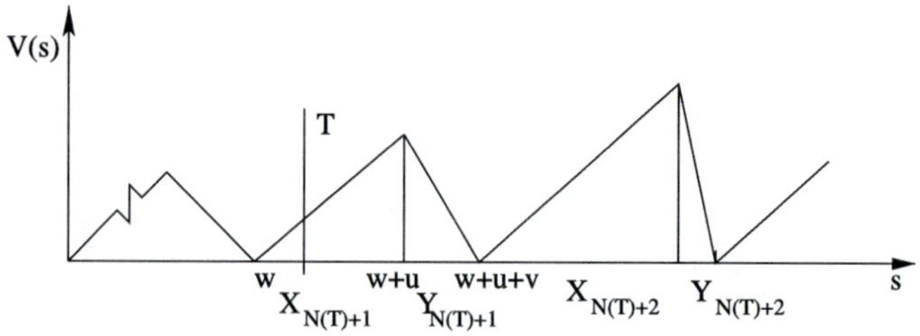

Fig. 3 $S_{N(T)} = w \neq 0$

Adding (5) and (6) evaluates $P(S_{N(T)} \leq t,$ on at $T)$. Then, the result of the theorem follows from the multiplication rule and (1). □

Theorem 3.4

$$P(S_{N(T)+1} + X_{N(T)+2} \leq t \,|\, \text{off at } T)$$
$$= \frac{1}{P(\text{off at } T)} \times \left(\int_0^T \int_{T-u}^{t-u} F_X(t-u-v)\, dF_Y(v)\, dF_X(u) \right.$$
$$\left. + \int_0^T \int_0^{T-w} \int_{T-w-u}^{t-w-u} F_X(t-w-u-v)\, dF_Y(v)\, dF_X(u)\, dm_{X+Y}(w) \right) \tag{7}$$

The proof of this theorem is similar to that given in Theorem 3.3, hence it is omitted.

3.3 *Warranty Cost Analysis*

Here we derive the expected warranty cost for non-renewing warranty over warranty period of length T and over the life cycle of length L. By assumption, the random variables C_i are independent and identically distributed and $E(C) = A + \delta E(Y)$.

3.3.1 Expected Costs Over $(0, T)$

Denote by $C(t)$ the total warranty cost accumulated up to time t. We have to distinguish two cases: first the warranty expires during an "off" time, then the total cost is accumulated over $N(T) + 1$ complete renewal cycles. Second, the warranty expires during an "on" time, so that only $N(T)$ complete renewal cycles contribute to the cost. Then

$$C(T) = \begin{cases} \sum_{i=1}^{N(T)} C_i, & \text{if the item is "on" at time T} \\ \sum_{i=1}^{N(T)+1} C_i, & \text{if the item is "off" at time T} \end{cases} \tag{8}$$

and the following result holds:

Theorem 3.5

$$E(C(T)) = (m_{X+Y}(T) + 1)\, E(C) - E(C_{N(t)+1}|\ \text{on at } T) P(\text{on at } T).$$

Proof: Using that $N(t) + 1$ is a stopping time for the sequence $\{C_i\}_1^\infty$ and Wald's equation (see [7]) we have

$$\begin{aligned} E(C(T)) &= E\left(\sum_{i=1}^{N(T)} C_i|\text{on at } T\right) P(\text{on at } T) + E\left(\sum_{i=1}^{N(T)+1} C_i|\text{off at } T\right) P(\text{off at } T) \\ &= E\left(\sum_{i=1}^{N(T)+1} C_i - C_{N(T)+1}|\text{on at } T\right) P(\text{on at } T) \\ &\quad + E\left(\sum_{i=1}^{N(T)+1} C_i|\text{off at } T\right) P(\text{off at } T) \\ &= E\left(\sum_{i=1}^{N(T)+1} C_i|\text{on at } T\right) P(\text{on at } T) + E\left(\sum_{i=1}^{N(T)+1} C_i|\text{off at } T\right) P(\text{off at } T) \\ &\quad - E(C_{N(T)+1}|\text{on at } T) P(\text{on at } T) \\ &= E\left(\sum_{i=1}^{N(T)+1} C_i\right) - E(C_{N(T)+1}|\text{on at } T)\, P(\text{on at } T) \\ &= (m_{X+Y}(T) + 1)\, E(C) - E(C_{N(t)+1}|\text{on at } T)\, P(\text{on at } T) \end{aligned}$$

□

We need to find $E(C_{N(t)+1}\ |\ \text{on at } T)\ P(\text{on at } T)$. The latter probability is given by (1). Since $C_{N(T)+1} = A + \delta Y_{N(T)+1}$, we need to evaluate $E(Y_{N(t)+1}\ |\ \text{on at } T)$. The following result holds:

Theorem 3.6

$$E(C(T)) = (A + \delta E(Y))\, (m_{X+Y}(T) + P(\text{off at } T))$$

The following lemma will be needed for the proof of the theorem:

Lemma 3.7

$$E(Y_{N(t)+1} \mid \text{on at } T) = E(Y)$$

Proof: By conditioning on $S_{N(T)}$, and using Theorem 3.1 and Corollary 3.2 we obtain

$$\begin{aligned}
&E(Y_{N(t)+1} | \text{on at } T) \\
&= E(Y_{N(t)+1} | S_{N(T)} = 0, \quad \text{on at } T)\, P(S_{N(T)} = 0 | \text{on at } T) \\
&+ \int_0^T E(Y_{N(t)+1} | S_{N(T)} = s, \text{ on at } T)\, dP(S_{N(T)} \le s | \text{on at } T) \\
&= E(Y_1 \mid X_1 > T) \frac{\bar{F}_X(T)}{\bar{F}_X(T) + \int_0^T \bar{F}_X(T-u)\, dm_{X+Y}(u)} \\
&+ \int_0^T E(Y_n \mid X_n > T - s) \frac{\bar{F}_X(T-s)\, dm_{X+Y}(s)}{\bar{F}_X(T) + \int_0^T \bar{F}_X(T-u)\, dm_{X+Y}(u)} \\
&= \int_0^T E(Y)\, dF_{S_{N(T)} \mid \text{on at } T}(s) = E(Y)
\end{aligned}$$

Using Lemma 3.7 it is easy to complete the proof of Theorem 3.6.

Proof: Indeed

$$\begin{aligned}
E(C(T)) &= E\left(\sum_{n=1}^{N(T)+1} C_i\right) - E(C_{N(T)+1} \mid \text{on at } T)\, P(\text{on at } T) \\
&= (m_{X+Y}(T) + 1)\, E(C) - E(C)\, P(\text{on at } T) \\
&= E(C)\, (m_{X+Y}(T) + P(\text{off at } T)).
\end{aligned}$$

□

3.3.2 Expected Costs Over (0, *L*)

Now we will focus on the evaluation of the expected warranty costs over the life cycle of an item. Let us consider the time between two consecutive purchases made by the consumer. Denote this time by ξ. It is a positive continuous random variable such that:

$$\xi = \begin{cases} S_{N(T)} + X_{N(T)+1}, & \text{if the item is "on" at time T} \\ S_{N(T)+1} + X_{N(T)+2}, & \text{if the item is "off" at time T} \end{cases}$$

Then, the expected costs over $(0, L)$ are expressed in terms of ξ in the following way:

$$E(C(L)) = E(N^*(L) + 1)\, E(C(T)),$$

where $N^*(t)$ is a renewal process with interevent time equal to ξ. Denote by $m^*_\xi(t)$ the renewal function of $N^*(t)$. Then

$$E(C(L)) = (m^*_\xi(L) + 1) E(C(T)). \tag{9}$$

In what follows we derive the distribution of the interevent time ξ.

Theorem 3.8

$$\begin{aligned}
P(\xi \le t) &= \bar{F}_X(T) - \bar{F}_X(t) \\
&\quad + \int_0^T (\bar{F}_X(T-u) - \bar{F}_X(t-u))\, dm_{X+Y}(u) \\
&\quad + \int_0^T \int_{T-u}^{t-u} F_X(t-u-v)\, dF_Y(v)\, dF_X(u) \\
&\quad + \int_0^T \int_0^{T-w} \int_{T-w-u}^{t-w-u} F_X(t-w-u-v)\, dF_Y(v)\, dF_X(u)\, dm_{X+Y}(w)
\end{aligned}$$

Proof:

$$\begin{aligned}
P(\xi \le t) &= P(S_{N(T)} + X_{N(T)+1} \le t \mid \text{on at } T) P(\text{on at } T) \\
&\quad + P(S_{N(T)+1} + X_{N(T)+2} \le t \mid \text{off at } T)\, P(\text{off at } T)
\end{aligned}$$

Applying Theorems 3.3 and 3.4 and using (1) and (2) completes the proof. □

3.4 Example

As an illustration of the ideas we will consider an example assuming that the lifetime of the item and the repair time are exponentially distributed random variables with parameters λ and μ.

3.4.1 Expected Costs Over $(0, T)$

In order to evaluate the expected warranty costs over $(0, T)$, we need to find the corresponding renewal function. Using Laplace transforms it can be shown that the renewal function for the renewal process with interevent time $X + Y$ is

Table 1 Expected warranty cost over a warranty period, $\mu = 92$

	T					
λ	0.5	1.0	1.5	2.0	2.5	3.0
2	2.958815	5.916262	8.873708	11.831155	14.788602	17.746049
3	4.392487	8.781961	13.171434	17.560908	21.950382	26.339855

Table 2 Expected warranty cost over a warranty period, $\mu = 122$

	T					
λ	0.5	1.0	1.5	2.0	2.5	3.0
2	2.968527	5.936269	8.904011	11.871752	14.839494	17.807236
3	4.417737	8.833737	13.249737	17.665737	22.081737	26.497737

$$m_{X+Y}(t) = \frac{\lambda\mu}{\lambda+\mu}\left(t - \frac{1}{\lambda+\mu}\left(1 - e^{-(\lambda+\mu)t}\right)\right).$$

Using (2), we get $P(\text{off at } t) = \frac{\lambda}{\lambda+\mu}(1 - e^{-(\lambda+\mu)T})$. Then, the expected warranty cost for non-renewing free replacement warranty policy with duration T is equal to

$$E(C(T)) = \left(A + \frac{\delta}{\mu}\right)\left(\frac{\lambda}{\lambda+\mu}\right)\left(\mu T + \frac{\lambda}{\lambda+\mu}\left(1 - e^{-(\lambda+\mu)T}\right)\right). \tag{10}$$

For selected values of T and λ and for $A = 3$ and $\delta = 2$, numerical values for the expected warranty cost are calculated and summarised in Tables 1 and 2.

The comparison between the two tables shows that it is better to have a longer average repair time (4 days for Table 1 against 3 days for Table 2). A possible reason for this result is the fact that for the fixed values of T and λ the value of μ will reflect on the number of renewal cycles per warranty period. Indeed, larger values of μ will increase the number of renewal cycles within the warranty period, which will increase the value of the expected warranty cost over $(0, T)$. Providing that the penalty cost δ is not too high, this is a reasonable strategy. On the other hand if δ is high and low expected warranty costs are targeted, it will require a reduction of the average repair time.

3.4.2 Expected Costs Over $(0, L)$

Even in this simple case of exponential lifetime and exponential repair time we encounter difficulties in evaluating the expected warranty cost over the life cycle of the item. The standard approach of finding the renewal function of the renewal process generated by the random variable ξ led to an expression with a limited value. The attempt to use MAPLE or MATHEMATICA to simplify the result was also not very successful. Hence, we used a numerical procedure. Based on the ideas of Xie [10], a

Table 3 Expected warranty cost over a life cycle, $\lambda = 2$, $\mu = 122$

	T					
L	0.5	1.0	1.5	2.0	2.5	3.0
3	10.75976	15.28904	17.36461	22.13644	24.21849	17.80759
5	16.69591	23.07293	26.60188	30.76411	29.57897	35.28823
7	22.66831	30.99828	35.74898	39.24450	43.15709	46.17403
10	31.53661	42.86935	49.25831	54.04590	57.49978	59.12158
15	46.36995	62.65530	71.51108	77.50511	82.34271	86.33264

Table 4 Expected warranty cost over a life cycle, $\lambda = 6$, $\mu = 122$

	T					
L	0.5	1.0	1.5	2.0	2.5	3.0
3	43.36922	51.46913	51.76007	68.92752	84.11078	51.75959
5	69.29022	83.65021	92.53079	102.91867	86.26324	103.51294
7	95.74140	112.94833	125.45428	135.87711	129.39141	154.36811
10	133.98574	156.33294	169.55625	172.45343	172.52479	203.51039
15	198.68439	230.64985	249.49196	255.03050	258.75241	258.7822

renewal equation solver has been written by Dr Richard Arnold in programming language R. The solver evaluates the renewal function under known cdf (in closed form), known pdf, or data for the renewal points. The last option is an extension of [10].

Using (9) and assuming $A = 3$ and $\delta = 2$, the expected warranty cost for selected values of L were evaluated. The comparison between Tables 3 and 4, with the given values of λ and μ, shows that the improvement of the reliability and quality of the product, reflecting on the increase of its average operating time, will highly reduce the expected warranty cost.

4 Renewing Warranty: Non-zero Repair Times

The alternating renewal process described in Sect. 3.2 is also assumed here. However, now we consider renewing warranty policy with perfect warranty repairs. Again, the cost of the ith repair is assumed to be $C_i = A + \delta Y_i$, and the random variables C_i are independent and identically distributed and their expected value is $A + \delta E(Y)$ (Fig. 4).

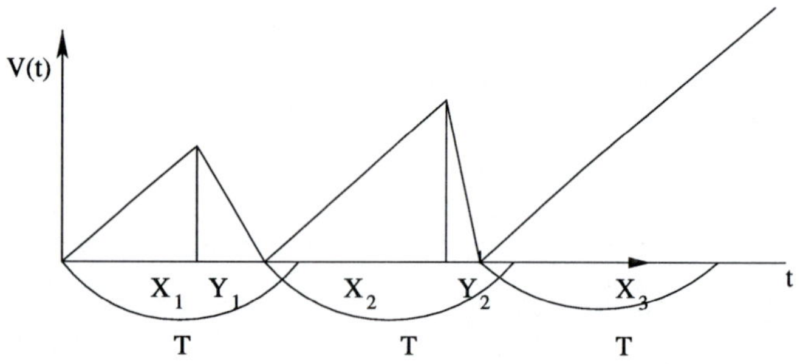

Fig. 4 Renewing warranty

4.1 The Model

We model the functioning of the item as in Sect. 3.1. Taking into account the impact of the renewing warranty, we adjust the model as follows: at the end of the repair time, the item is warranted anew for a period of length T, i.e. after each repair the item is assumed to be as good as new. If the warranty period ends during an operating period, the cost of the following repair is not incurred by the warranter and the warranty coverage expires. Here we will distinguish between warranty coverage W_T, which is a random variable, and warranty period, which is a predetermined constant T.

4.2 Warranty Cost Analysis

Here we derive the expected warranty cost for renewing warranty over warranty period of length T and over the life cycle of length L.

4.2.1 Expected Cost Under Renewing Warranty Coverage

Due to the mechanism of the renewing warranty, W_T is equal to:

$$W_T = \begin{cases} T, & \text{if } X_1 > T \\ T + \sum_{i=1}^{n}(X_i + Y_i), & \text{if } X_1 \le T, \cdots, X_n \le T, X_{n+1} > T \text{ for some } n. \end{cases}$$

Then, the warranty cost $C(W_T)$ over the warranty coverage is a random variable and its distribution is:

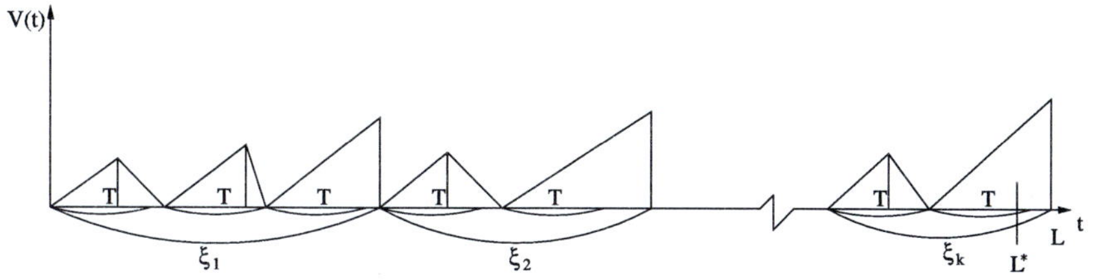

Fig. 5 Life cycle over (0, L)

$$C(W_T) = \begin{cases} 0, & \text{with } 1 - F_X(T) \\ C_1, & \text{with } F_X(T)(1 - F_X(T)) \\ \vdots & \\ \sum_{i=1}^{n} C_i, & \text{with } F_X(T)^n(1 - F_X(T)). \\ \vdots & \end{cases} \tag{11}$$

Thus, $C(W_T)$ has a geometric distribution with parameter $(1 - F_X(T))$ and

$$E(C(W_T)) = \frac{F_X(T)}{1 - F_X(T)}(A + \delta E(Y)). \tag{12}$$

Therefore, provided that the distributions of X and Y are known, using (12), the expected cost under renewing warranty coverage can easily be evaluated. Otherwise, (12) should be used with appropriate estimations of $F_X(T)$ and $E(Y)$.

4.2.2 Expected Costs Under Renewing Warranty Coverage Over Life Cycle

Let L^* be a prespecified time during which a product is considered to be contemporary and competitive with similar products in the market. Let L be the time of the first off warranty failure of the product after L^*. Then, we call (0, L) the life cycle of the item. The idea of life cycle and the relationship between L and L^* are represented in Fig. 5.

In what follows we evaluate the expected warranty costs over (0, L), where the value of L^* is known. Let us consider the continuous positive random variable, ξ, representing the time between two consecutive product purchases. By definition,

$$\xi = \begin{cases} X_1 & \text{if } X_1 > T \\ \sum_{i=1}^{n}(X_i + Y_i) + X_{n+1} & \text{if } X_1 \le T, \ldots, X_n \le T, X_{n+1} > T \text{ for some } n. \end{cases}$$

Then, the expected costs over (0, L), denoted by $E(C(L))$, are expressed in terms of ξ in the following way:

$$E(C(L)) = (m_\xi^*(L) + 1)\, E(C(W_T)),$$

where $m_\xi^*(t)$ is the renewal function of the renewal process generated by ξ.

Now, let us introduce the age parameter for ξ denoted by τ, i.e. τ is the time origin where ξ is measured from. We will derive the probability density function (pdf) of ξ, $g_\xi(\tau, t)$, given τ. The following theorem holds:

Theorem 4.9 *The pdf, $g_\xi(\tau, t)$, satisfies the following integral equation:*

$$g_\xi(\tau,t) = \begin{cases} f_X(t) + \int_0^{t-T} \int_0^{t-T-u} g_\xi \\ \quad (\tau + u + v, t - u - v) f_Y(v) f_X(u)\, dv du & \text{if } T < t < 2T \\ f_X(t) + \int_0^{T} \int_0^{t-T-u} g_\xi \\ \quad (\tau + u + v, t - u - v) f_Y(v) f_X(u)\, dv du & \text{if } t \geq 2T. \end{cases} \tag{13}$$

Proof: Using the definition of pdf, namely,

$$g_\xi(\tau, t)\Delta t \approx P(\xi \in (t, t + \Delta t))$$

for small Δt, we will condition on X_1. There are two scenarios under which the event $\{\xi \in (t, t+\Delta t)\}$ can occur. Their pictorial representations are given in Figs. 6 and 7.

- Scenario 1 (Fig. 6)
  $X_1 > T$, thus $\xi = X_1$. Then $\{\xi \in (t, t + \Delta t)\} \equiv \{X_1 \in (t, t + \Delta t)\}$.
- Scenario 2 (Fig. 7)
  $X_1 \leq T$.

Here our main idea is to find a relationship between $g_\xi(\tau, t)$ and $g_\xi(\tau + s, t - s)$, where s is the point of the first warranty renewal. We need to consider two cases:

1. $T < t < 2T$.
   Due to the definition of ξ, $0 < s < t - T$. (Recall that ξ will terminate only if $X_k > T$ for some k.) Then,

Fig. 6 $X_1 > T$

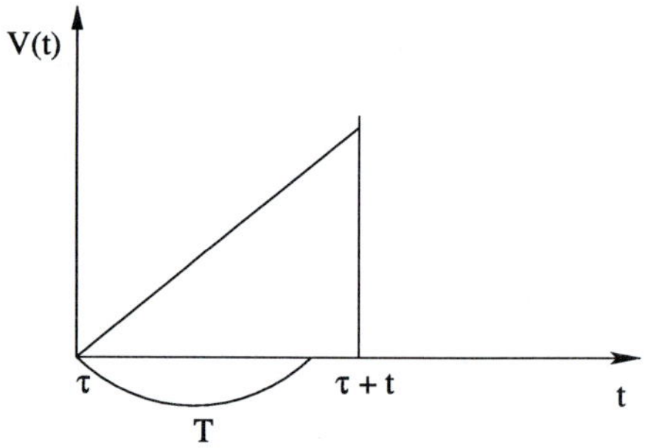

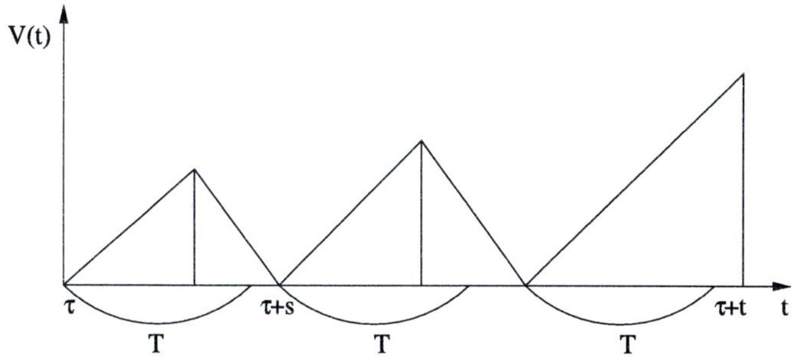

Fig. 7 $X_1 < T$

Fig. 8 $T < t < 2T$, $t - T < T$

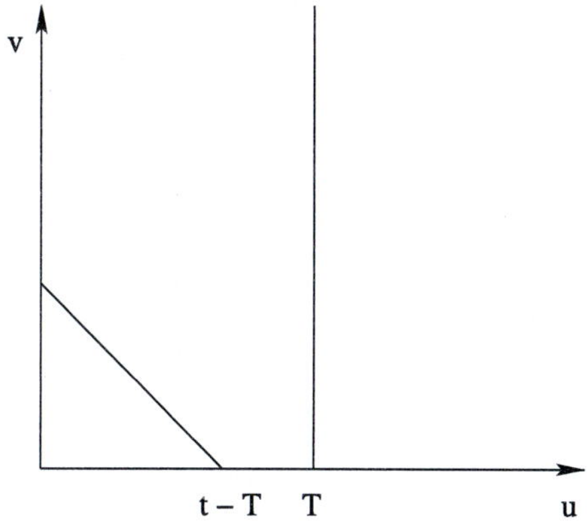

$$0 < s < t - T < 2T - T = T \Rightarrow 0 < s < T$$

and (see Fig. 8)

$$\{X_1 + Y_1 \in (s, s + \Delta s)\} \subset \{X_1 \in (u, u + \Delta u)\} \text{ for any } 0 < u \le s < T.$$

Hence, using Scenario 1, we obtain

$$g_\xi(\tau, t) = f_X(t) + \int_0^{t-T} \int_0^{t-T-u} g_\xi(\tau + u + v, t - u - v) f_Y(v) f_X(u)\, dv\, du.$$

2. $t > 2T$.
   Again, $0 < s < t - T$, but now $t - T > T$, (see Fig. 9), $\{\xi \in (t, t + \Delta t)\}$ is equivalent to the event: there is a failure at time u (measured from the origin τ) and $u < T$ (i.e. failure within the warranty period) which occurs with probability $f_X(u)du$, and repair lasting v, which occurs with probability $f_Y(v)dv$ and $\{\xi \in (t - u - v, t - u - v + \Delta(t - u - v))\}$ with initial age $\tau + u + v$, which occurs with probability $g_\xi(\tau + u + v, t - u - v)\Delta(t - u - v)$. Then, taking into account

Fig. 9 $t - T > T$

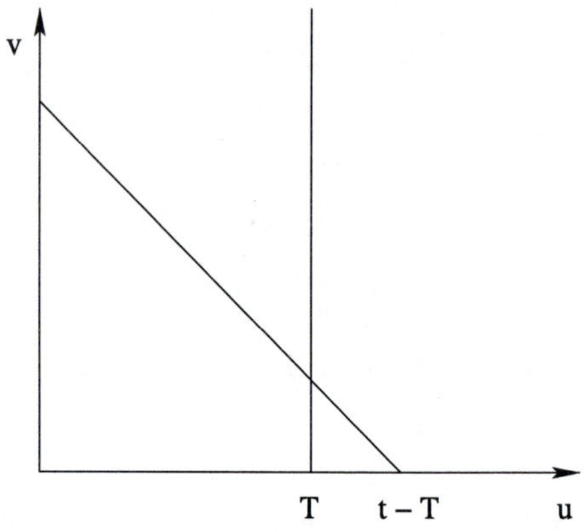

Scenario 2, we have

$$g_\xi(\tau, t) = f_X(t) + \int_0^T \int_0^{t-T-u} g_\xi(\tau + u + v, t - u - v) f_Y(v) f_X(u)\, dv\, du,$$

which completes the proof of the theorem.

4.2.3 Numerical Procedure for Calculating the pdf of ξ

Let us denote $S = X_1 + Y_1$. It is easy to notice that:

1. The support of S is $(0, t - T)$ for any t.
2. The differences in the limits of integration in (13) are due to the restriction $X_1 < T$.

Equation (13) can be rewritten in terms of S in the following way:

$$g_\xi(\tau, t) = f_X(t) + \int_0^{t-T} g_\xi(\tau + s, t - s) f_T(s) ds, \quad t > T, \tau \geq 0 \qquad (14)$$

where $f_T(s) = \int_0^{T \wedge s} f_X(u) f_Y(s - u) du$. The sub-density $f_T(s)$ reflects the comments at the beginning of this section.

Let us consider a grid of step h in the two-dimensional plane (τ, t). Let NL be the number of points on both τ and t axes. Note that, for convenience, the count of the points starts from 1.

Using the definition of Riemann–Stiltjes integral, $g_\xi(\tau, t)$ can be approximated by:

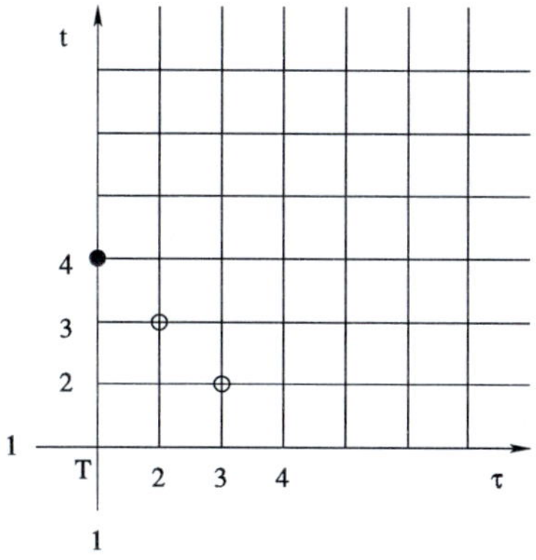

Fig. 10 Grid in (τ, t) plane

$$\begin{cases} g_{i1} = 0, \quad i = 1, 2, \cdots, NL; \\ g_{i2} = f_X(T + h); \\ g_{1j} = f_X(T + (j-1)h) + h \sum_{k=1}^{j-2} g_{1+k,j-k} f_T(kh), \quad j = 3, 4, \ldots, NL; \\ g_{ij} = g_{1j}, \quad i = 2, 3, \ldots, NL. \end{cases}$$

The reasoning for this algorithm is the following: The sum approximating the integration in (14) consists of only values of $g_\xi(\cdot, \cdot)$ calculated over the diagonals of the grid, i.e. if g_{ij} is to be calculated, then the previous values of $g_\xi(\cdot,\cdot)$ needed are only those $g_{i+k,j-k}$ for $k = 1, \ldots, j-k$. These values are calculated at points located on the diagonal consisting of (i, j). In each step of the procedure, once g_{1j} is evaluated, the remaining values $g_{ij}, i = 2, 3, \ldots, NL$ are assigned to equal to g_{1j}. This is because the meaning of the first parameter is the age and the distribution of ξ is independent of the age. The parameter τ was introduced only for convenience in an attempt to simplify the notations and the reasoning of the computational procedure (Fig. 10).

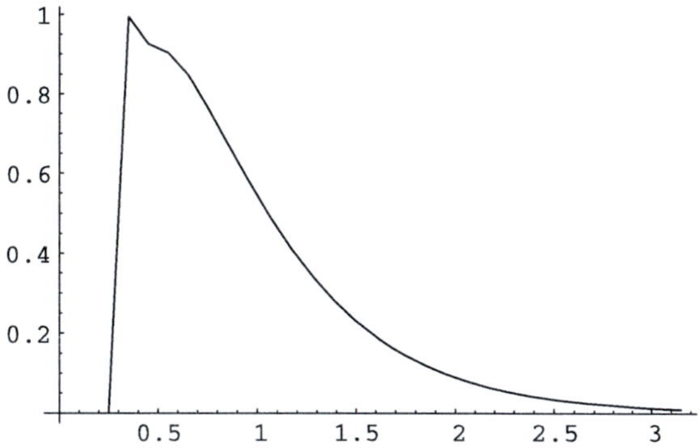

Fig. 11 $\lambda = 2, \mu = 10, T = 0.25$

Fig. 12 $\lambda = 0.5, \mu = 52, T = 0.25$

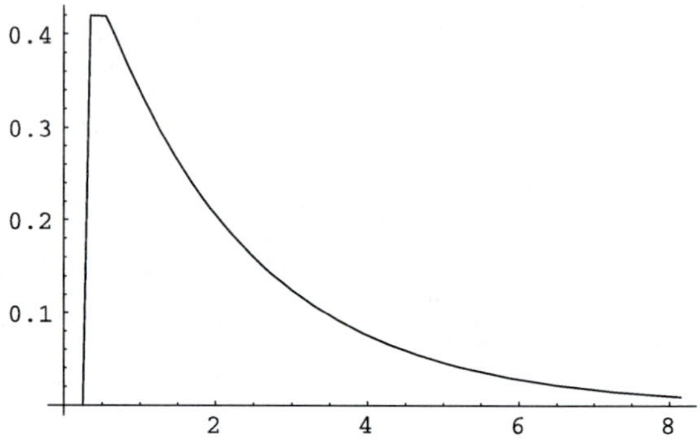

We used MATHEMATICA to write a code for the numerical procedure for obtaining the density $g_\xi(t)$. The graphs of the pdf of ξ show that the distribution of ξ is skewed to the right. It is unimodal and the value of its maximum decreases when T increases. For fixed value of T, the shape of the density depends on λ, μ (Figs. 11–12). For more details on $g_\xi(t)$, see [3].

4.2.4 Estimating the Renewal Function of ξ

Our next step is to use the suggested numerical procedure to evaluate the renewal function generated by ξ. Again, we used the renewal equation solver written by Dr. Richard Arnold.

4.3 Example

As an illustration we consider an example assuming that the lifetime of the item and the repair time are exponentially distributed random variables with parameters λ and μ. The same procedure is valid for general distribution of the operating time X and the repair time Y.

4.3.1 Expected Costs Over $(0, W_T)$

Using (12) for selected values of T and λ and for $A = 3$ and $\delta = 2$, numerical values for the expected warranty cost were calculated and summarised in the following two tables. We measure the time in years.

In the first row of Table 5 $\lambda = 2$, which means that on average there is a failure of the product every 6 months. The length of the following repair, again on average, is set to be equal to approximately four days. As expected the expected warranty cost is an increasing function of the length of the warranty period. The second row represents the values of the expected warranty cost for $\lambda = 4/3$, which means that on average, there is a failure of the product every nine months. The values of the remaining parameters are kept the same as for the first row. By comparing the rows in Table 5, it is easy to see that the shorter mean operating time leads to higher expected warranty cost, which is in agreement with our intuition. These conclusions are similar to the ones we have reached for non-renewing warranty in [2].

In Table 6, the length of the repair, again on average, is set to be equal to approximately 1 week. The same as for Table 5 comments apply for Table 6 which is: the expected warranty cost is an increasing function of the length of the warranty period T, i.e. row-wise increasing values of the expected warranty cost and it is a decreasing function of the mean operating time, i.e. column-wise increasing values of the expected warranty cost.

The comparison between Tables 5 and 6 shows that it is better to have a shorter average repair time (four days for Table 5 against 1 week for Table 6). This conclusion is opposite to the one we reached in [2]. This is due to the differences between renewing and non-renewing warranty policies. The comparison between Tables 6 and 7, (all parameters are kept the same, only the value of δ is different) does not lead to surprising conclusions. The expected warranty cost is an increasing function of δ. In [2], the expected warranty cost over a warranty period as a function of δ had a maximum. This is because the warranty coverage W_T is a random variable, against fixed length T of the warranty in a non-renewing scenario.

4.3.2 Expected Costs Over $(0, L)$

Using the computational procedure for (13) and assuming $A = 3$ and $\delta = 2$, the expected warranty cost for selected values of L were evaluated. The comparison between these values shows that the improvement of the reliability and quality of the product, reflecting on the increase of its average operating time, will highly reduce the expected warranty cost. For numerical results and graphical summary, the reader is referred to [3].

5 Maintenance: Non-zero Periodic Preventive Repairs

Next, we summarise the results presented in Wang and Zhang [9]. The authors consider a simple deteriorating system. After a failure the system is replaced at a high

Table 5 Expected warranty cost over a warranty period, $\mu = 122$, $\delta = 2$

	T			
λ	1/12	1/4	1/2	1
2	0.547054	1.9568	5.18301	19.2719
4/3	0.354484	1.19332	2.85874	8.4268
1	0.262137	0.856732	1.9568	5.18301

Table 6 Expected warranty cost over a warranty period, $\mu = 52, \delta = 2$

	T			
λ	1/12	1/4	1/2	1
2	0.551057	1.97111	5.22093	19.4129
4/3	0.357077	1.20205	2.87965	8.48845
1	0.264055	0.863	1.97111	5.22093

Table 7 Expected warranty cost over a warranty period, $\mu = 52$ and $\delta = 22$

	T			
λ	1/12	1/4	1/2	1
2	0.620811	2.22062	5.88181	21.8702
4/3	0.402277	1.35421	3.24417	9.56294
1	0.297479	0.972241	2.22062	5.88181

cost. To extend the operating lifetime and to reduce the operating cost, at the time the system lifetime reaches a constant level B, the system could be repaired preventively, through an imperfect repair. The following scenario is considered: the successive operating times of the system after preventive repair form a stochastically decreasing geometric process, while the consecutive non-zero preventive repair times of the system form a stochastically increasing geometric process. The objective of this study is to determine an optimal bivariate replacement policy such that the average cost rate (the long-run average cost per unit time) is minimised.

5.1 The Model

The model is constructed under the following assumptions:

- At the beginning, a new system with preventive repairs (PR) is installed. At some point of time the system will be replaced by a new one and the replacement time is negligible.
- The PR will be adopted as soon as the operating time of the system reaches level B, and the PR is imperfect. Henceforth, the following notations will be used:
  - X_n - the operating time of the system after the $(n-1)$th PR with cdf $F_n(t) = F(a^{n-1}t), a \geq 1$; $EX_1 = \lambda$. $\{X_n\}, n = 1, 2, 3, \ldots$ form a stochastically decreasing geometric process with ratio a.
  - Y_n - the repair time of the system in the nth cycle with cdf $G_n(t) = G(b^{n-1}t)$, $0 < b \leq 1$; $\mu = EY_1$. $\{Y_n\}, n = 1, 2, 3, \ldots$ form a stochastically increasing geometric process with ratio b.
  - $\{X_n\}$ and $\{Y_n\}, n = 1, 2, 3, \ldots$ are independent.
  - A bivariate replacement maintenance policy (B, N) is adopted, i.e. B is a fixed period of time between consecutive PR and N is the number of PR's before the system is replaced. In other words, if the system is free of failure until the

$(N+1)^{st}$ PR, then it is replaced instead of performing a PR. At failure the system brings failure cost η and it is instantaneously replaced.
- The time between two consecutive system replacements $\tau_{\{.\}}$ is called a renewal cycle. $\{\tau_1, \tau_2, \tau_3, \ldots\}$ form a renewal process, where τ_1 is the time to first replacement.

5.2 The Average Cost Rate

All results and their derivations follow the presentation in [9]. Before the main result regarding the average cost rate $C(B, N)$ is provided, we need a list of preliminary results, so as to facilitate the understanding and utilisation of the main result.

1. We start with the distribution of M, the number of PR before system replacement. It is easy to see that:

$$P(M=0) = F(B) \quad \text{and} \quad P(M=k) = \prod_{i=0}^{k-1} \bar{F}(a^i B) F(a^k B). \tag{15}$$

2. The system total operating time $T(B, N)$ before renewal can be expressed as follows:

$$T(B, N) = \begin{cases} MB + \{X_{M+1} | X_{M+1} \le B\}, & \text{if } M \le N \\ (N+1)B, & \text{if } M > N. \end{cases} \tag{16}$$

3. The total PR time $S(B, N)$ in a renewal cycle is:

$$S(B, N) = \begin{cases} Y_1 + Y_2 + \ldots + Y_M, & \text{if } M \le N \\ Y_1 + Y_2 + \ldots + Y_N, & \text{if } M > N. \end{cases} \tag{17}$$

4. The total cost function $\Phi(B, N)$ in a renewal cycle is given by:

$$\Phi(B, N) = \left(-c_w (MB + X_{M+1} | X_{M+1} \le B) + c_r \sum_{k=1}^{M} Y_k + \eta \right) I_{\{M \le N\}} + \left(-c_w (N+1)B + c_r \sum_{k=1}^{N} Y_k \right) I_{\{M > N\}} + c, \tag{18}$$

where $I_{\{.\}}$ is an indicator function, c_w is the system's working reward rate, c_r is the system's PR cost rate, c is the system's replacement cost, η is the system's invalidation cost.

Thus, now having the expressions (16), (17) and (18) and noticing that

$$E[X_k|X_k \le B] = \frac{1}{F(a^{k-1}B)} \int_0^B x dF(a^{k-1}x),$$

allow to derive the expectations of the above random variables as follows:

1. $$E[T(B,N)] = \int_0^B x dF(x) + \sum_{k=1}^{N} \left[kB + \frac{1}{F(a^k B)} \int_0^B x dF(a^k x) \right] \times \prod_{i=0}^{k-1} \bar{F}(a^i B) F(a^k B) + (N+1)B \prod_{i=0}^{N} \bar{F}(a^i B); \tag{19}$$

2. $$E[S(B,N)] = \sum_{k=1}^{N} \left[F(a^k B) \prod_{i=0}^{k-1} \bar{F}(a^i B) \sum_{i=1}^{k} \frac{\mu}{b^{i-1}} \right] + \sum_{i=1}^{N} \frac{\mu}{b^{i-1}} \prod_{k=0}^{N} \bar{F}(a^k B); \tag{20}$$

3. $$\begin{aligned} E[\Phi(B,N)] = &-c_w \sum_{k=1}^{N} \left[kB + \frac{1}{F(a^k B)} \int_0^B x dF(a^k x) \right] \\ &\times (-c_w) \left[\prod_{i=0}^{k-1} \bar{F}(a^i B) F(a^k B) + \int_0^B x dF(x) \right] \\ &+ F(b) E\eta + \sum_{k=1}^{N} \left[c_r \sum_{i=1}^{k} \frac{\mu}{b^{i-1}} + E\eta \right] \\ &\times \prod_{i=0}^{k-1} \bar{F}(a^i B) F(a^k B) - c_w (N+1) B \\ &\times \prod_{i=0}^{N} \bar{F}(a^i B) + c_r \sum_{i=1}^{N} \frac{\mu}{b^{i-1}} \prod_{i=0}^{N} \bar{F}(a^i B) + c. \end{aligned} \tag{21}$$

For more details on the derivation of (19), (20) and (21) the reader is referred to [9]. Now, using these results it can be shown that the average cost rate of the system $C(B,N)$ is given by:

$$\begin{aligned} C(B,N) &= \frac{\text{E[costs in renewal cycle]}}{\text{E[length of a renewal cycle]}} = \frac{E[\Phi(B,N)}{E[T(B,N)+S(B,N)} \\ &= \frac{-c_w \phi_1 + c_r \phi_2 + \phi_3 E\eta + c}{\phi_1 + \phi_2}, \end{aligned} \tag{22}$$

where

- $$\phi_1 = \int_0^B x dF(x) + \sum_{k=1}^{N} \left[kB + \frac{1}{F(a^k B)} \int_0^B x dF(a^k x) \right] \times \prod_{i=0}^{k-1} \bar{F}(a^i B) F(a^k B) + (N+1)B \prod_{i=0}^{N} \bar{F}(a^i B);$$
- $$\phi_2 = \sum_{k=1}^{N} \left[F(a^k B) \prod_{i=0}^{k-1} \bar{F}(a^i B) \sum_{i=1}^{k} \frac{\mu}{b^{i-1}} \right] + \sum_{i=1}^{N} \frac{\mu}{b^{i-1}} \prod_{k=0}^{N} \bar{F}(a^k B);$$
- $$\phi_3 = F(B) + \sum_{k=1}^{N} \left[F(a^k B) \prod_{i=0}^{k-1} \bar{F}(a^i B) \right].$$

Hence, the next step is to identify the optimal replacement policy (B^*, N^*) that minimises the average cost rate of the system $C(B, N)$ given in (22).

5.3 Example

The following example is taken from [9]. Let us assume that the distribution of the nth operating time X_n is Weibull with parameters β and α, i.e.

$$F_n(t) = 1 - e^{-\left(\frac{a^{n-1}t}{\beta}\right)^\alpha}, \text{ for } \ t > 0.$$

Then, the average cost rate of the system $C(B, N)$ simplifies to

$$C(B, N) = \frac{c_r \mu l_1 + l_2 E\eta - c_w l_3 + c}{l_1 + l_3},$$

where

- $$l_1 = \sum_{k=1}^{N} \sum_{i=1}^{k} \frac{1}{b^{i-1}} e^{-\sum_{i=0}^{k-1} \left(\frac{a^i B}{\beta}\right)^\alpha} - \sum_{k=1}^{N-1} \sum_{i=1}^{k} \frac{1}{b^{i-1}} e^{-\sum_{i=0}^{k} \left(\frac{a^i B}{\beta}\right)^\alpha};$$
- $$l_2 = 1 + \sum_{k=2}^{N} e^{-\sum_{i=0}^{k-1} \left(\frac{a^i B}{\beta}\right)^\alpha} - \sum_{k=1}^{N} e^{-\sum_{i=0}^{k} \left(\frac{a^i B}{\beta}\right)^\alpha};$$
- $$l_3 = \int_0^B e^{-(\frac{x}{\beta})^\alpha} dx + \sum_{k=1}^{N} e^{-\sum_{i=0}^{k-1} \left(\frac{a^i B}{\beta}\right)^\alpha} \int_0^B e^{-(\frac{a^k x}{\beta})^\alpha} dx.$$

Assigning specific values to the parameters, such as $a = 1.05$, $b = 0.95$, $\mu = 8$, $E\eta = 1500$, $c_r = 20$, $c_w = 50$, $c = 2000$, $\beta = 1000$ and $\alpha = 2$, and after using a numerical procedure, it is shown that the optimal strategy is $(B^*, N^*) = (380, 10)$. In other words, for a system with characteristics, as given in the example, the optimal fixed period of time between consecutive PR is $B^* = 380$ and if the system is free of failure until the 11th PR, then at the time scheduled for this PR, the system has to be replaced. This maintenance strategy assures the minimum average cost rate of the system of $C(B^*, N^*) = -47.5977$. For more comments and details on the example, please see [9].

6 Maintenance: Markovian Model for Non-zero Preventive Repair Times

In this section, we summarise the Markovian approach proposed in Fang and Liu [4] to model for non-zero preventive repair times. The main objective of this study is to design a maintenance policy (B, N), so that the steady-state profit rate of the system is maximised, with B being the interval of preventive maintenance (repairs) and N being the number of failure-free preventive repairs to system replacement. The parameters of this strategy have the same meaning as in [9]. Also, the settings considered here are close to the settings in Sect. 5, but have some specifics and we discuss these below.

6.1 The Model

The model is built-up upon the following assumptions:

- At the beginning, a new system with preventive repairs (PR) is installed. At failure the system is repaired and the repair is imperfect with non-zero repair time.
- The times between two consecutive system failures are called cycles.
- The system failure in cycle N is catastrophic and the system is replaced by a new, identical system. The replacement requires negligible time.
- The PR will be adopted as soon as the operating time of the system reaches level B, and within a cycle the PR is perfect and the PR times are i.i.d. Moreover, the imperfect failure repair affects the first lifetime of the follow-up cycle and the lifetimes within a cycle are i.i.d. Henceforth, the following notations will be used:
  - $X_i^{(n)}$ - the operating time of the system after the nth PR within the ith cycle with cdf $H_i(x)$, pdf $h_i(x)$, failure rate function $a_i(x)$, and $E[X_i^{(n)}] = \lambda_i, \quad i = 1, 2, \ldots; \quad n = 0, 1, 2, \ldots$. Moreover, $\{X_i^{(0)}\}$ form a decreasing stochastic process.

- $Z_i^{(n)}$ - the preventive repair time of the system in cycle i after n PR with cdf $F_i(z)$, pdf $f_i(z)$, hazard function $b_i(z)$, and $E[Z_i^{(n)}] = b_i, \quad i = 1, 2, \ldots; \quad n = 0, 1, 2, \ldots$. Moreover, $\{Z_i^{(0)}\}$ form an increasing stochastic process.
- Y_i - the failure repair time of the system in cycle i cdf $G_i(y)$, pdf $g_i(y)$, hazard function $\mu_i(y)$, and $E[Y_i] = \mu_i, \quad i = 1, 2, \ldots; \quad n = 0, 1, 2, \ldots$. Moreover, $\{Y_i\}$ form a monotonically increasing stochastic process.
- $\{X_i^{(n)}\}$, $\{Z_i^{(n)}\}$ and $\{Y_i\}$, $\quad i = 1, 2, \ldots; \quad n = 0, 1, 2, \ldots$ are independent.

- The working reward per unit time is C_1, failure repair cost per unit time is C_2, preventive repair cost per unit time is C_3, and the system replacement cost is C.

The state of the system is modelled as follows:

- $(i, 0, n)$ - the system is working after the nth PR in cycle i, $n = 0, 1, 2, \ldots$; $i = 1, 2, \ldots, N$.
- $(i, 1, n)$ - the system is under PR after the nth PR in cycle i, $n = 0, 1, 2, \ldots$; $i = 1, 2, \ldots, N$.
- $(i, 2)$ - the system is under failure repair in cycle i, $i = 1, 2, \ldots, (N-1)$.

The modelling is reduced to a vector Markov process, so that it allows for the derivation of the state probability density equations. For more details on the modelling and results see [4].

6.2 Steady-State PR-Replacement Policy

Next we summarise the results regarding the steady-state performance measures of the system and use them to identify the optimal steady-state maintenance strategy as described in Sect. 6.1.

The authors show that:

- the steady-state replacement frequency M_r is given by

$$M_r = \frac{1}{\sum_{i=1}^{N} \frac{\int_0^B \bar{H}(x)dx}{H_i(B)} + \sum_{i=1}^{N} \frac{b_i \bar{H}(B)}{H_i(B)} + \sum_{i=1}^{N-1} \mu_i}; \tag{23}$$

- the steady-state availability A is equal to

$$A = M_r \sum_{i=1}^{N} \frac{\int_0^B \bar{H}(x)dx}{H_i(B)}; \tag{24}$$

- the steady-state PR frequency M_1 is given by

$$M_1 = M_r \sum_{i=1}^{N} \frac{b_i \bar{H}(B)}{H_i(B)}; \tag{25}$$

- the steady-state failure repair probability P is

$$P = M_r \sum_{i=1}^{N-1} \mu_i. \tag{26}$$

Therefore, using (23), (24), (25) and (26) the steady-state average profit rate $C(B, N)$ of the system is obtained to be equal to:

$$C(B, N) = \frac{C_1 \sum_{i=1}^{N} \frac{\int_0^B \bar{H}(x)dx}{H_i(B)} - C_2 \sum_{i=1}^{N-1} \mu_i - C_3 \sum_{i=1}^{N} \frac{\bar{H}(B)}{H_i(B)} - C}{\sum_{i=1}^{N} \frac{\int_0^B \bar{H}(x)dx}{H_i(B)} + \sum_{i=1}^{N} \frac{b_i \bar{H}(B)}{H_i(B)} + \sum_{i=1}^{N-1} \mu_i}. \tag{27}$$

It is easy to see that the steady-state average profit rate $C(\infty, N)$ of the system without PR is given by

$$C(B, N) = \frac{C_1 \sum_{i=1}^{N} \lambda_i - C_2 \sum_{i=1}^{N-1} \mu_i - C}{\sum_{i=1}^{N} \lambda_i + \sum_{i=1}^{N-1} \mu_i}. \tag{28}$$

Therefore, it is worth to perform PR only if $C(\infty, N) < C(B^*, N^*)$, where (B^*, N^*) are the parameters of the optimal maintenance strategy. A possible approach in finding the parameters of the optimal strategy is first, to find B_N^* for every N, so that $C(B_N^*, N)$ reaches maximum for $N = 1, 2, 3, \ldots$ and second, find the maximum among these values to determine $C(B_N^*, N^*)$, so that (B_N^*, N^*) are the parameters of the optimal maintenance policy. For more on this approach, see [11].

6.3 Example

As in [4], assume that $C_1 = 4,900$, $C_2 = 2,100$, $C_3 = 20,000$ and $C = 2,200,000$. Also let $H_i(x) = 1 - e^{(0.0001 \times 1.04^{i-1} x)^2}$, for $x \geq 0$, $i = 1, 2, \ldots, N$. Moreover, $b_i = 5 \times 1.05^{i-1}$, $i = 1, 2, \ldots, N$ and $\{Y_i, i = 1, 2, \ldots, N-1\}$ is a geometric process with $\mu_i = 150 \times 1.1^{i-1}$, $i = 1, 2, \ldots, N-1$. For these parameter values, it is shown that

$$C(B, N) = 4900 - \frac{A}{B}, \tag{29}$$

where

-

$$A = \sum_{i=1}^{N} \frac{(24500 \times 1.05^{i-1} + 20000)e^{-(0.0001 \times 1.04^{i-1} B)^2}}{1 - e^{-(0.0001 \times 1.04^{i-1} B)^2}}$$
$$+ 10500000(1.1^{N-1} - 1) + 2200000$$

- $$B = \sum_{i=1}^{N} \frac{\int_0^B e^{-(0.0001 \times 1.04^{i-1} x)^2} dx}{1 - e^{-(0.0001 \times 1.04^{i-1} B)^2}}$$
$$+ \sum_{i=1}^{N} \frac{5 \times 1.05^{i-1} e^{-(0.0001 \times 1.04^{i-1} B)^2}}{1 - e^{-(0.0001 \times 1.04^{i-1} B)^2}} + 1500(1.1^{N-1} - 1).$$

By using numerical computations, the parameters of the optimal maintenance policy are found to be equal to: $B^* = 1,727.343$ and $N^* = 3$ with maximum steady-state average profit reaching $C(B^*, N^*) = 4,847.148$ per unit time. For more details and comments, see [4].

7 A Case Study: Maintenance Optimisation for Age-Based Replacement Policy

In this section, a case study of maintenance optimisation introduced by Pintelon, van Puyvelde, and Gelders [6] is summarised. An age-based replacement model, which allows for non-zero (preventive and corrective) repair times is used to determine an optimal replacement policy. This study sheds some light on problems that need to be dealt with when mathematical models are applied to solve practical problems.

7.1 Description of the Case Study

In the case study of Pintelon et.al. [6], an optimal age-based maintenance policy is sought for the bottleneck machine of a manufacturing plant of beverage cans. Cans are produced through several phases of the production lines. The bottleneck phase of the production lines is associated with the cupper by which each cup is formed with sheets of metal. The cupper capacity influences the output level of the production heavily.

The company was using a classical block replacement policy under which preventive maintenance was conducted every ten days in addition to corrective maintenance at failure. A new replacement policy is desired to efficiently maintain the equipment by incorporating the data collected by its maintenance information system and making use of mathematical models.

7.2 The Model

Here, an age-based model with a modification of non-negligible maintenance times is applied. It is referred to as an extended age-based model. The following list summarises notations and assumptions with some justification.

- At the beginning, the cupper is new. When the operation time of the cupper reaches T_a (in days), a preventive maintenance (PM) with cost p [in Belgian Franc (BF)] is carried out. In addition, at each failure (before the operation time reaches T_a) a corrective maintenance (CM) with cost c (in BF) is executed.
- The duration of CM is fixed at t_r (in days). Likewise, the duration of PM is prespecified at t_m (in days). The classical age-based model assumes negligible maintenance time, but in the settings of this case study non-zero maintenance times are appropriate. Moreover, the property of production process justifies deterministic durations of CM and PM times.
- The times between two consecutive maintenance completion times, either corrective or preventive, is said to be a cycle.
- Let T denote the time to failure of the cupper in each cycle with the cumulative distribution function $F(t)$, density function $f(t)$ and failure rate function $z(t)$.
- Single component machine: Since cupper failures are mostly caused by one component, this assumption is appropriate.
- The system has two states ("on" or "off"): The production process is required to be with high speed and high accuracy and allows for no deterioration. Hence, it is either working denoted by "on" or not working denoted by "off".
- Failure-based versus use-based maintenance: An optimal balance between the frequencies of corrective and preventive maintenance is sought.
- As-good-as-new repairs: In each maintenance action, the cupper is repaired to be as-good-as-new.
- Model approach: (1) continuous time, (2) infinite horizon, (3) stochastic model
- Continuous production process: The cupper is working continuously.
- Failure distribution is not known: The classical age-based model assumes a known failure distribution, but in this case study, it was not the case. A few appropriate failure distributions are applied for a sensitivity analysis.
- Maintenance times: No consensus was formed in regard to independence of maintenance times. Two scenarios are considered. (1) optimistic (maintenance times are independent) (2) pessimistic (maintenance times are dependent).
- Optimisation of objective functions: (1) minimisation of the long-run maintenance cost per unit time; (2) maximisation of the average availability of the cupper (this would not be an objective function for the model with negligible maintenance time).

7.3 The Extended Age-Based Policy-Objective Functions and Properties

Using the renewal reward arguments (Barlow and Hunter [1], Tijms [8]), the following objective functions are obtained in [6].

- the average availability of the cupper is equal to

$$E(\text{availability}) = \frac{E(\text{on time in a cycle})}{E(\text{cycle length})} = \frac{\int_0^{T_a} t\,dF(t) + T_a(1 - F(T_a))}{\int_0^{T_a} t\,dF(t) + t_r F(T_a) + (T_a + t_m)(1 - F(T_a))} \tag{30}$$

- the long-run maintenance cost per unit time is given by

$$E(\text{cost}) = \frac{E(\text{cost per cycle})}{E(\text{cycle length})} = \frac{cF(T_a) + p(1 - F(T_a))}{\int_0^{T_a} t\,dF(t) + t_r F(T_a) + (T_a + t_m)(1 - F(T_a))} \tag{31}$$

The former is to be maximised, while the latter is to be minimised.

Some properties of the extended age-based model given below are discussed in [6].

- If the maintenance times are negligible, then the model concerned reduces to a classical age-based model for minimising the long-run maintenance cost per unit time.
- Under the assumption that the failure time T is exponentially distributed, the optimal T_a for both objective functions become infinite, i.e. preventive maintenance is unnecessary.

7.4 Example

Numerical methods and results presented in [6] are summarised in this subsection. The preventive maintenance time is set at $t_m = 3\,\text{h}$. As for the corrective maintenance time, two values are used: (1) $t_r = 10\,\text{h}$ (pessimistic scenario) and (2) $t_r = 5\,\text{h}$ (optimistic scenario). A global cost of 125,000 BF/h (approximately 4,000 dollars/h) including wages and materials is assumed. The case makes both maintenance times and global costs predictable, so using deterministic values for them is justified.

Based on the data collected by the company, the mean time between failures (MTBF) for the cupper is 12 days. For the sake of sensitivity analysis, a few different failure distributions are used with the MTBF of 12 days. The two-parameter Weibull distribution is selected as the failure distribution with $f(t) = \alpha\tau(\tau t)^{\alpha-1}e^{-(\tau t)^{\alpha}}$,

$F(t) = 1 - e^{-(\tau t)^{\alpha}}$ and $z(t) = \alpha\tau(\tau t)^{\alpha-1}$, where α and τ are shape and scale parameters, respectively. The managers' knowledge suggests that $\alpha = 4.0$. Using the expression for the first moment μ of T, given by

$$\mu = \frac{1}{\tau}\Gamma\left(1 + \frac{1}{\alpha}\right)$$

together with MTBF of 12 days (i.e., 1.714286 wk) and $\alpha = 4.0$, the corresponding τ can be obtained.

With the Weibull failure distribution the objective functions (30) and (31) contain integrals which cannot be evaluated analytically. If $\tau T_a < 1$, then (30) and (31) can be computed via an appropriate numerical method. Otherwise, a simulation method can be used to evaluate them. For details, see [6]

Under the pessimistic scenario $t_r = 10$ h, both objective functions (30) and (31) are optimised at $T_a = 1.190$ wk [corresponding to 8 days and a shift (8h)] with values 0.979713 and 426020 BF/wk, respectively. It is observed that as T_a increases the CM cost increases, whereas the PM cost decreases.

A comparison is made between the current model, i.e. the block-based model with preventive maintenance conducted every $T_b = 10$ days (1.428571 wk), and the extended age-based model. At optimality the former has average availability of 0.976234, while the latter has 0.979713. Contrary to the intuition, this difference may lead to a significant increase in income due to the fact that the cupper is a bottleneck machine. For details, the reader is referred to [6].

Under the optimistic scenario $t_r = 5$ h, the optimal preventive maintenance interval T_a is 1.571 wk with average availability of 0.984698 and the long-run cost of per unit time 321237 BF/wk.

A sensitivity analysis is carried out for the pessimistic scenario. In addition to the case $\alpha = 4.0$, α is set at $1, 1.5, 2.0$, and 2.5 and optimal values of T_a, average availability and long-run maintenance cost per unit time are compared. It is observed that (1) the higher the shape parameter α the shorter the T_a , but the higher the optimal average availability; (2) the higher the α , the clearer the optimum; (3) it is confirmed that when $\alpha = 1$ (the exponential failure distribution), the average availability increases without bound as $T_a \to \infty$.

8 A Simulation Model

In what follows we propose a simple simulation warranty model. We extend our study [2] by assuming imperfect warranty repairs. We model the "on" times of the system using a decreasing geometric process with parameters (X_1, a) and the "off" times (the warranty repair times) by an increasing geometric process with parameters (Y_1, b). Our goal is not only to estimate the expected warranty cost over a prespecified warranty period, but also to formulate and solve an optimisation problem regarding the length of the warranty period and provide some sensitivity analysis on the results.

- **The expected warranty cost**
  The evaluation of the expected warranty cost is straightforward and follows the standard approach. We assign a cost $C_i = A + cY_i$ to the "off" times, as in sect. 3.3. First, we generate the two geometric processes, the "ON" process and the "OFF" process, each with prespecified parameters. Based on the "OFF" process and the parameter values of the cost function, for a fixed value of the warranty period T, taking into account whether the warranty ends in an "on" or "off" period, the warranty cost is computed. For a fixed value of T at least 100 realisations of the "ON" and "OFF" processes are considered and the warranty cost for these realisations are averaged to obtain the expected warranty cost for the chosen value of T.
- **Optimisation problem on** T
  Next, we aim to formulate and justify an optimisation problem for determining the optimal warranty period for our model. Our objective function is the probability of product's sale $P(T)$ and we aim to maximise it. We assume that $P(T)$ is an increasing function of the difference $D(T)$, which has the following representation:

$$D(T) = v\,\{\text{total "on" time in } T\} - c\,\{\text{total "off" time in } T\},$$

where $c \geq 0$ and $v \geq 0$. Of course, the probability $P(T)$ might depend on other factors, but in this study we focus only on the above difference. What could be the interpretation of the parameters v and c ? One possible interpretation is as follows: the parameter v could be thought of as the rate of customer satisfaction due to the proper product functioning, and c as the rate of customer dissatisfaction due to the product failure. Next, let

$$r = \frac{v}{c}$$

be the ratio of the two rates. Now, if the warranty expires in an "off" period, i.e. the last "off" period is included in the warranty period, and the warranty coverage consists of total of d complete cycles, our optimisation criterion becomes

$$\begin{aligned} \max\ D(T) &= v \sum_{i=0}^{d} X_i - c \sum_{i=0}^{d} Y_i = v\,(T - \sum_{i=0}^{d} Y_i) - c \sum_{i=0}^{d} Y_i \\ &= v\,T - (v + c) \sum_{i=0}^{d} Y_i = c\,(r\,T - (1 + r) \sum_{i=0}^{d} Y_i). \end{aligned} \tag{32}$$

If the warranty expires in an "on" period, i.e. there is an incomplete cycle at the end of the warranty with d complete cycles before it, our optimisation criterion becomes:

$$\max \quad D(T) = v\,(T - \sum_{i=0}^{d} Y_i) - c\,\sum_{i=0}^{d} Y_i = v\,T - (v+c)\,\sum_{i=0}^{d} Y_i$$
$$= c\,(r\,T - (1+r)\,\sum_{i=0}^{d} Y_i). \tag{33}$$

Therefore, according to (32) and (33), the difference $D(T)$ is expressed equivalently in both cases. Of course, in the simulation we need to keep track whether the warranty expires during "on" or "off" time.
Next we present several illustrations of the model. In these illustrations the "on" times follow a geometric process with $F_1(x) = 1 - e^{-\left(\frac{t}{\beta}\right)^{\alpha}}$, for $t > 0$, i.e. the underlying distribution is Weibull with parameters $(\alpha_{on}, \beta_{on})$ and $a > 1$, and the "off" times follow a geometric process with $G_1(x)$, which is also Weibull with parameters $(\alpha_{off}, \beta_{off})$ and $0 < b < 1$. In Fig. 13, the remaining model parameters have the following values: $A = 0$, $r = 0.01$, $(\alpha_{on}, \beta_{on}) = (2, 1500)$ and $a = 1.05$, $(\alpha_{off}, \beta_{off}) = (2, 10)$ and $b = 0.95$, and the optimal value of the warranty period is $T^* = 5900$. In Fig. 14, the dependence of $D(T)$ on r is depicted for $r = 0.01$; 0.0075; 0.005, with corresponding optimal values $T^* = 5900$, 2000, 650. As expected, T^* also decreases as r decreases.
In Figs. 15 and 16, we vary the ratio r and obtain the two limiting cases $T^* = 0$ and $T^* = \infty$. As expected, when r is very small, i.e. the dissatisfaction rate is much higher that the satisfaction rate, the warranty period is zero, which will lead to $P(T) = 0$. Hence, the product has to be significantly improved before being introduced into the market. On the other hand, if r is relatively high, so that the two rates are comparable, the warranty period could be large and the probability for product sale will tend to one.

- **sensitivity analysis**
  Figures 13–16 provide an insight that the optimal value of T, if it exists, depends on the ratio $r = \frac{v}{c}$. Figure 17 depicts $D(T)$'s (and the optimal value of T) dependence on the parameter β_{off} of G_1 with the values of all of the remaining parameters as in Fig. 13. The upper curve shows $D(T)$ from Fig. 13 and the lower curve is $D(T)$ for $\beta_{off} = 15$, which leads, as expected, to a lower optimal value of $T^* = 1050$.

Fig. 13 $r = 0.01; T^* = 5900$

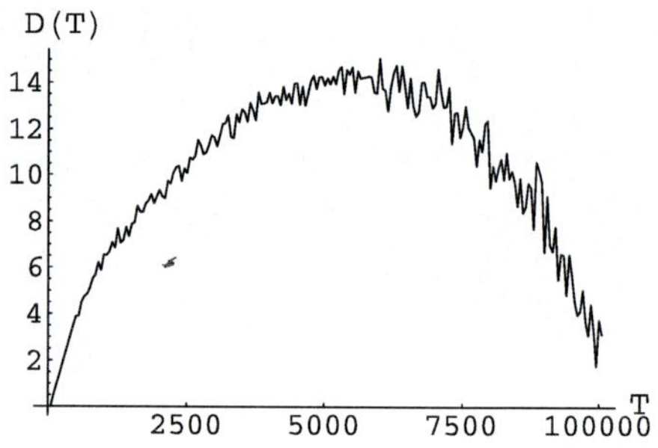

Fig. 14 $r = 0.01; 0.0075; 0.005$

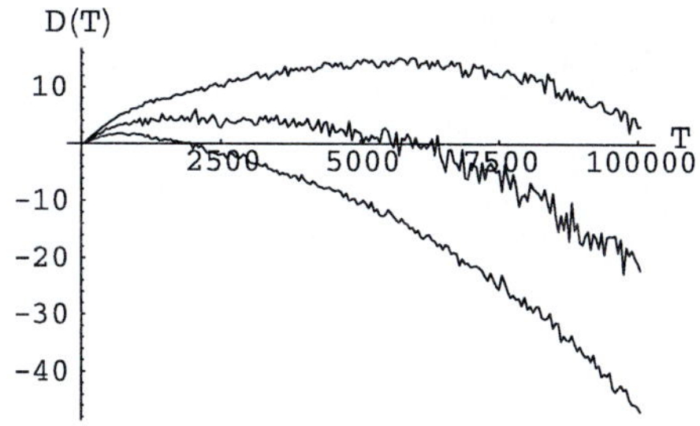

Fig. 15 $r = 0.001; T^* = 0$

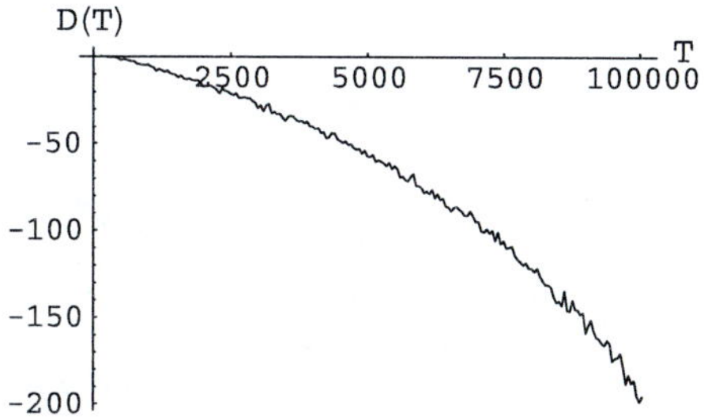

Fig. 16 $r = 0.1; T^* = \infty$

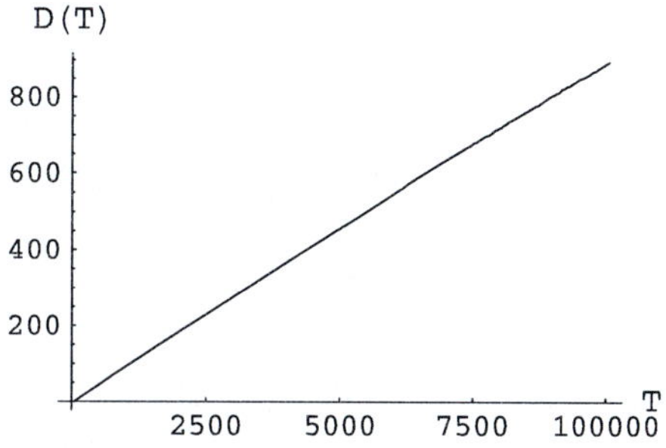

Fig. 17 $(\beta_{1;off}, \beta_{2;off}) = (10, 15)$

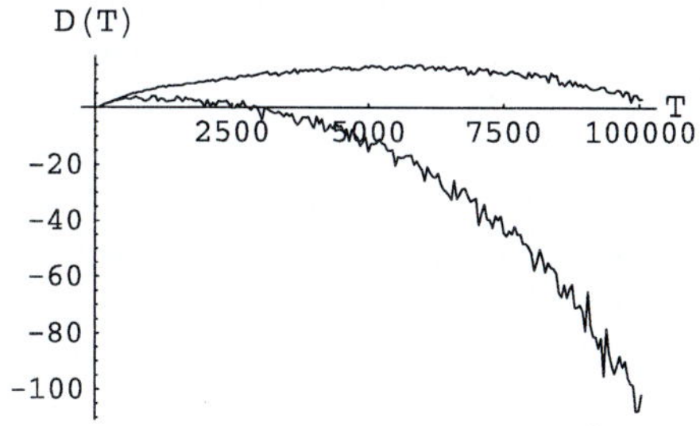

Lastly, Fig. 18 depicts $D(T)$'s (and the optimal value of T) dependence on the parameter b of the "off" times geometric process, keeping all remaining parameters as in Fig. 13. The upper curve is appropriately scaled curve from Fig. 13 and the

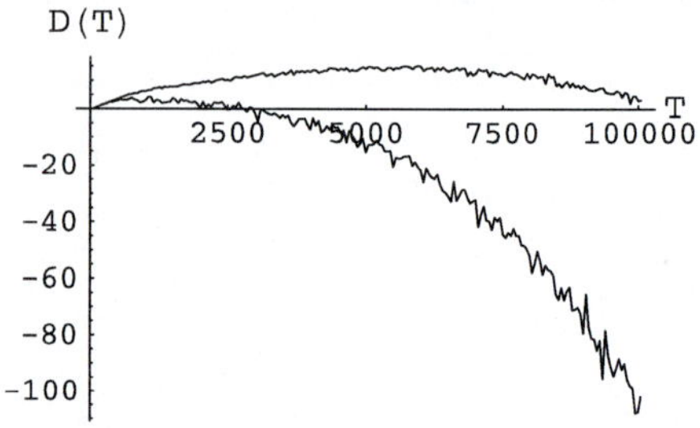

Fig. 18 $(b_1, b_2) = (1.05, 1.25)$

lower curve is $D(T)$ for $b = 0.80$, which leads, again as expected, to a lower optimal value of $T^* = 2600$.

Currently, we are working on the extension of the periodic preventive repair-replacement model presented in [9] (see Sect. 5). In this new simulation model, we introduce product warranty, and aim to solve an optimisation problem that will result in an optimal maintenance-warranty strategy with parameters (B^*, N^*, T^*). The detailed description and illustration of this model will be presented elsewhere.

9 Conclusions

In this chapter, we have reviewed several published studies with a common theme to emphasise the importance of taking into account the non-zero length of rectification actions. Our goal was to show that while modelling the product performance and related cost analysis, it is important to include in the model the non-zero times of warranty repairs, as well as the preventive/corrective maintenance repairs and the “cost” associated with them. In most situations it is acceptable to consider the repairs to be instantaneous, especially if they are not associated with high penalties, losses, or dissatisfaction. At the same time, it is well known that the harm to the producer/manufacturer’s reputation due to one dissatisfied customer is much higher than the positive impact of this reputation due to a group of satisfied customers. A faulty product could lead to a high customer dissatisfaction and could have a significant negative impact on the producer’s market standing. Hence, if the rate of this dissatisfaction, i.e. the “cost” of the “off” times, is taken into account, then better maintenance/warranty strategies from manufacturers’ as well as customers’ point of view could be designed.

Acknowledgments We would like to dedicate this study to Professor Shunji Osaki, in appreciation of his significant contributions to reliability theory and related fields, which inspired us in our effort to advance our own research. In addtion, we thank Professor Estate Khmaladze, Dr Mark Johnston, and Dr Richard Arnold from the School of Mathematics, Statistics and Operations Research, Victoria University of Wellington, New Zealand, for their support during the preparation of this manuscript.

References

1. Barlow RE, Hunter LC (1960) Optimal preventive maintenance policies. Operations Research 8:90–100
2. Chukova S, Hayakawa Y (2004) Warranty cost analysis: non-zero repair time. Applied Stochastic Models in Business and Industry 20:59–71
3. Chukova S, Hayakawa Y (2004) Warranty cost analysis: renewing warranty with non-zero repair time. Int J Reliab Qual Saf Eng 1:1–20
4. Fang YT, Liu BY (2006) Preventive repair policy and replacement policy of repairable system taking non-zero preventive repair time. J Zhejiang Univ SCI A 7:207–212
5. Pham H, Wang H (1996) Imperfect maintenance. Eur J Oper Res 94:425–438
6. Pintelon LMA, Van Puyvelde FLB, Gelders LF (1995) An age-based replacement policy with non-zero repair times for a continuous production process. Int J Prod Res 3:2111–2123
7. Ross S (1996) Stochastic Processes. John Wiley & Sons
8. Tijms H (1986) Stochastic Modeling and Analysis. Wiley, New York
9. Wang GJ, Zhang YL (2006) Optimal periodic preventive repair and replacement policy assuming geometric process repair. IEEE Transactions on Reliability 55:118–122
10. Xie M (1989) On the solution of renewal-type integral equations. Communications in Statistics - Simulation 18:281–293
11. Zhang YL (1994) A bivariate optimal replacent policy for repairable systems. J Appl Probab 31(4):1123–1127

Repair-Time Limit Replacement Policies

Won Young Yun and Naoto Kaio

Abstract This article concerns repair-limit replacement problems and review the existing stochastic models in which repair times are random variables. If a system fails, we should decide whether we repair the failed system (repair option) or replace it by new one (replacement option with a lead time). We classify the existing repair-time limit models based on available information amount of repair times (perfect, partial, and no information), repair type (perfect and imperfect repair), and objective functions (expected cost and profit with and without discounting). We summarize the modeling assumptions and explain how to obtain the optimal repair-limit replacement policies. Finally, we propose some interesting topics for future studies.

1 Introduction

For repairable systems, the maintenance plan during life cycle is important and affects the life cycle cost. Usually we can repair the failed system or sometimes replace it by new one. The maintenance engineer estimates repair cost (time) and if the repair cost (time) is relatively cheap (short), the failed system is repaired. Otherwise, it is replaced by new one. A lot of papers deal with replacement problems based on repair limit. In the existing literature, the proposed models may be classified into two main types: repair-cost limit and repair-time limit models. In the repair-cost limit models, when a unit fails, the repair cost is estimated and repair is undertaken if the

W. Y. Yun (✉)
Department of Industrial Engineering, Pusan National University, 30 Changjeon-Dong, Kumjeong-Ku, Busan 609-735, Korea
e-mail: wonyun@pusan.ac.kr

N. Kaio
Department of Economic Informatics, Hiroshima Shudo University, 1-1-1 Ozukahigashi, Asaminami-ku, Hiroshima 731-3195, Japan
e-mail: kaio@shudo-u.ac.jp

T. Dohi and T. Nakagawa (eds.), *Stochastic Reliability and Maintenance Modeling*, Springer Series in Reliability Engineering 9, DOI: 10.1007/978-1-4471-4971-2_5,

estimated cost is less than a pre-specified cost limit. Otherwise, the unit is replaced (refer Wang and Pham [46]). In the repair-time limit models, a unit is repaired at failure: if the repair is completed within a pre-specified time, it is put into operation again. Otherwise, it is replaced by a new one and the new one is used.

This chapter concerns mainly repair-time limit replacement problems and reviews the existing stochastic models. The repair-limit replacement problems are considered by Drinkwater and Hastings [18]. Hastings [19–21] and Lambe [33] formulate the repair-limit replacement problems by applying the dynamic programming. Love et al. [35] and Love and Guo [36] consider repair-limit problems in vehicle replacement cases. Nakagawa and Osaki [40], Kaio and Osaki [24, 25], Muth [37], and Nguyen and Murthy [41, 42] derive the optimal repair-time limits minimizing the expected cost rates with and without discounting. Osaki and Okumoto [43], Kapil and Sinha [27], and Kapur and Kapoor [28] introduce the repair-limit suspension policies for two-unit systems, and discuss the similar maintenance optimization problems to the repair-limit replacement ones. Kapur et al. [29, 30] consider the combined models with the repair-limit policy and the other maintenance options, and propose extended repair replacement policies. L'Ecuyer and Haurie [34], White [47], and Segawa and Ohnishi [44] develop the Markov and semi-Markov decision processes in repair-limit models. Jiang [22, 23] proves the optimality of repair-cost limit replacement policies and consider repair-limit replacement problems under general repair models.

Dohi et al. [1–4, 6, 8–10, 12–15, 17] deal with cost and profit models in repair-time limit problems with and without discounting. In particular, Dohi et al. [1, 4, 12, 13] consider stochastic profit models recently under an earning rate criterions. Dohi et al. [1, 6, 8, 12–16] introduced the concept of subjective repair-time distribution and considered graphical optimization problems to minimize the expected cost in estimation of the optimal decision. Dohi et al. [2, 3, 8, 10, 13, 15] also study imperfect repair models. For an excellent survey of repair-time limit replacement problems, see Dohi and Kaio [5]. In the most models related to repair-time limit policies, the exact repair times are not known before completing the repair. Kim and Yun [31] consider a repair-time limit model with estimation error.

This chapter concerns repair-time limit replacement problems and reviews the existing stochastic models. We describe the modeling assumptions and derive the objective functions (cost and profit functions) in the repair-time limit models. Basically, we consider a single unit system and assume that we have two options to recover the failed system; repair and replacement. Repair time is a random variable and replacement needs a new unit provided only by an order after a lead time. When the unit has failed, we decide to start repair or order a new unit. In this review chapter, we classify the existing repair-time limit models and explain the models and optimization problems. Finally, we propose some promising research topics in this area.

The remaining part of this chapter is organized as follows. The perfect repair models are summarized in Sect. 2. The imperfect repair models are studied in Sect. 3. Some miscellaneous topics are dealt in Sect. 4. Finally, Sect. 5 concludes this chapter.

Notation

t_0: Repair-time limit
$F(t), f(t)$: cdf, pdf of time to failure of a part
$F_{ir}(t)$: cdf of time to failure after imperfect repair
$G(t), g(t), r(t)$: cdf, pdf, and failure rate function of repair time
k: Penalty cost per unit time when the production machine is in down state
e_0: Earning rate per unit operation time
e_r: Per unit time
c: Fixed cost associated with the ordering of a new unit
h: Holding cost per unit time
L: Lead time for delivery of a new unit
Z: Order time point
β: Discount rate
$\overline{\varphi}(\cdot) = 1 - \varphi(\cdot)$

2 Perfect Repair Models

In this section, we consider perfect repair models in which the unit after repair is same as the new one. Consider a simple production machine with a part where each failed part is repairable but may be provided after a lead time L if it is replaced by a new one. The machine starts to operate at time 0. The time to failure of each part X is a non-negative random variable having the mean life μ_f whose distribution function and probability density function are $F(x)$ and $f(x)$, respectively. Once the machine failed, the maintenance engineers wish to decide whether to repair it or order a new unit. Basically, the repair times are different and can be considered as a random variable.

2.1 No Information Case

In this basic model, we assume that we have no information about repair time and know only general information (distribution function of repair time) about repair time. When the unit has failed, the repair is started immediately. If the repair is completed up to the time limit for repair t_0 (repair time limit), then the unit is installed at that time. It is assumed that the unit once repaired is presumed as good as new (perfect repair). However, if the repair time is greater than t_0, *i.e.,* the repair is not completed after the time t_0, then the failed unit is scrapped, and a new unit is ordered immediately and delivered after the lead time L. It is assumed that the time required for replacement is negligible. The repair time for each unit has an arbitrary distribution $G(t)$ with the density $g(t)$ and finite mean μ_r. Under these model assumptions, we consider the interval from the start of the operation to the next start as one renewal cycle. For an infinite planning horizon, it is appropriate to adopt the expected cost per unit time in the steady-state (expected cost rate) as an optimization criterion. The total expected cost for one cycle is given by the following three costs:

(1) The expected repair cost is

$$e_r \left\{ \int_0^{t_0} t dG(t) + t_0 \overline{G}(t_0) \right\} = e_r \int_0^{t_0} \overline{G}(t) dt$$

(2) The expected shortage cost is

$$k \left\{ \int_0^{t_0} t dG(t) + (t_0 + L)\overline{G}(t_0) \right\} = k \left\{ \int_0^{t_0} \overline{G}(t) dt + L\overline{G}(t_0) \right\}$$

(3) The expected ordering cost is $c\overline{G}(t_0)$.

The expected duration of one cycle is given by

$$\mu_f + \int_0^{t_0} \overline{G}(t) dt + L\overline{G}(t_0).$$

The total expected cost per unit time in the steady-state(expected cost rate) is ,

$$TC_{p1}(t_0) = \frac{(k + e_r) \int_0^{t_0} \overline{G}(t) dt + (c + kL)\overline{G}(t_0)}{\mu_f + \int_0^{t_0} \overline{G}(t) dt + L\overline{G}(t_0)} \tag{1}$$

Two special cases are the following

$$TC_{p1}(0) = \frac{c + kL}{\mu_f + L}, \quad TC_{p1}(\infty) = \frac{(k + e_r)\mu_r}{\mu_f + \mu_r}.$$

Theorem 1: Suppose that $e_r L < c$.

(1) If the repair-time distribution $G(t)$ is increasing hazard rate (IHR), the optimal repair-time limit is 0 (always repair case) or infinite (always ordering case).
(2) If the repair-time distribution $G(t)$ is strictly decreasing hazard rate (DHR), there exists a finite and unique optimal repair-time limit t_{01}^* $(0 < t_{01}^* < \infty)$ under some conditions (refer to Dohi and Kaio [5]) and the corresponding minimum expected cost rate is given by

$$TC_{p1}(t_{01}^*) = \frac{e_r + k - (kL + c)r(t_{01}^*)}{1 - Lr(t_{01}^*)} \tag{2}$$

Theorem 1 gives a sufficient condition for existence of the finite and unique optimal repair time limit. As a special case (L=0), Eq. 1 is equal to the expected cost rate in Nakagawa and Osaki [40].

2.2 Perfect Information Case

In this subsection, we assume that we can estimate the repair time perfectly. When the system is failed, the repair time can be estimated and known. In this case, we consider the following repair and replacement policy. If the estimated repair time is less than t_0, we start to repair the failed system. Otherwise, we order a new unit and replace the system after the unit is delivered after a lead time. Then

(1) The expected cost of a renewal cycle is

$$e_r \int_0^{t_0} t dG(t) + k \left\{ \int_0^{t_0} t dG(t) + L\overline{G}(t_0) \right\} + c\overline{G}(t_0).$$

(2) The expected duration of one cycle is given by

$$\mu_f + \int_0^{t_0} t dG(t) + L\overline{G}(t_0).$$

Then, the expected cost rate is

$$TC_{p2}(t_0) = \frac{(e_r + k) \int_0^{t_0} t dG(t) + (c + kL)\overline{G}(t_0)}{\mu_f + \int_0^{t_0} t dG(t) + L\overline{G}(t_0)} \tag{3}$$

In a similar way in Theorem 1, we can obtain the optimal repair-time limit t_{02}^*. If there exists a finite and unique optimal repair-time limit $t_{02}^* (0 < t_{02}^* < \infty)$, the corresponding minimum expected cost rate is given by

$$TC_{p2}(t_{02}^*) = \frac{(e_r + k)t_{02}^* - (kL + c)}{t_{02}^* - L} \tag{4}$$

The optimal cost rate (4) with perfect information of repair times is not greater than one without information. The difference between Eqs. 2 and 4 is the expected value of perfect information per unit time (EVPI) and

$$\begin{aligned} EVPI &= TC_{p2}(t_{02}^*) - TC_{p1}(t_{01}^*) \\ &= \frac{e_r + k - (kL + c)r(t_{01}^*)}{1 - Lr(t_{01}^*)} - \frac{(e_r + k)t_{02}^* - (kL + c)}{t_{02}^* - L} \end{aligned} \tag{5}$$

Thus, the cost to get the perfect information should be less than EVPI to use the perfect information.

2.3 Partial Information Case

In this subsection, we assume that we can estimate the actual repair time T_a but there is the estimation error in repair-time estimation. Once the machine is failed (failure time; X), we can estimate the repair time but the estimation error occurs. If the estimated repair time is greater than a pre-specified limit t_0, then we order a new unit. After the new unit is delivered after the lead time L, the failed unit is replaced by the new one and the machine starts operating at time $t = X + L$ again, where the replacement time can be negligible. Otherwise we repair the failed part and the machine operates at $t = X + T_a$. In case that the actual repair time is less than t_0 but the repair time is estimated greater thant_0, a replacement is carried out. In the reverse case, a repair is performed over t_0. After the completion of repair operation, the repaired part becomes as good as new (perfect repair). Thus, the time interval from the start of operation to the next starting point can be defined as one cycle. The estimated repair time, T_e is a function of the actual repair time and estimation error, i.e., $T_e = T_a + \boldsymbol{\varepsilon}$. Given the actual repair time $T_a = t$, the estimated repair time is $T_e|t$, whose conditional probability density function is denoted as $h(u|t)$.

To obtain the expected duration of a renewal cycle, we first consider conditional expected duration and cost, and then obtain the unconditional ones. If the actual repair time is given as $\boldsymbol{t}$, then the estimated value of repair time for a given actual repair time is a random variable with conditional distribution function, $\Pr\{T_e \leq y\,|T_a = t\} = \int_{-\infty}^{y} h(u\,|t\,)du = H(y\,|t\,)$. Since the starting point of machine operation can be regarded as a renewal point, the expected duration of a renewal cycle for a given actual repair time is given by

$$\mu_f + tH(t_0\,|t\,) + L\overline{H}(t_0\,|t\,)$$

Also, the expected cost of a renewal cycle for a given actual repair time becomes

$$(e_1 + k)tH(t_0\,|t\,) + (kL + c)\overline{H}(t_0\,|t)$$

Thus, the unconditioned expected duration and cost of a renewal cycle are given by

$$\begin{aligned} T_{L1}(t_0) &= \mu_f + \int_0^{\infty} \left[tH(t_0\,|t\,) + \overline{H}(t_0\,|t\,)L\right]dG(t) \\ &= \mu_f + L + \int_0^{\infty} (t - L)H(t_0\,|t\,)dG(t) \\ V_{L1}(t_0) &= \int_0^{\infty} \left[(e_r + k)tH\,(t_0\,|t\,) + (kL + c)\overline{H}(t_0\,|t)\right]dG(t) \\ &= (kL + c) + \int_0^{\infty} [(e_r + k)t - (kL + c)]\,H(t_0\,|t\,)dG(t) \end{aligned} \tag{6}$$

From the renewal reward argument, the expected cost rate is given by $TC_{p3}(t_0)$=V_{L1} (t_0)/$T_{L1}(t_0)$. The problem is to derive the optimal repair-time limit, t_{03}^* satisfying

$$TC_{p3}\left(t_{03}{}^*\right) = \max_{0\leq t_0<\infty} TC_{p3}\left(t_0\right). \tag{7}$$

In general case, it is difficult to obtain analytically the optimal solutions of the Eq. 7. We consider a special case where the expected cost rate over an infinite time horizon can be derived as a closed form. The estimated repair time, T_e is a function of the actual repair time and estimation error, i.e., $T_e = T_a + \boldsymbol{\varepsilon}$ where the estimation error assumed to be an independent and normally distributed random variable with mean 0 and variance σ_e^2. We also assume that given the actual repair time $T_a = t$, the estimated repair time $T_e|t$, has an independent identical normal distribution with mean t and variance σ_e^2, i.e. $T_e|t \sim N(t,\sigma_e^2)$. This assumption is reasonable in practice, as first the estimated repair times are usually symmetrically distributed around the actual repair time t and second the estimated repair times are typically accurate enough to be close to the actual repair time with higher probability. Second, the actual repair times are assumed to be independent and follow approximately a normal distribution. In addition, we assume that the probability that the repair time has negative value is too small and can be negligible. Thus, the actual repair time, T_a follows a normal distribution with mean, μ_a and variance, σ_a^2. To obtain the expected cost rate we use the following results;

(1) $\int_{-\infty}^{\infty} H(t_0\,|t)\; dG\,(t) = \Phi(\frac{t_0-\mu_a}{\sqrt{\sigma_a^2+\sigma_e^2}})$

(2) When $U \sim N(\mu_a, \sigma_a^2 + \sigma_e^2)$, $m(t) = \int_{-\infty}^{t_0} ug(u)du = \mu_a\Phi(\frac{t_0-\mu_a}{\sqrt{\sigma_a^2+\sigma_e^2}}) - \sqrt{\sigma_a^2 + \sigma_e^2}\phi(\frac{t_0-\mu_a}{\sqrt{\sigma_a^2+\sigma_e^2}})$

(3) $\int_{-\infty}^{\infty} tH(t_0\,|t)dG(t) = \mu_a\Phi(\frac{t_0-\mu_a}{\sqrt{\sigma_a^2+\sigma_e^2}}) - \frac{\sigma_a^2}{\sqrt{\sigma_a^2+\sigma_e^2}}\phi(\frac{t_0-\mu_a}{\sqrt{\sigma_a^2+\sigma_e^2}})$

The expected duration and cost of a renewal cycle are given by

$$T_{L1}(t_0) = \mu_f + L + (\mu_a - L)\Phi(\frac{t_0-\mu_a}{\sqrt{\sigma_a^2+\sigma_e^2}}) - \frac{\sigma_a^2}{\sqrt{\sigma_a^2+\sigma_e^2}}\phi(\frac{t_0-\mu_a}{\sqrt{\sigma_a^2+\sigma_e^2}})$$

$$V_{L1}(t_0) = (kL+c) - (kL+c-(e_1+k)\mu_a)\Phi(\frac{t_0-\mu_a}{\sqrt{\sigma_a^2+\sigma_e^2}}) - \frac{(e_1+k)\sigma_a^2}{\sqrt{\sigma_a^2+\sigma_e^2}}\phi(\frac{t_0-\mu_a}{\sqrt{\sigma_a^2+\sigma_e^2}})$$

For detail derivation, refer Kim and Yun [31]. It is difficult to obtain the optimal repair time limit to minimize the expected cost rate but we can find the approximate optimal solutions numerically.

3 Imperfect Repair Models

In Sect. 2, we studied perfect repair models and in this section we consider imperfect repair models. When the unit fails, then we should determine to start repair or to order new unit. The mean failure times after repair and replacement are $\mu_f$and μ_{ir}, respectively. If the new one unit is ordered, then it is delivered after the lead time L. it is assumed that the replacement time is negligible.

3.1 No Information Case

In this subsection, we assume that we have no information about repair time and we know only general information (distribution function of repair time) about repair time. When the unit fails, the repair is started immediately. If the repair is completed within the repair-time limit t_0, then the unit start to operate again. On the other hand, if the repair is not completed after the time t_0, then the failed unit is scrapped, and the new unit is ordered immediately. For this repair-time limit model, the total expected cost for one cycle is given as follows;

The expected repair cost is $e_r \left\{ \int_0^{t_0} t dG(t) + t_0 \overline{G}(t_0) \right\} = e_r \int_0^{t_0} \overline{G}(t)dt$

The expected shortage cost is $k \left\{ \int_0^{t_0} t dG(t) + (t_0 + L)\overline{G}(t_0) \right\} = k \left\{ \int_0^{t_0} \overline{G}(t)dt + L\overline{G}(t_0) \right\}$

The expected ordering cost is $c\overline{G}(t_0)$.

The expected duration of one cycle is given by

$$\int_0^{t_0} (\mu_{ir} + t)dG(t) + \int_{t_0}^{\infty} (\mu_f + t_0 + L)dG(t) = \mu_{ir} + \int_0^{t_0} \overline{G}(t)dt + (L + \mu_f - \mu_{ir})\overline{G}(t_0).$$

The expected cost rate is,

$$TC_{p4}(t_0) = \frac{(k + e_r)\int_0^{t_0} \overline{G}(t)dt + (c + kL)\overline{G}(t_0)}{\mu_{ir} + \int_0^{t_0} \overline{G}(t)dt + (L + \mu_f - \mu_{ir})\overline{G}(t_0)}. \quad (8)$$

Theorem 2: Suppose that $e_r L + (k + e_r)(\mu_r - \mu_{ip}) < c$.

(3) If the repair-time distribution $G(t)$ is IHR, the optimal repair-time limit is 0 (always repair case) or infinite (always ordering case).
(4) If the repair-time distribution $G(t)$ is strictly DHR, there exists a finite and unique optimal repair-time limit $t_{04}^*(0 < t_{04}^* < \infty)$ under some conditions(refer to Dohi and Kaio [5]) and the corresponding minimum expected cost rate is given by

$$TC_{p4}(t_{04}^*) = \frac{(e_r + k) - (kL + c)r(t_{04}^*)}{1 - (L + \mu_r - \mu_{ir})r(t_{04}^*)} \tag{9}$$

From Eqs. 2 and 9, the expected profit of perfect repair per unit time is given by

$$\begin{aligned} EPPR &= TC_{p4}(t_{04}^*) - TC_{p1}(t_{01}^*) \\ &= \frac{(e_r + k) - (kL + c)r(t_{04}^*)}{1 - (L + \mu_r - \mu_{ir})r(t_{04}^*)} - \frac{e_r + k - (kL + c)r(t_{01}^*)}{1 - Lr(t_{01}^*)} \end{aligned}$$

3.2 Perfect Information Case

In this subsection, we assume that we can estimate the repair times perfectly. Thus if the system is failed, we estimate the repair time. If the estimated time is less than t_0, we repair the failed system. Otherwise, we order the new unit and replace the system. Then, the expected cost of a renewal cycle is

$$e_r \int_0^{t_0} \overline{G}(t)dt + k\left\{\int_0^{t_0} tdG(t) + L\overline{G}(t_0)\right\} + c\overline{G}(t_0)$$

The expected duration of one cycle is given by

$$\mu_f + \int_0^{t_0} tdG(t) + L\overline{G}(t_0)$$

Then, the expected cost rate is

$$TC_{p5}(t_0) = \frac{(e_r + k)\int_0^{t_0} tdG(t) + (c + kL)\overline{G}(t_0)}{\mu_{ir} + \int_0^{t_0} tdG(t) + (L + \mu_r - \mu_{ir})\overline{G}(t_0)}. \tag{10}$$

In a similar way in Theorem 2, we can obtain the optimal repair-time limit t_{05}^*. If there exists a finite and unique optimal repair-time limit $t_{05}^*(0 < t_{05}^* < \infty)$, the corresponding minimum expected cost rate is given by

$$TC_{p5}(t_{05}^*) = \frac{(e_r + k)t_{05}^* - (kL + c)}{t_{05}^* - (L + \mu_r - \mu_{ir})}. \tag{11}$$

3.3 Partial Information Case

In this subsection, we consider the imperfect repair model and all assumptions about partial information in repair estimation are same as in Sect. 2.3. Once the machine is failed, we estimate the repair time but the estimation error occurs. If the estimated imperfect repair time is greater than a pre-specified limit t_0, then we order a new unit. After the new unit is delivered after the lead time L, the failed part is replaced by the new one and the machine starts operating at time again. Otherwise we repair the failed part and the machine operates after imperfect repair.

The expected duration of a renewal cycle for a given actual repair time is given by

$$(t + \mu_{ir})H(t_0\,|t) + (L + \mu_f)\overline{H}(t_0\,|t).$$

Also, the expected cost of a renewal cycle for a given actual repair time is same as

$$(e_1 + k)tH(t_0\,|t) + (kL + c)\overline{H}(t_0\,|t).$$

Thus the unconditioned expected duration and cost of a renewal cycle are given by

$$\begin{aligned}
T_{L2}(t_0) &= \int_0^\infty \left[(t + \mu_{ir})H(t_0\,|t) + (L + \mu_f)\overline{H}(t_0\,|t)\right]dG(t) \\
&= \mu_f + L + \int_0^\infty (t + \mu_{ir} - L - \mu_f)H(t_0\,|t)dG(t) \qquad (12) \\
V_{L2}(t_0) &= \int_0^\infty \left[(e_1 + k)tH\,(t_0\,|t) + (kL + c)\overline{H}(t_0\,|t)\right]dG(t) \\
&= (kL + c) + \int_0^\infty \left[(e_1 + k)t - (kL + c)\right] H(t_0\,|t)dG(t).
\end{aligned}$$

From the renewal reward argument, the expected cost rate is given by $TC_{p6}(t_0)$=V_{L2} (t_0)/$T_{L2}(t_0)$. The problem is to derive the optimal repair-time limit, t_{06}^* satisfying

$$TC_{p6}\left(t_{06}{}^*\right) = \max_{0 \le t_0 < \infty} TC_{p6}\,(t_0) \qquad (13)$$

As special cases, we can obtain the closed form of the expected cost rate and find the optimal solution numerically.

4 Miscellaneous Models

In this section, we consider some extended and modified models in repair-time limit problems. In particular, some studies consider different optimization criteria from the expected cost rate. Stochastic profit and discounting models are derived in Sects. 4.1 and 4.2. Most of existing models in repair-time limit problems consider the ordering option after failure. In Sect. 4.3, we consider a simple preventive ordering model. Finally, we review simply the estimation problem of model parameters.

4.1 Profit Models

In Sect. 3, we considered three cost terms (repair, shortage, and ordering costs) to determine the optimal repair-time limit. In this subsection, we consider earning rate during operating period and derive the expected profit models.

Perfect Repair Case

In this subsection, we consider an expected profit model with perfect repair and all model assumptions are same as in Sect. 2.1 The total profit for a cycle consists of the expected repair cost and total earning amount. The expected cost is same as

$$(k + e_r)\int_0^{t_0} \overline{G}(t)dt + (c + kL)\overline{G}(t_0).$$

The total earning amount for a cycle is $\mu_f e_0$.

The expected duration of one cycle is same as

$$\mu_f + \int_0^{t_0} \overline{G}(t)dt + L\overline{G}(t_0).$$

The total expected profit per unit time in the steady-state is,

$$TP_{p1}(t_0) = \frac{\mu_f e_0 - (k + e_r)\int_0^{t_0} \overline{G}(t)dt - (c + kL)\overline{G}(t_0)}{\mu_f + \int_0^{t_0} \overline{G}(t)dt + L\overline{G}(t_0)}. \tag{14}$$

In a similar way in Theorem 1, we can obtain the optimal repair-time limit t_{07}^*. If there exists a finite and unique optimal repair-time limit $t_{07}^*(0 < t_{07}^* < \infty)$, the corresponding minimum expected profit per unit time is given by

$$TP_{p1}(t_{07}^*) = \frac{-(e_r + k) + (kL + c)r(t_{07}^*)}{1 - Lr(t_{07}^*)} \tag{15}$$

As a similar model, the expected profit model with perfect repair and perfect information is studied in Dohi et al. [12].

Imperfect Repair Case

In this subsection, we consider an expected profit model with imperfect repair and all model assumptions are same as in Sect. 3.1 The total profit for a cycle consists of the expected repair cost and total earning amount. The expected cost is same as

$$(k + e_r)\int_0^{t_0} \overline{G}(t)dt + (c + kL)\overline{G}(t_0).$$

The total earning amount for a cycle is $\left(e_0\left(\mu_{ir}G(t_0)+\mu_f\overline{G}(t_0)\right)\right)$.
The expected duration of one cycle is same as

$$\mu_{ir}+\int_0^{t_0}\overline{G}(t)dt+(L-\mu_f-\mu_{ir})\overline{G}(t_0).$$

The total expected profit per unit time in the steady-state is,

$$TP_{p2}(t_0)=\frac{e_0\mu_f-(k+e_r)\int_0^{t_0}\overline{G}(t)dt-(e_0(\mu_f-\mu_{ir})+c+kL)\overline{G}(t_0)}{\mu_{ir}+\int_0^{t_0}\overline{G}(t)dt+(L+\mu_f-\mu_{ir})\overline{G}(t_0)} \tag{16}$$

In a similar way in Theorem 2, we can obtain the optimal repair-time limit t_{08}^*. If there exists a finite and unique optimal repair-time limit $t_{08}^*(0<t_{08}^*<\infty)$, the corresponding minimum expected profit per unit time is given by

$$TP_{p2}(t_{08}^*)=\frac{-(e_r+k)+\left(e_0(\mu_f-\mu_{ir})+kL+c\right)r(t_{08}^*)}{1-(L+\mu_{ir}-\mu_f)r(t_{08}^*)} \tag{17}$$

As a similar model, the expected profit model with imperfect repair and perfect information is studied in Dohi et al. [4].

Partial Information Model

In this subsection, we consider an expected profit model with perfect repair and all model assumptions are same as in Sect. 2.3 The expected profit for a cycle consists of the expected cost for a cycle and total earning amount. Thus, the unconditioned expected duration and profit of a renewal cycle are given by

$$\begin{aligned}T_{L3}(t_0)&=\mu_f+\int_{-\infty}^{\infty}\left[tH(t_0\,|t)+\overline{H}(t_0\,|t)L\right]dG(t)\\&=\mu_f+L+\int_{-\infty}^{\infty}(t-L)H(t_0\,|t)dG(t)\\V_{L3}(t_0)&=e_0\mu_f-\int_{-\infty}^{\infty}\left[(e_r+k)tH\,(t_0\,|t)+(kL+c)\overline{H}(t_0\,|t)\right]dG(t)\\&=e_0\mu_f-(kL+c)-\int_{-\infty}^{\infty}\left[(e_r+k)t-(kL+c)\right]H(t_0\,|t)dG(t)\end{aligned} \tag{18}$$

From the renewal reward argument, the expected profit per unit time is given by

$$TP_{p3}\left(t_{09}{}^*\right)=\max_{0\le t_0<\infty}TP_{p3}\left(t_0\right)=\frac{V_{L3}(t_0)}{L_{L3}(t_0)}. \tag{19}$$

As special cases, we can obtain the closed form of the long-run average profit function by similar way in Sect. 2.3

4.2 Discounting Models

In this subsection, we consider some typical cases in which the cost is discounted with the discount rate over an infinite horizon.

Perfect Repair Model

The perfect repair model in Sect. 2.1 is studied again but we consider the repair-limit replacement problem under discounted cost criterion. The present value of a unit cost for one cycle is

$$\delta_1(t_0) = \int_0^\infty \int_0^{t_0} e^{-\beta(t+x)} dG(t)dF(t) + \int_0^\infty \int_{t_0}^\infty e^{-\beta(t+x+L)} dG(t)dF(t).$$

The expected total discounted cost for one cycle is

$$\begin{aligned} V_1(t_0) = &\int_0^\infty \int_0^{t_0} \int_0^t c_r e^{-\beta(t+x)} dy dG(t)dF(t) \\ &+ \int_0^\infty \int_{t_0}^\infty \int_0^{t_0} c_r e^{-\beta(t+x)} dy dG(t)dF(t) \\ &+ \int_0^\infty \int_0^{t_0} \int_0^t k e^{-\beta(t+x)} dy dG(t)dF(t) \\ &+ \int_0^\infty \int_{t_0}^\infty \int_0^{t_0+L} k e^{-\beta(t+x)} dy dG(t)dF(t) \\ &+ \int_0^\infty \int_{t_0}^\infty c e^{-\beta(t_0+x+L)} dG(t)dF(t) \end{aligned}$$

Then the expected total discounted cost for an infinite time horizon is given by

$$TDC_1(t_0) = \sum_{n=0}^{\infty} V_1(t_0)\delta_1(t_0)^n = \frac{V_1(t_0)}{\overline{\delta}(t_0)} \tag{20}$$

Thus, we should obtain the optimal repair-time limit minimizing the Eq. 20.

Theorem 3: Suppose that$e_r \int_0^L e^{-\beta t} dt < c e^{-\beta L}$.

If the repair-time distribution $G(t)$ is IHR, the optimal repair-time limit is 0 (always repair case) or infinite (always ordering case).

If the repair-time distribution $G(t)$ is strictly DHR, there exists a finite and unique optimal repair-time limit $t_{010}^*(0 < t_{010}^* < \infty)$ under some conditions(refer to Dohi and Kaio [5]) and the corresponding minimum total discounted cost is given by

$$TDC_1(t_{010}^*) = \frac{e_r + k - \frac{k}{\beta}(1 - e^{-\beta L})\left(\beta + r(t_{010}^*)\right)}{\left(\beta + r(t_{010}^*)\right) e^{-\beta L} - r(t_{010}^*)} \tag{21}$$

Theorem 3 gives a sufficient condition for existence of the finite and unique optimal repair-time limit.

Imperfect Repair Model

In this subsection, we formulate the imperfect repair model with discounting under the expected total discounted cost over an infinite time horizon. The present value of a unit cost for one cycle is

$$\delta_2(t_0) = \int_0^{t_0}\int_0^{\infty} e^{-\beta(t+x)} dF_{ir}(x)dG(t) + \int_{t_0}^{\infty}\int_0^{\infty} e^{-\beta(t_0+x+L)} dF(x)dG(t)$$

The expected total discounted cost is given by

$$V_2(t_0) = \int_0^{t_0}\int_0^{t} (c_r + k)e^{-\beta x} dx dG(t) + \int_{t_0}^{\infty}\int_0^{t_0} (c_r + k)e^{-\beta x} dx dG(t)$$
$$+ \overline{G}(t_0)\left[\int_0^{L} ke^{-\beta(t_0+x)}dx + ce^{-\beta(t_0+L)}\right]$$

Then the expected total discounted cost for an infinite time horizon is given by

$$TDC_2(t_0) = \sum_{n=0}^{\infty} V_2(t_0)\delta_2(t_0)^n = \frac{V_2(t_0)}{\overline{\delta}_2(t_0)} \tag{22}$$

Thus, we obtain the optimal repair-time limit minimizing the Eq. 22.

Theorem 4: Suppose that

$$(e_r + k)e^{-\beta L}\left[\int_0^{\infty} e^{-\beta t} dF(t) - \int_0^{\infty} e^{-\beta t} dF_{ir}(t)\right]$$
$$- e_r[1 - e^{-\beta L}]\int_0^{\infty} e^{-\beta t} dF_{ir}(t) + c\beta e^{-\beta L}\int_0^{\infty} e^{-\beta t} dF_{ir}(t) > 0$$

(1) If the repair-time distribution $G(t)$ is IHR, the optimal repair time limit is 0 (always repair case) or infinite (always ordering case).

(2) If the repair-time distribution $G(t)$ is strictly DHR, there exists a finite and unique optimal repair-time limit $t_{011}^{*}(0 < t_{011}^{*} < \infty)$ under some conditions(refer Dohi and Kaio [5]) and the corresponding minimum total discounted cost is given by

$$TDC_2(t_{011}^{*}) = \frac{e_r + ke^{-\beta L} - \frac{k}{\beta}\left(1 - e^{-\beta L}\right) r(t_{011}^{*}) - ce^{-\beta L}\left[\beta + r(t_{011}^{*})\right]}{\int_0^{\infty} e^{-\beta t} dF(t) e^{-\beta L}\left[\beta + r(t_{011}^{*})\right] - \int_0^{\infty} e^{-\beta t} dF_{ir}(t) r(t_{011}^{*})} \tag{23}$$

Theorem 4 gives a sufficient condition for existence of the finite and unique optimal repair-time limit.

Partial Information Case

We consider the partial information case where the profit is discounted with the discount rate $\boldsymbol{\beta}$ over an infinite time horizon. To find the expected total discounted cost over an infinite time horizon, we obtain the expected total discounted cost during one cycle firstly. Then, the expected total discounted cost during one cycle is given by

$$\begin{aligned} V_3(t_0) &= \int_0^{\infty}\int_0^{\infty} [H(t_0 \left| t\right.) \int_0^{t} (e_r + k)e^{-\beta(x+y)} dy \\ &\quad + \overline{H}(t_0 \left| t\right.)[ce^{-\beta(x+L)} + \int_0^{L} ke^{-\beta(x+y)} dy\,]]dG(t)dF(x) \\ &= \frac{1}{\beta}\int_0^{\infty} e^{-\beta x}[\int_0^{\infty} [H(t_0 \left| t\right.)(e_r + k)(1 - e^{-\beta t}) \\ &\quad + \overline{H}(t_0 \left| t\right.)[c\beta e^{-\beta L} + k(1 - e^{-\beta L})]dG(t)]dF(x)] \end{aligned}$$

Since the expected present value of the unit cost just after one cycle is given by

$$\begin{aligned} \delta_3(t_0) &= \int_0^{\infty}\int_0^{\infty} [H(t_0 \left| t\right.)e^{-\beta(x+t)} + \overline{H}\ t_0 \left| t\right.)e^{-\beta(x+L)}\,]dG(t)dF(x) \\ &= \int_0^{\infty} e^{-\beta x}[\int_0^{\infty} [H(t_0 \left| t\right.)e^{-\beta t} + \overline{H}(\ t_0 \left| t\right.)e^{-\beta L}\,]dG(t)]dF(x). \end{aligned}$$

Thus the expected total discounted cost over an infinite time horizon becomes

$$TDC_3(t_0) = \sum_{j=0}^{\infty} V_3(t_0)\delta_3(t_0)^j = \frac{V_3(t_0)}{\overline{\delta_3}(t_0)} \tag{24}$$

The problem is to derive the optimal repair-time limit, t_{012}^{*} satisfying

$$TDC_3\left(t_{012}^{*}\right) = \max_{0 \le t_0 < \infty} TDC_3\left(t_0\right). \tag{25}$$

As a special case, all model assumptions are same as in Sect. 2.3 We also assume that the failure time follows an exponential distribution with rate λ=1/μ_f. Using the below result,

$$\int_{-\infty}^{\infty} e^{-\beta t} H(t_0\,|t)dG(t) = \exp\left[-\frac{2\beta\mu_a - \beta^2\sigma_a^2}{2}\right]\Phi\left(\frac{t_0 - (\mu_a - \beta\sigma_a^2)}{\sqrt{(\sigma_a^2 + \sigma_e^2)}}\right)$$

The expected total discounted cost during one cycle and the expected present value of the unit cost just after one cycle are given by

$$\begin{aligned}\beta V_D(t_0) &= \int_0^{\infty} e^{-\beta x}\Bigg[(e_r + k)\Bigg[\Phi\left(\frac{t_0 - \mu_a}{\sqrt{\sigma_a^2 + \sigma_e^2}}\right) \\ &\quad - \exp\left[-\frac{2\beta\mu_a - \sigma_a^2\beta^2}{2}\right]\Phi\left(\frac{t_0 - (\mu_a - \sigma_e^2\beta)}{\sqrt{(\sigma_e^2 + \sigma_a^2)}}\right)\Bigg] \\ &\quad + \left[c\beta e^{-\beta L} + k(1 - e^{-\beta L})\right]\left[1 - \Phi(\frac{t_0 - \mu_a}{\sqrt{\sigma_a^2 + \sigma_e^2}})\right]\Bigg]dF(x) \\ &= \Bigg[(e_r + k)\Bigg[\Phi\left(\frac{t_0 - \mu_a}{\sqrt{\sigma_a^2 + \sigma_e^2}}\right) \\ &\quad - \exp\left[-\frac{2\beta\mu_a - \sigma_a^2\beta^2}{2}\right]\Phi\left(\frac{t_0 - (\mu_a - \sigma_a^2\beta)}{\sqrt{(\sigma_e^2 + \sigma_a^2)}}\right)\Bigg] \\ &\quad + [c\beta e^{-\beta L} + k(1 - e^{-\beta L})]\left[1 - \Phi\left(\frac{t_0 - \mu_a}{\sqrt{\sigma_a^2 + \sigma_e^2}}\right)\right]\Bigg]\frac{\lambda}{\lambda + \beta}\end{aligned}$$

$$\begin{aligned}\delta(t_0) &= \int_0^{\infty}\int_0^{\infty}\left[H(t_0\left|t\right)e^{-\beta(x+t)} + \overline{H}(\,t_0\left|t\right)e^{-\beta(x+L)}\right]dG(t)dF(x) \\ &= \Bigg[\exp\left[-\frac{2\beta\mu_a - \sigma_a^2\beta^2}{2}\right]\Phi\left(\frac{t_0 - (\mu_a - \sigma_a^2\beta)}{\sqrt{(\sigma_e^2 + \sigma_a^2)}}\right) \\ &\quad + e^{-\beta L}\left(1 - \Phi(\frac{t_0 - \mu_a}{\sqrt{\sigma_a^2 + \sigma_e^2}}\right)\Bigg]\frac{\lambda}{\lambda + \beta}\end{aligned}$$

For detail derivation, refer Kim and Yun [31]. Thus, we can obtain the closed form of the expected total discounted profit over an infinite time horizon.

4.3 Preventive Order Model

In previous models, after the system is failed, we consider whether to order or repair. In this subsection, we consider a preventive order of a new unit before failure. If the system does not fail up to a pre-specified time Z, the order for a spare is made at the time Z and after a lead time L the spare unit is delivered. In this case, if the unit fails after Z, we replace the unit by the spare unit delivered. On the other hand, if the unit is failed before Z, the repair is started immediately. If the repair is completed up to the time limit for repairt_0(repair time limit), then the unit is installed at that time. It is assumed that the unit once repaired is presumed as good as new (perfect repair). However, if the repair time is greater than t_0, *i.e.* the repair is not completed after the time t_0, then the failed unit is scrapped, and the spare unit is ordered immediately and delivered after the lead time L. It is assumed that the time required for replacement is negligible. Under these model assumptions, we consider the interval from the start of the operation to the next start as one renewal cycle. The total expected cost for one cycle is given by the following three costs:

The expected cost of a renewal cycle is

$$\int_0^Z [(k+e_r)\int_0^{t_0} \overline{G}(t)dt + (c+kL)\overline{G}(t_0)]dF(x)$$
$$+\int_z^{Z+L} [c+k(Z+L-x)]dF(x)$$
$$+\int_{Z+L}^{\infty} [c+(t-Z-L)h]\,dF(x).$$

The expected duration of one cycle is given by

$$\int_0^Z \left[x+\int_0^{t_0} \overline{G}(t)dt + L\overline{G}(t_0)\right]dF(x) + \int_z^{Z+L}(Z+L)dF(x) + \int_{Z+L}^{\infty} x dF(x).$$

The total expected cost rate is given by,

$$TC_{p7}(t_0,Z) = \frac{\begin{array}{l}\int_0^Z \left[(k+e_r)\int_0^{t_0}\overline{G}(t)dt + (c+kL)\overline{G}(t_0)\right]dF(x)\\ \quad +\int_z^{Z+L}[c+k(Z+L-x)]dF(x)\\ \quad +\int_{Z+L}^{\infty}[c+(t-Z-L)h]\,dF(x)\end{array}}{\begin{array}{l}\mu_f+\int_0^Z\left[\int_0^{t_0}\overline{G}(t)dt + L\overline{G}(t_0)\right]dF(x)\\ \quad +\int_z^{Z+L}(Z+L-x)dF(x)\end{array}}. \quad (26)$$

Under this model, we should obtain the optimal repair-time limit and ordering time minimizing the expected cost rate.

Remark: This subsection has dealt with a simple order and repair/replacement problem. These models can be extended from various viewpoints. Thomas and Osaki [45] and Dohi et al. [7] presented continuous models with stochastic lead times.

4.4 Estimation Problem

In repair-time limit models, we should know the distributions of repair, failure distribution, and cost terms to obtain the optimal repair-time limit minimizing the expected cost rate or maximizing the expected total profit. Accurate statistical estimation of cost terms and or repair-time distributions is definitely needed to execute the repair-limit replacement program. Dohi et al. [1–4, 6, 8, 12–15] and Koshimae et al., [32] proposed graphic optimization methods and non-parametric estimation methods in the optimal repair-limit replacement policies. Under the assumption that the repair-time distribution is unknown but the complete repair data are available, non-parametric estimators of the optimal repair-time limits or repair-cost limits minimizing the expected cost and profit functions have been obtained. The basic idea is to utilize the total time on test (TTT) concept and Lorenz transform approach. TTT transform and plot have been applied to maintenance optimization problems by some researchers in reliability area (refer Nakagawa[38]) but Lorenz transform approach has been used to optimize the repair-time limits recently by Dohi et al. [3, 9, 12, 13]. Additionally, the graphical methods and non-parametric estimation with the complete repair data showed that the resulting estimator were strongly consistent, i.e., the estimates asymptotically approach to the real (but unknown) optimal solutions as the number of failure data increases. The property on consistency seems to be very attractive because it is not so easy to guarantee the goodness-of-fit of probability distributions estimated from the field data. In other words, engineers may skip statistical procedures of parameter estimation, goodness-of-fit test, etc., if the resulting estimate of the optimal repair-limit replacement policy is satisfactory. Dohi et al., [1–3, 6, 8, 14] and Koshimae et al., [32] investigated asymptotic properties of the estimated repair-limit replacement policies through Monte Carlo simulations and showed that the reasonable data size such as 30–50 was enough to estimate the minimum expected cost with higher accuracy.

5 Conclusions

In this chapter, we have reviewed repair-time limit replacement problems. In repair-time limit models, we have two options to recover the system failure; repair and replacement. The repair time is a random variable but the replacement time with a new unit is negligible. The new unit should be ordered and is delivered after a lead time L. If a system fails, we should decide to start to repair or to order a new unit. Based on the available information of repair times, we can choose one among repair and replacement options and we estimate the repair time to recover the system failure. If the estimated repair time is less than the pre-specified limit, then we start to repair the failed system. Otherwise we order the new item and finally replace the system. We considered three models based on available information amount (no, perfect and partial information) and two types of repair (perfect and imperfect repairs).

As modified models, profit and discounting models were also reviewed. Additionally, an ordering model with inventory cost was studied. Finally, we discussed estimation problems. For further studies, the following topics will be promising ones in repair-time limit models;

(1) Repair-time limit models with different repair types: we can consider minimal repair or other imperfect repair models, for example, age reduction and failure reduction models (refer Nakagawa [38] and Wang and Pham [46]).
(2) Repair-time limit models with preventive maintenance: age-based preventive maintenance (PM) can be also made before failure and PM interval will be also another decision variable (refer Kapur et al. [30], Wang and Pham [46]).
(3) Repair-time limit models with finite time horizon: We can study the repair-time limit optimization problems with finite or random time horizon (refer Nakagawa and Mizutani [39]).
(4) Integrated models with repair time and cost limits: Repair time and cost limits can be determined to minimize the expected cost rate together.
(5) Repair-time limit models with general inventory policies: General inventory policies, for example, (Q,r) policy, can be applied to the repair-time limit models (refer Dohi et al. [16]).

Acknowledgments This research was supported by Basic Science Research Program through the National Research Foundation of Korea (NRF) funded by the Ministry of Education, Science and Technology (No.2010-0025084). The authors are grateful to Professor S. Osaki who stimulated their interest in maintenance optimization area through many papers and edited books in reliability and maintenance.

References

1. Dohi T, Aoki T, Kaio N, Osaki S (1998) Non-parametric preventive maintenance optimization models under earning rate criteria. IIE Trans 30:1099–1108
2. Dohi T, Aoki T, Kaio N, Osaki S (2001) Optimization the repair time limit replacement schedule with discounting and imperfect repair. J Qual Maintenance Eng 7:71–84
3. Dohi T, Aoki T, Kaio N, Osaki S (2003) The optimal repair-time limit replacement policy with imperfect repair: Lorenz transform approach. Math Comput Modell 38:1169–1176
4. Dohi T, Aoki T, Kaio N, Osaki S (2006) Statistical estimation algorithms for some repair-limit replacement scheduling problems under earning rate criteria. Comput Math Appl 51:345–356
5. Dohi T, Kaio N (2005) Repair-limit replacement program in industrial maintenance-renewal reward process modeling. In: Tokimasa T, Hiraki S, Kaio N (eds) Applied economic informatics and systems sciences. Kyushu University Press, Fukuoka, pp 157–172
6. Dohi T, Kaio N, Osaki S (1995) Solution procedure for a repair limit problem using the TTT concept. IMA J Math Appl Bus Ind 6:101–111
7. Dohi T, Kaio N, Osaki S (1998) On the optimal ordering policies in maintenance theory-survey and applications. Appl Stochast Models Data Anal 14:309–321
8. Dohi T, Kaio N, Osaki S (2000) A graphical method to repair cost limit replacement policies with imperfect repair. Math Comput Modell 31:99–106
9. Dohi T, Kaio N, Osaki S (2001) Determination of optimal repair-cost limit on the Lorenz curve. J Oper Res Soc Jpn 44:207–219

10. Dohi T, Kaio N, Osaki S (2003) A new graphical method to estimate the optimal repair-time limit with incomplete repair and discounting. Comput Math Appl 46:999–1007
11. Dohi T, Kaio N, Osaki S (2003) Preventive maintenance models: replacement, repair, ordering and inspection. In: Pham H (ed) Springer handbook of reliability. Springer, London, pp 349–366
12. Dohi T, Kaio N, Osaki S (2007) Stochastic profit models under repair-limit replacement program. In: Proceedings of international workshop on recent advances in stochastic, operations research II:27–34
13. Dohi, T. Kaio, N. and Osaki, S. (2010), A stochastic profit model under repair-limit replacement program with imperfect repair. In: Proceedings of 4th Asia-Pacific, International Symposium(APARM 2010), pp 153–160
14. Dohi T, Koshimae H, Kaio N, Osaki S (1997) Geometrical interpretations of repair cost limit replacement policies. Int J Reliab Qual Safe Eng 4:309–333
15. Dohi T, Matsushima N, Kaio N, Osaki S (1996) Nonparametric repair limit replacement policies with imperfect repair. Eur J Oper Res 96:260–273
16. Dohi H, Okamura H, Osaki S (1999) Optimal order-limit policies for an (r, Q) inventory system. IMA J Math Appl Bus Ind 10:127–145
17. Dohi T, Takeita K, Osaki S (2000) Graphical methods for determining/estimating optimal repair-limit replacement policies. Int J Reliab Qual Safe Eng 7:43–60
18. Drinkwater RW, Hastings NAJ (1967) An economical replacement model. Oper Res Q 18:121–138
19. Hastings NAJ (1968) Some notes on dynamic programming and replacement. Oper Res Q 19:453–464
20. Hastings NAJ (1969) The repair limit replacement method. Oper Res Q 20:337–349
21. Hastings NAJ (1970) Equipment replacement and the repair limit method. In: Jardine AKS (ed) Operational research in maintenance. Manchester University Press, Barnes & Noble Inc., New York, pp 100–118
22. Jiang X, Cheng K, Makis V (1998) On the optimality of repair-cost-limit polices. J Appl Probab 35:936–949
23. Jiang X, Makis V, Jardine AKS (2001) Optimal repair/replacement policy for a general repair model. Adv Appl Probab 33:206–222
24. Kaio N (1981) Optimum repair limit policies with cost constraint. Microelectron Reliab 21:597–599
25. Kaio N, Osaki S (1982) Optimum repair limit policies with a time constraint. Int J Syst Sci 13:1345–1350
26. Kaio N, Dohi T, Osaki S (2002) Classical replacement models. In: Osaki S (ed) Stochastic models in reliability and maintenance. Springer, Berlin, pp 65–87
27. Kapil DVS, Sinha SM (1978) Repair limit suspension policies for a 2-unit redundant system with 2-phase repairs. IEEE Trans Reliab R- 30:90
28. Kapur PK, Kapoor KR (1978) A note on repair limit suspension policies for a 2-unit standby redundant system with two phase repairs. Microelectron Reliab 17:591–592
29. Kapur PK, Kapoor KR, Kapil DVS (1980) Joint optimum preventive-maintenance and repair-limit replacement policies. IEEE Trans Reliab R-29:276–279
30. Kapur PK, Garg RB, Butani NL (1989) Some replacement policies with minimal repairs and repair cost limit. Int J Syst Sci 20:267–279
31. Kim HG, Yun WY (2010) A repair time limit replacement policy with estimation error. Commun Stat Theory Methods 39:1–14
32. Koshimae H, Dohi T, Kaio N, Osaki S (1996) Graphical/statistical approach to repair limit replacement problem. J Oper Res Soc Jpn 39:230–246
33. Lambe TA (1974) The decision to repair or scrap a machine. Oper Res Q 25:99–110
34. L'Ecuyer P, Haurie A (1987) The repair vs replacement problem: a stochastic control approach. Optimal Control Appl Methods 8:219–230
35. Love CE, Rodger R, Blazenko G (1982) Repair limit policies for vehicle replacement. INFOR 20:226–237

36. Love CE, Guo R (1996) Utilizing Weibull failure rates in repair limit analysis for equipment replacement/preventive maintenance decisions. J Oper Res Soc 47:1366–1376
37. Muth EJ (1977) An optimal decision rule for repair vs replacement. IEEE Trans Reliab R-26:179–181
38. Nakagawa T (2006) Maintenance theory of reliability. Springer, Berlin
39. Nakagawa T, Mizutani S (2009) A summary of maintenance policies for a finite interval. Reliab Eng Syst Safe 94:89–96
40. Nakagawa T, Osaki S (1974) The optimum repair limit replacement policies. Oper Res Q 25:311–317
41. Nguyen DG, Murthy DNP (1980) A note on the repair limit replacement policy. J Oper Res Soc 31:1103–1104
42. Nguyen DG, Murthy DNP (1981) Optimal repair limit replacement policies with imperfect repair. J Oper Res Soc 32:409–416
43. Osaki S, Okumoto K (1977) Repair limit suspension policies for a two unit standby redundant system with two phase repairs. Microelectron Reliab 16:41–45
44. Segawa Y, Ohnishi M (2000) The average optimality of a repair-limit replacement policy. Math Comput Modell 31:327–334
45. Thomas LC, Osaki S (1978) An optimal ordering policy for a spare unit with lead time. Eur J Oper Res 2:409–419
46. Wang H, Pham H (2005) Reliability and optimal maintenance. Springer, Berlin
47. White DJ (1989) Repair limit replacement. OR Spektrum 11:143–149

Repair Strategies in an Uncertain Environment: Stochastic Game Approach

Y.-H. Kim and Lyn C. Thomas

Abstract This chapter deals with Repair strategies for stand-by equipment which maximises the time until failure when there is a vital need for the equipment, and it is unable to respond. We model conflict situations where the operating environment is controlled by an opponent. We develop stochastic game models to determine the form of the optimal Maintenance/Repair policy under these conditions and present numerical examples.

1 Introduction

A cold stand-by redundancy is where a unit is only brought into operation when there is a vital need for it. Hospital emergency power supplies, emergency response vehicles, and many military weapon systems are typical examples of standby unit. The cost of such failures is large compared with all other costs and so a cost criterion is inappropriate. Instead, we maximise the time until a catastrophic event occurs (when the equipment is needed and is unable to function) for a standby unit in an uncertain environment. The uncertainty in the environment is reflected in the frequency with which initiating events (to which the equipment needs to respond) occur. In other research, changes in the environment and hence the frequency of the initiating events were modelled as a random process ([7]), but here the environment is controlled by an opponent and so the solution is modelled as a stochastic game.

When on duty in peace keeping roles countering terrorist threats, troops and their equipment cannot remain on perpetual standby. The troops have to be given rest and relaxation, and even if replaced by other forces there will be a learning period when the new forces will not be able to respond as rapidly as their predecessors. The equipment has to receive regular maintenance, and where appropriate, repair.

Y.-H. Kim · Lyn C. Thomas (✉)
School of Management, University of Southampton, Southampton, UK
e-mail: L.Thomas@soton.ac.uk

T. Dohi and T. Nakagawa (eds.), *Stochastic Reliability and Maintenance Modeling*, Springer Series in Reliability Engineering 9, DOI: 10.1007/978-1-4471-4971-2_6,

The terrorists or warring parties wish to initiate events which will require the troops or equipment to respond. It is assumed that the readiness of the terrorists to initiate events in the next period of time is partially known by the authorities and is reflected in their state of alertness level (such as the U.S. DEFCOM levels). The terrorist player decides how active they will be in the next period, which then determines the alertness level. This is equivalent to saying that the terrorist player chooses what the alertness level will be. One also assumes that the terrorists have a good knowledge of the state of the standby "equipment" or troops, both by calculating how long they have been on standby and also by open or clandestine inspection of the equipment. This is then a maintenance model involving two players and such situations can be modeled as stochastic games.

The literature on Maintenance, Repair and Replacement policies for deteriorating equipment is long and distinguished. It started with the work of [1], and as the surveys and bibliographies of Refs. [5, 10, 12, 19, 21, 22, 24] and Wang [25] indicate, it has continued apace to the present day. Almost all the literature concentrates on policies which minimise the average discounted cost criterion. The idea of using a catastrophic event criterion to overcome the problem that failure will result in unquantifiably large cost was suggested first by Thomas et al. [23], with other instances being considered by Kim and Thomas [7]. In all these cases, the background environment and hence the probability of an initiating event is either fixed or follows a random Markovian process. Other authors such as Refs. [2–4], [9, 20] and [17, 18] have looked at maintenance in a random environment but in those cases the unit is always in use so the changes in the environment age the equipment at different rates, but do not affect when it is needed. Refs. [8, 26–28] and [6] study protective systems, such as circuit breakers, alarms, and protective relays with non-self-announcing failures where the rate of deterioration is governed by a random environment. We, on the other hand, allow the deterioration of the equipment to be independent of the environment, but the environment affects the need for the equipment. Yeh [29] studied an optimal maintenance model for a standby system but focused on availability and reliability as the criteria. Modelling the maintenance process as a game where the opponent is able to set the environment conditions has not been discussed before. In fact the application of game theory in the maintenance problem is restricted to warranty contracts [13, 14]. Here, we model the situation using stochastic games which were first introduced by Shapley (1953).

It is clear that there has to be some constraint on the activity of the "terrorist" and hence on the alertness level. Otherwise, the game is trivial—the "terrorist" will always force the activity level to its highest (most dangerous) state. This then reduces to a problem with one decision maker and no variation in the external state, which was the problem considered in [23].

In Sect. 2, we define our notation, set up the basic unconstrained game and confirm that in such a game it is optimal for the terrorist player to keep the state of alertness at its highest level. In Sect. 3, we consider the situations where there are constraints on the frequency with which the terrorist can be sufficiently active to force the alertness index to its highest level. For ease of notation, we will concentrate on the game where there are only two alertness states—Peaceful or Dangerous—but the results apply in more complicated situations. We investigate two constraints. The first type of constraint is on the average frequency of dangerous states in the game played so far. The second constraint discounts the activity of the terrorist, so what he was doing in the last period is much more important than his activity (or lack of it) several periods ago. In Sect. 4, we produce numerical examples and in Sect. 5 we draw conclusions on how the maintenance/ recuperation strategy depends on the interaction between the state of the equipment and the alertness level. We believe these models are a useful step in estimating Repair and Maintenance policies for standby equipment (and staff) which is used to combat the events initiated by intelligent and malevolent opponents.

2 Unconstrained Stochastic Game Model

We assume throughout that Player I, is the owner of the standby capability (hereafter called the equipment) and Player II is the one who seeks to create a catastrophic event–that is initiates an event to which the equipment fails to respond. The parameters of the model are

$i = 1, 2, \ldots, N$—the state of the equipment where N is the failed state;

P_{ij}—probability of equipment moving from state i to state j in one period of time, if no Repair action is performed.

This is independent of whether it is "used" or not that period. The standby unit is inspected regularly each period and this gives information on the operational state of the equipment to Player I. We assume that either through open inspection or by clandestine means, Player II is also aware of the state of the equipment.

Assume $\sum_{j=1}^{N} P_{ij} = 1$, $P_{NN} = 1$ and the Markov chain is such that there exists a $T = \min\{n \geq 0;\ P_{iN}^{n} > 0 \text{ for all } i\}$ so that within T periods, the chance of the equipment failing is positive from all starting states, i.e. $(P)_{iN}^{T} > 0$ for all i (equivalent to $\min_i (P)_{iN}^{T} = p > 0$)..

This ensures that without some maintenance of the equipment it is bound to fail eventually. The "ordering" of the intermediate states of the equipment reflects increasing pessimism about their future operability. This corresponds to P_{ij} satisfying a first-order stochastic condition namely

$$\sum_{j<k} P_{ij} \geq \sum_{j<k} P_{i+1,j} \quad \text{for all} \quad i = 1, \ldots, N-1,\ k = 1, \ldots, N..$$

This means if one considers states lower than k to be the "good" ones , one is more likely to move to a good state from i then from $i + 1$. The preventive Maintenance/Repair action (the former if equipment is in state $i = 1, \ldots, N - 1$, the latter if the state is N) takes one time period, during which the equipment cannot be used, if required. Such actions return the equipment to state 1-the good as new state. The subsequent results also hold if the maintenance action is not perfect, and returns the equipment to state i with probability r_i, but we will not complicate the notation by describing this case. $a = 1, 2, \ldots, M$ is the level of alertness of Player I but is really a decision by Player II on how active he intends to be in the next period. Both sides know that Player I has sufficient information sources to be able to correctly identify what this activity level will be. When Player II decides on his activity level, this corresponds to him choosing the "environment" for the next period. b_a is the probability of an initiating event occurring when the environment is a where $b_1 \leq b_2 \leq \ldots \leq b_M$ since the higher the alertness level the more likely that Player II will seek to initiate an event.

In the basic game, Player I has to decide at each period whether to undertake preventive Maintenance or Repair on his standby equipment, and Player II has to decide what the threat level of the environment should be. The game is played repeatedly until there is a catastrophic event when Player I cannot respond to an initiating event either because the equipment is being preventively maintained or because it has failed. Thus, Player I wants a Repair/Maintenance strategy that maximises the expected time until a catastrophic event, while Player II wishes to choose effort levels (environments) to minimise this expected time.

	$\Gamma_i, i \neq N$	II		
		Environmental level 1	Environmental level a	Environmental level M
I	Do nothing	$1 + \sum_{j=1}^{N} P_{ij}\Gamma_j$	$1 + \sum_{j=1}^{N} P_{ij}\Gamma_j$	$1 + \sum_{j=1}^{N} P_{ij}\Gamma_j$
	Repair	$(1 - b_1)(1 + \Gamma_1)$	$(1 - b_a)(1 + \Gamma_1)$	$(1 - b_M)(1 + \Gamma_1)$

(1)

Thus the basic game Γ) is a two person zero sum stochastic game consisting of N subgames Γ_i, $i = 1, 2, ..N$, where Γ_i is the game starting in the situation when the equipment is in state i. Player I decides whether to perform a maintenance action or Do Nothing for the next period while Player II decides what the environment will be. This defines the probability that an initiating event will occur during the period, and hence if the equipment is down or being repaired, whether there is a catastrophic event. If the equipment is in state $i(\Gamma_i)$ and no maintenance is carried out, it will move to state $j(\Gamma_j)$ for the next period with probability P_{ij}. The payoff matrix when the game is in subgame Γ_i is given by

	Γ_N	II		
		Environrnental level 1	Environmental level a	Environmental level M
	Do nothing	$(1 - b_1)(1 + \Gamma_N)$	$(1 - b_a)(1 + \Gamma_N)$	$(1 - b_M)(1 + \Gamma_N)$
	Repair	$(1 - b_1)(1 + \Gamma_1)$	$(1 - b_a)(1 + \Gamma_1)$	$(1 - b_M)(1 + \Gamma_1)$

(2)

The deterioration assumption guarantees that there is a probability p that the equipment will be down or in repair every T periods. In that period any initiating event will become a catastrophic event, and the least chance of an initiating event in any period is b_1. Thus, the time until an initiating event is bounded above by T/b_1^p. So Γ is a two person zero-sum stochastic game with a finite number of subgames, each of which has only a finite number of pure strategies ($2 \times M$) and where the total reward to each player is bounded above. Mertens and Neyman [11] proved that such games have a solution. The value of the game v(i) starting with equipment in state i satisfies the following

$$v(i) = val\begin{bmatrix} 1+\sum_{j=1}^{N} P_{ij}v(j) & \dots & 1+\sum_{j=1}^{N} P_{ij}v(j) \\ (1-b_1)(1+v(1)) & \dots & (1-b_M)(1+v(1)) \end{bmatrix} \quad \text{for} \quad i \neq N \qquad (3)$$

$$v(N) = val\begin{bmatrix} (1-b_1)(1+v(N)) & \dots & (1-b_M)(1+v(N)) \\ (1-b_1)(1+v(1)) & \dots & (1-b_M)(1+v(1)) \end{bmatrix} \qquad (4)$$

where val means the value of the game whose payoff matrix follows. Moreover, this game can be solved using a value iteration approach where the nth iterate $v_n(i)$ (which corresponds to value if only n periods were allowed) satisfies $v_0(i) = 0$ for all i and then

$$v_n(i) = val\begin{bmatrix} 1+\sum_{j=1}^{N} P_{ij}v_{n-1}(j) & \dots & 1+\sum_{j=1}^{N} P_{ij}v_{n-1}(j) \\ (1-b_1)(1+v_{n-1}(1)) & \dots & (1-b_M)(1+v_{n-1}(1)) \end{bmatrix} \quad \text{for} \quad i \neq N \qquad (5)$$

with a similar equation based on (4) for $v_n(N)$. This allows us to solve the game with help of the following results.

Theorem 1

(i). $v_n(i)$ is non-deceasing in n and non-increasing in i and converges to $v(i)$.
(ii). $v(i)$ is non-increasing in i.
(iii). The optimal strategy in the unconstrained game is: for Player II always to choose the most dangerous environment (level M); for Player I to Do Nothing in states $i < i^*$, where $i^* \leq N$, and perform maintenance/repair in state i^* to N.

Proof.

(i). The non-decreasing result in n follows since $v_1(i) \geq v_0(i) = 0$ and then by induction. Since $v_{n-1}(i) \geq v_{n-2}(i)$ for all i, the terms in the payoff matrix for $v_n(i)$ are greater than or equal to the terms in the matrix for $v_{n-1}(i)$. Hence $v_n(i) \geq v_{n-1}(i)$ and the induction step is proved.
Similarly $0 = v_0(i+1) \leq v_0(i) = 0$ for all i, so the hypothesis of v(i) non-increasing in i holds for $n = 0$. Assume true for $v_{n-1}(i)$ then the stochastic ordering plus the monotonicity of $v_{n-1}(i)$ implies $\sum_{j=1}^{N} P_{i+1,j}v_{n-1}(j) \leq \sum_{j=1}^{N} P_{i,j}v_{n-1}(j)$. Each entry in (5) of $v_n(i)$ is as large if not larger than the corresponding terms for $v_n(i+1)$, so $v_n(i+1) \leq v_n(i)$ for $i = 1, \dots, N-1$. The same result holds for $v_n(N) \leq v_n(N-1)$ since for $v_n(N)$ it is clear that Repair dominates do nothing because $v_{n-1}(N) \leq v_{n-1}(i)$. Hence $v_n(N) = \min\{(1-b_1)(1+v_{n-1}(i)), (1-b_2)(1+v_{n-1}(i))\} \leq v_n(N-1)$ and the induction step holds.

(ii). Trivially since $v_n(i) \leq v_{n+1}(i)$, $v_n(\cdot)$ converges to $v(\cdot)$ because $v_n(i)$ is a bounded increasing function, bounded above by T/pb_1. The monotonicity of $v_n(i)$ then guarantees the monotonicity of $v(i)$.

(iii). Player II's strategy is obvious since the values for the most dangerous environmental choice (M) always dominates the other strategies. Since v(1) $\geq v(N)$, the repair strategy (for the dangerous environment) is as good if not better than the do nothing strategy for state *N*. The monotonicity of $v(j)$ together with the stochastic ordering of P_{ij} implies $\Sigma P_{ij}\ v(j)$ is non-increasing in *i* and so once $\Sigma P_{ij} v(j)$ goes below $(1 - b_M(1 + v(1))$ (the definition of i^* it will remain below it for all higher states *i*.

So the unconstrained game is solved by the terrorist player always being at the highest state of activity. This is both unrealistic and reduces the problem to a single decision maker problem such as that in [7]. In the next section, we look at a more realistic assumption, namely that there is some limit on the terrorist's activity and hence on the frequency the environment is at its highest danger level. To keep the situation clear, we will hereafter assume there are only two levels of alertness—which we will label Dangerous (level 2) and Peaceful (level 1).

3 Models with Constraints on Effort

One reason an enemy cannot continuously create a dangerous environment, is that it needs time to regroup, plan and rest its forces—which we facetiously describe as "sleep". One possible assumption is that in stage *n* of the game, the enemy can only have created a dangerous environment for a proportion *c* of these stages. Thus if it has created a dangerous situation in *d* of the n periods that the game has been running, $d \leq cn$ then $s = cn - d$ is a measure of the "sleep index". This "sleep index" relates to how many consecutive periods of dangerous environment the enemy can create before it has to rest. If the sleep index is *s* and at the next period Player II chooses a Peaceful environment, the index will move to $s + c$, while if he chooses a dangerous environment, the index will move to $s + c - 1 = s - (1 - c)$. In this model, the effect of the rest induced by a peaceful environment will endure undiminished throughout all the future. An alternative view is that the *c* value that the restful period adds to the "sleep" index should diminish to αc next period, αc^2 the period thereafter and so on. In this case, if the current sleeping index is *s*, and Player II chooses a Peaceful environment this period, the index will move to $\alpha s + c$, while if Player II chooses to make the environment dangerous the index will move to $\alpha s - (1 - c)$.

We will prove results for the two cases $\alpha = 1$ (undiscounted) and $\alpha < 1$ (discounting of the index) in the same model though in the former case the sleep index could be infinite, while in the latter case it is bounded above by $c/(1 - \alpha$. In order to ensure a finite set of subgames, we will always assume in the undiscounted case that the index cannot exceed S. So the stochastic game Γ model of this situation consists of a series of subgame $\Gamma_{i,s}$ where $i = 1, - - -, N$ and $0 \leq$ s $\leq$ min $S, c/(1 - \alpha)$.

Although the sleep index set appears continuous, it is in fact countably infinite, and in fact finite if only r stages are allowed. If the index starts with s_0 then after r stages, the value can only be $\alpha^r s_0 + c(1-\alpha^r)/(1-\alpha) - \sum_{i=1}^{r} Z_i \alpha^{r-i}$ where $Z_i = 1$ or 0 depending on where Player II played Dangerous or Peaceful at the ith stage.

Let $v(i, s)$ be the value of the game Γ starting in $\Gamma_{i,s}$, where the equipment is in state i and the sleep index is s, then the values satisfy the equations

$$v(i,s) = val \begin{bmatrix} (1-\delta_N(i)b_1)(1+\sum\limits_{j=1}^{N} P_{ij}v(j,\alpha s + c)) & (1-\delta_N(i)b_2)(1+\sum\limits_{j=1}^{N} P_{ij}v(j,\alpha s + c - 1)) \\ (1-b_1)(1+v(1,\alpha s + c)) & (1-b_2)(1+v(1,\alpha s + c - 1)) \end{bmatrix} \quad (6)$$

where $\delta_N(i) = 1$ if $i = N, 0$ otherwise

One can solve this problem as in the previous section using value iteration. The iterates $v_n\ (i,\ s)$ satisfy an equation like (6) but with $v(i, s)$ replaced by $v_n\ (i,\ s)$ on the left hand side of (6) and $v(i,\ s)$ replaced by $v_{n-1}(i, s)$ on the right hand side of (6).

As in Sect. 2, in order to prove results about the optimal policies for the game, Γ, one proves results about $v_n(i, s)$ and hence $v(i, s)$.

Lemma 1

(i). $v_n(i, s)$ is non-deceasing in n and non-increasing in i and s.
(ii). $v(i, s)$ is non-increasing in i and s.

Proof.

(i). All the results follow by induction and the fact that if $W_1 = val \begin{bmatrix} a_1 & b_1 \\ c_1 & d_1 \end{bmatrix}$ and $W_2 = val \begin{bmatrix} a_2 & b_2 \\ c_2 & d_2 \end{bmatrix}$ then if $a_1 \geq a_2, b_1 \geq b_2, c_1 \geq c_2, d_1 \geq d_2, W_1 \geq W_2$.

(ii). Since $v_n(i, s)$ is non-deceasing in n and bounded above by T/pb_1, then $v_n(i, s)$ is a monotonic bounded sequence and so converges to $v(i,s)$. So the properties, $v_n(i+1, s) \leq v_n(i, s)$, $v_n(i, s') \leq v_n(i, s)$ if $s \leq s'$ hold for the limit function $v(i, s)$.

This allows one to describe features of the optimal strategies . If the item is "down (in state N)" then Player I will want to Repair it, while Player II will want to make the environment dangerous if they can. This ability to make the environment dangerous can only occur if αs- $(1-c) \geq 0$ or $s \geq (1-c)/\alpha$. Since if Player II starts with a sleep index of 0, the maximum the index can be is $s < c/(1-\alpha)$. Player II can only play the Dangerous strategy if $c/(1-\alpha) > (1-c)/\alpha$, i.e. $\alpha + c > 1$. So if $\alpha + c \leq 1$, the resultant game becomes trivial with Player II only able to play Peaceful and the results of the 1-player situation in [7], holding.

Theorem 2 Provided $\alpha + c \geq 1$, then in state N

1. if s satisfies $s \geq (1-c)/\alpha$, the optimal strategies are "Repair vs Dangerous"
2. if s satisfies $s < (1-c)/\alpha$, the optimal strategies are "Repair vs Peaceful"

Proof. The payoff matrix in the subgame $\Gamma_{N,s}$ is

$\Gamma_{N,s}$	Making peaceful situation	Making dangerous situation
Do nothing	$(1-b_1)(1+v(N,\alpha s+c))$	$(1-b_2)(1+v(N,\alpha s+c-1))$
Repair	$(1-b_1)(1+v(1,\alpha s+c))$	$(1-b_2)(1+v(1,\alpha s+c-1))$

Since by Lemma 1, $v(N,s) \leq v(1,s)$, it is trivial that the Repair strategy dominates the Do Nothing strategy for Player I. If $s < (1-c)/\alpha$, then Player II can only play the Peaceful strategy and so "Repair versus Peaceful" is optimal. If $s \geq (1-c)/\alpha$, we need to show that it is better for Player II to play Dangerous than Peaceful at the first occasion the system is in state N. Assume the system is currently down and let π^* be the policy that chooses to play "peaceful" at this period and plays optimally thereafter so $v^{\pi_P^*}(N,s) = (1-b_1)(1+v(1,\alpha s+c))$. Let π_1 be the policy that plays "peaceful" in the current period when $i = N$, and is the same as π^* except that at the next down situation it will choose the dangerous environment. Since playing Dangerous rather than Peaceful cannot increase the time until a catastrophic event $v^{\pi_1}(N,s) \leq v^{\pi_P^*}(N,s)$. Let π_2 be the policy that plays Dangerous now and Peaceful at the next down event, but otherwise chooses the same actions as π^*. Let K be the expected time between now and the next time when $i = N$ under π^*. Let T be the expected time from the next time $i = N$ to when a catastrophic event occurs under the π^* policy, conditional on there being a next down time when $i = N$. Then $v^{\pi_1}(N,s) = (1-b_1)(K+(1-b_2)T) > (1-b_2)(K+(1-b_1)T) = v^{\pi_2}(N,s)$

If π_D^* is the optimal policy for Player II to play against the optimal policy of Player I provided he chooses dangerous for his period, then $v^{\pi_D^*}(N,s) \leq v^{\pi_2}(N,s) \leq v^{\pi_1}(N,s) \leq v^{\pi_P^*}(N,s)$. So it is best for Player II to choose the "dangerous" environment as the best response to Player I's optimal policy.

If the standby system is working, then one can have any of the four combinations of pure strategies being chosen or even mixed strategies. What one can show though is that if the sleeping index is so low, that Player II cannot provoke a dangerous environment either this period or next period; then Player I will do nothing if the system is working.

Theorem 3 If $s < (1-\alpha c-c)/\alpha^2$, then Player I will do nothing in state (i,s) when i is a working state $(i < N)$.

Proof. The condition on s means that Player II must let the environment be peaceful for the next two periods. Consider the possible strategies for Player I over these next two periods,

strategy 1 : Repair in both periods
strategy 2 : Repair in period 1 and Do Nothing in period 2
strategy 3 : Do Nothing in period 1 and Repair in period 2

Let W_1, W_2, W_3 be the respective expected times until a catastrophic event if the optimal policy is used after the first two periods. Then

$$W_1 = (1 - b_1)(1 + (1 - b_1) + (1 - b_1)\, v\,(1, \alpha^2 s + \alpha c + c)$$

$$W_2 = (1 - b_1)(1 + 1 + \sum_{j=1}^{N} P_{ij}\, v\,(1, \alpha^2 s + \alpha c + c)$$

$$W_3 = 1 + (1 - b_1) + (1 - b_1) + (1 - b_1)\, v\,(1, \alpha^2 s + \alpha c + c)$$

and trivially $W_3 \geq W_1$and $W_3 \geq W_2$ since $v(1, s) \geq v(j, s)$ for all j and s. Hence, the Do Nothing now policy dominates the policies that Repair now and the result holds.

It need not be the case that it is optimal to Do Nothing even if one is in the new state $i = 1$ because one may recognise that an opponent has to play peacefully this period if the sleep index is s where $\alpha s + c - 1 < 0$. Repairing keeps the item in state 1, while it could degrade under the Do Nothing strategy. This result will be found in an example in the next section ($s = 0.6$ in Table 1). Before doing that we will show that in the undominated case if the system is working, and if s is large enough, then the players will either play "do nothing" against "peaceful" or they will play mixed strategies where Player I has a very high chance of playing "do nothing". To do that, we need the following limit result.

Lemma 2 *In the case $\alpha = 1$, as $s \to \infty$, v_n (i,s) and v(i,s) converge, respectively, to $v_n(i)$ and v(i) where*

$$v_n(i) = \max\left\{1 + \sum_{j=1}^{N} P_{ij} v_{n-1}(j),\ (1 - b_2)(1 + v_{n-1}(1))\right\}$$

and

$$v(i) = \max\left\{1 + \sum_{j=1}^{N} P_{ij} v(j),\ (1 - b_2)(1 + v(1))\right\}$$

These equations correspond to the situation where Player II is choosing the dangerous environment all the time.

Proof. From Lemma 1, v_n (i, s) and v (i, s) are non-increasing sequence in s, and as they are bounded above, they must converge. In the limit since $b_1 < b_2$, Player II's Dangerous strategy dominates its Peaceful one, since the payoffs against Do Nothing are the same, and against repair $(1 - b_2)(1 + v_{n-1}(1)) < (1 - b_1)(1 + v_{n-1}(1))$.

We are now in a position to describe what happens in the game when the sleep index gets very large.

Theorem 4 In the game with $\alpha = 1$, if the equipment is in a working state i, then for any $\varepsilon > 0$, $\exists S$ so that for $s \geq S$, the optimal strategies are either a) Do Nothing versus Peaceful, or *b*) mixed strategies where Player I plays Do Nothing with a probability at least $1 - \varepsilon$.

Proof. Consider the payoff matrix in the subgame $\Gamma^n_{i,s}$ of the game with n periods to go,

$\Gamma^n_{k \neq N,s}$	Making Peaceful Situation	Making Dangerous Situation
Do Nothing	$1 + \sum_{j=k}^{N} P_{kj} v_{n-1}(j, s+c) : A_n$	$1 + \sum_{j=k}^{N} P_{kj} v_{n-1}(j, s+c-1) : B_n$
Repair	$(1-b_1)(1 + v_{n-1}(1, s+c)) : C_n$	$(1-b_2)(1 + v_{n-1}(1, s+c-1)) : D_n$

and let A, B, C, D be the comparable values in $\Gamma_{i,s}$ when v_{n-1} is replaced by v. From Lemma 1 and the stochastic ordering property it follows that $B > A$. We also can prove $B > D$. By convergence, we can choose a N and a S so that $|v_n(j,s) - v(j,s)| < \varepsilon$ for all j, s if $n \geq N$ and provided $s > S$ we can choose $|v(j,s) - v(j)| < \varepsilon$ for all j where $v_n(j)$ is defined in Lemma 3.2. Then,

$$\begin{aligned}
1 + \sum_{j=k}^{N} P_{kj} v(j,s) &\geq 1 + \sum_{j=k}^{N} P_{kj} v_{n+1}(j,s) - \varepsilon \\
&\geq 1 + \sum_{j=k}^{N} P_{kj} v_{n+1}(j) - 2\varepsilon \geq 1 + \sum_{j=k}^{N} P_{kj}(1-b_2)(1+v_n(1)) - 2\varepsilon \\
&= 1 + (1-b_2)(1+v_n(1)) - 2\varepsilon \geq 1 + (1-b_2)(1+v_n(1,s)) - 3\varepsilon \\
&\geq 1 + (1-b_2)(1+v(1,s)) - 4\varepsilon \geq (1-b_2) + (1-b_2)v_n(1,s)
\end{aligned}$$

provided $\varepsilon < 1/4$

Hence $B > D$.

If $A \geq C$, then the fact $B > D$, means Do Nothing dominates Repair for Player I and $A < B$ means that Peaceful dominates Dangerous for Player II. Thus, Do Nothing versus Peaceful is optimal.

In the case $A < C$, note that as $b_2 > b_1$, then for s large enough $C > D$ since

$$\begin{aligned}
(1-b_1)(1+v(1,s+c)) &\geq (1-b_1)(1+v(1)) - \varepsilon \\
&\geq (1-b_2)(1+v(1)) + \varepsilon \geq (1-b_2)(1+v(1,s+c-1))
\end{aligned}$$

Hence with $C > A$, $C > D$, $B > A$, $B > D$, the optimal strategy is a mixed one with Player I playing $\left(\frac{C-D}{C+B-A-D}, \frac{B-A}{C+B-A-D}\right)$. For any $\delta > 0$ choose ε so that $\delta > 2\varepsilon/(b_2 - b_1)$ and $\varepsilon < \frac{b_2-b_1}{2(1-b_2)}$. Then the convergence of $v(j,s)$ in s means one can choose a $S*$ so for $s \geq S*$ $|v(j, s+c-1) - v(j, s+c)| < \varepsilon$ for all $s, \geq S*$ and all j. For such s

$$0 \leq B - A = \sum_{j=k}^{N} P_{kj}[v(j, s + c - 1) - v(j, s + c)] < \varepsilon$$

$$\begin{aligned} C + B - A - D \geq C - D &= [(1 - b_1)v(1, s + c) - (1 - b_2)v(1, s + c - 1)] \\ + b_2 - b_1 &\geq ((1 - b_1) - (1 - b_2)v(1, s + c)) - (1 - b_2)\varepsilon + (b_2 - b_1) \\ &\geq (b_2 - b_1) - (1 - b_2)\varepsilon > (b_2 - b_1)/2 \end{aligned}$$

Then Player I plays Repair with probability

$$\frac{B - A}{C + B - A - D} \leq \frac{\varepsilon}{(b_2 - b_1)/2} < \delta$$

and the result holds.

4 Numerical Examples

The actual policies in specific case can be obtained by value iteration calculations. The following examples have three equipment states-1 (new), 2 (used) and 3 (failed)—and doing nothing gives the following transition probabilities,

$$P = \begin{pmatrix} 0.3 & 0.4 & 0.3 \\ 0 & 0.4 & 0.6 \\ 0 & 0 & 1 \end{pmatrix}$$

The first examples are the non-discounted cases when $\alpha = 1$. Assume the constraint is that $c = 0.3$ so Player II can only create a dangerous environment 30 % of the time.

Tables 1 and 2 give the results in the new state ($i = 1$) first when $b_1 = 0.1$ and $b_2 = 0.5$ so there is a large difference between the Peaceful and Dangerous states (Table 1), and then when $b_1 = 0.4$ and $b_2 = 0.5$ (Table 2) so there is little difference between the two states. Notice in all cases, Player II can only choose the Peaceful environment if the sleep index s is less than 0.7. Theorem 3 says that for $s < 0.4$, Player I does nothing but notice in Table 1 at $s = 0.6$; Player I will Repair, even though (perhaps because) Player II can only ensure a Peaceful environment at this period but at the next period, could move the environment to the dangerous level.

Looking at Table 1, when the b_1, b_2 are quite different, the optimal strategies are mixed as s increases, though Player I's is getting more and more likely to Do Nothing. When s is large enough, Theorem 4 applies and in Table 1 an ε mixed strategy is optimal. In Table 2 when b_1, b_2 are similar, then Do Nothing versus Peaceful is optimal at all sleep index values since there is no point in repairing equipment in the best state since the impact of the environment is so small.

Table 1 The result for $\boldsymbol{i} = 1$(new), $\boldsymbol{b}_1 = 0.1$, $\boldsymbol{b}_2 = 0.5$

s	Value				GV	I		II	
	DN vs P	*DN vs D*	*R vs P*	*R vs D*	v	*DN*	*R*	*P*	*D*
0.0	8.545694	0	7.901159	0	8.545694	*DN*	-	*P*	-
0.1	7.991002	0	7.244038	0	7.991002	*DN*	-	*P*	-
0.2	7.846352	0	7.063115	0	7.846352	*DN*	-	*P*	-
0.3	7.779409	0	7.001468	0	7.779409	*DN*	-	*P*	-
0.4	7.049254	0	6.966484	0	7.049254	*DN*	-	*P*	-
0.5	6.848216	0	6.845989	0	6.848216	*DN*	-	*P*	-
0.6	6.771734	0	6.799698	0	6.799698	-	*R*	*P*	-
0.7	6.734876	9.258600	6.745477	4.772648	6.740826	0.44	0.56	0.97	0.03
0.8	6.549694	8.756620	6.663171	4.495315	6.606939	0.5	0.5	0.97	0.03
0.9	6.468176	8.610763	6.599401	4.422995	6.533275	0.5	0.5	0.97	0.03
1.0	6.429325	8.545694	6.562935	4.389533	6.495242	0.51	0.49	0.91	0.09
2.0	5.980036	6.429325	6.257458	3.646075	6.020760	0.85	0.15	0.89	0.11
3.0	5.845124	5.980036	6.152195	3.476366	5.859863	0.95	0.05	0.88	0.12
4.0	5.794211	5.845124	6.111484	3.417886	5.800096	0.98	0.02	0.88	0.12
5.0	5.774245	5.794211	6.095469	3.395269	5.776603	0.99	0.01	0.88	0.12
15.0	5.761531	5.761532	6.085306	3.380726	5.761531	$1-\varepsilon$	ε	0.88	0.12
27.0	5.761531	5.761531	6.085306	3.380726	5.761531	$1-\varepsilon$	ε	0.88	0.12
35.0	5.761531	5.761531	6.085306	3.380726	5.761531	$1-\varepsilon$	ε	0.88	0.12

i	s=0			s=35
1	*DN vs P*	*R vs P*	*Mixed*	(1-ε, ε) *vs Mixed*
2	*DN vs P*	*R vs P*	*Mixed*	(1-ε, ε) *vs Mixed*
3	*R vs P*		*R vs D*	

Fig. 1 Simple Form of the Result in Tables 1, 3 and 4

Tables 3 and 4 are the policies for the used and failed states in the case when $b_1 = 0.1$ and $b_2 = 0.5$ (which are the same parameters as in Table 1 for the new state). In state 2, one has Do Nothing vs Peaceful for $s < 0.4$ (no dangerous environments for at least two periods), then one has Repair vs Peaceful, at $0.4 \leq s < 0.7$. The mixed strategies are optimal as s increases and as $s \to \infty$ Player I tends to Do Nothing with probability $1-\varepsilon$ while Player II tends to (0.62, 0.38). Table 4 confirms the results of Theorem 2 that when the unit is down it must be repaired and the enemy will seek to make the environment dangerous if he can.

Figure 1 summarises the results of Tables 1, 3 and 4. If the equipment has failed one must repair it and the enemy will try to ensure a dangerous environment if its sleep index is high enough to allow it to. If the equipment is working then for a low sleep index, the solution is Do Nothing against Peaceful. As the sleep index increases so the enemy will be able to be dangerous in the next period, the equipment is repaired ready for that. If the sleep index is high enough that the enemy can ensure a dangerous environment this period, both sides play a mixed strategy with Player I more and more likely to Do Nothing and Player II being slightly more likely to play

Table 2 The result for $i = 1$ (new, $b_1 = 0.4, b_2 = 0.5$)

s	Value				GV	I		II	
	DN vs P	*DN vs D*	*R vs P*	*R vs D*	v	*DN*	*R*	*P*	*D*
0.0	6.231987	0	4.212781	0	6.231987	*DN*	-	*P*	-
0.1	6.079537	0	4.128808	0	6.079537	*DN*	-	*P*	-
0.2	6.036238	0	4.100696	0	6.036238	*DN*	-	*P*	-
0.3	6.021390	0	4.090763	0	6.021390	*DN*	-	*P*	-
0.4	5.881433	0	4.087023	0	5.881433	*DN*	-	*P*	-
0.5	5.834579	0	4.073457	0	5.834579	*DN*	-	*P*	-
0.6	5.818203	0	4.066845	0	5.818203	*DN*	-	*P*	-
0.7	5.811789	6.394473	4.064020	3.615947	5.811789	*DN*	-	*P*	-
0.8	5.789178	6.278914	4.062832	3.539723	5.789178	*DN*	-	*P*	-
0.9	5.778158	6.244166	4.060517	3.518074	5.778158	*DN*	-	*P*	-
1.0	5.773448	6.231987	4.059103	3.510651	5.773448	*DN*	-	*P*	-
2.0	5.761807	5.773448	4.056922	3.382586	5.761807	*DN*	-	*P*	-
5.0	5.761531	5.761531	4.056871	3.380726	5.761531	*DN*	-	*P*	-
15.0	5.761531	5.761531	4.056871	3.380726	5.761531	*DN*	-	*P*	-
27.0	5.761531	5.761531	4.056871	3.380726	5.761531	*DN*	-	*P*	-
35.0	5.761531	5.761531	4.056871	3.380726	5.761531	*DN*	-	*P*	-

Table 3 The result for $i = 2$ (not new, but working)

s	Value				GV	I		II	
	DN vs P	*DN vs D*	*R vs P*	*R vs D*	v	*DN*	*R*	*P*	*D*
0.0	8.312322	-	7.901159	-	8.312322	*DN*	-	P	-
0.3	7.779409	-	7.001468	-	7.779409	*DN*	-	*P*	-
0.4	6.458816	-	6.966484	-	6.966484	-	*R*	*P*	-
0.5	6.214748	-	6.845989	-	6.845989	-	*R*	*P*	-
0.6	6.138669	-	6.779698	-	6.779698	-	*R*	*P*	-
0.7	6.103182	7.065249	6.745477	4.772648	6.488706	0.40	0.60	0.87	0.13
1.0	5.708000	8.312322	6.562935	4.389533	6.174224	0.45	0.55	0.82	0.18
2.0	5.265155	5.708540	6.257458	3.646075	5.409183	0.85	0.15	0.68	0.32
5.0	5.059901	5.079677	6.095469	3.395269	5.067430	0.99	0.01	0.62	0.38
15.0	5.047299	5.047299	6.085306	3.380726	5.047299	$1-\varepsilon$	ε	0.62	0.38
27.0	5.047299	5.047299	6.085306	3.380726	5.047299	$1-\varepsilon$	ε	0.62	0.38
35.0	5.047299	5.047299	6.085306	3.380726	5.047299	$1-\varepsilon$	ε	0.62	0.38

Dangerous (but is still likely to play Peaceful most of the time because of the "sleep" restrictions).

Looking at the same problem $b_1 = 0.1, b_2 = 0.5$ but in the discounted case with α= 0.8 and c = 0.4 (not 0.3) leads to Tables 5, 6 and 7.

Again Table 6 confirms the results of Theorem 2, since Player II can only play Dangerous if $s \geq 0.75$, while Table 5 shows as the sleep index increase the strategies change from Do Nothing versus Peaceful to Repair versus Peaceful and then to mixed

Table 4 The result for $i = 3$ (down)

s	Value				GV	*I*		*II*	
	DN vs P	*DN vs D*	*R vs P*	*R vs D*	v	*DN*	*R*	*P*	*D*
0.0	7.201043	-	7.901159	-	7.901159	-	R	P	
0.1	7.169558	-	7.244038	-	7.244038	-	R	P	
0.4	5.195171	-	6.966484	-	6.966484	-	R	P	
0.5	4.945585	-	6.845989	-	6.845989	-	R	P	
0.6	4.880504	-	6.779698	-	6.779698	-	R	P	
0.7	4.850395	4.450395	6.745477	4.772648	4.772648	-	R		D
1.0	4.400580	4.000580	6.562935	4.389533	4.389533	-	R		D
2.0	4.112814	2.444766	6.257458	3.646075	3.646075	-	R		D
5.0	3.952436	2.204379	6.095469	3.395269	3.395269	-	R		D
15.0	3.942610	2.190339	6.085306	3.380726	3.380726	-	R		D
27.0	3.942610	2.190339	6.085306	3.380726	3.380726	-	R		D
35.0	3.942610	2.190339	6.085306	3.380726	3.380726	-	R		D

Table 5 The result for $i = 1$ (new), $c = 0.4$, $\alpha = 0.8$

s	Value				GV	*I*		*II*	
	DN vs P	*DN vs D*	*R vs P*	*R vs D*	v	*DN*	*R*	*P*	*D*
0.0	9.119139	0	8.433221	0	9.119139	*DN*	-	*P*	-
0.1	8.499747	0	7.655538	0	8.499747	*DN*	-	*P*	-
0.4	8.390626	0	7.551563	0	8.390626	*DN*	-	*P*	-
0.5	7.517332	0	7.513618	0	7.517332	*DN*	-	*P*	-
0.6	7.390844	0	7.492803	0	7.492803	-	*R*	*P*	-
0.7	7.332755	0	7.409354	0	7.409354	-	R	*P*	-
0.8	7.327393	9.760024	7.403906	5.030051	7.366117	0.49	0.51	0.98	0.02
1.1	7.196503	9.206344	7.263204	4.719848	7.225783	0.56	0.44	0.99	0.01
1.2	7.021900	9.120808	7.231623	4.685983	7.116675	0.55	0.45	0.95	0.05
1.5	6.928468	8.474689	7.122899	4.237344	6.996304	0.65	0.35	0.96	0.04
1.6	6.872290	8.392972	7.038725	4.196486	6.930300	0.65	0.35	0.96	0.04
1.7	6.743334	7.525998	6.983073	4.175824	6.795611	0.78	0.22	0.93	0.07
1.8	6.713779	7.396593	6.963334	4.167040	6.762757	0.80	0.20	0.93	0.07
1.9	6.682847	7.351381	6.944154	4.130449	6.733013	0.81	0.19	0.92	0.08
2.0	6.680185	7.330340	6.942264	4.115207	6.729240	0.81	0.19	0.92	0.08

i	*S*=0		*s*=2.0
1	*DN vs P*	*R vs P*	Mixed
2	*DN vs P*	*R vs P*	*Mixed*
3	*R vs P*		*R vs D*

Fig. 2 Simple form of the result in Table 5, 6 and 7

strategies. Note that 2 is the greatest value the sleep index can be when $c = 0.4$ and $\alpha = 0.8$, and in this case both players are playing a mixed strategy.

The results of Tables 5, 6 and 7 are summarised in Fig. 2. The results are very similar to the undominated case. The only difference is that because discounting

Table 6 The result for $i = 2$ (not new, but working), $c = 0.4$, $\alpha = 0.8$

s	Value				GV	*I*		*II*	
	DN vs P	*DN vs D*	*R vs P*	*R vs D*	v	*DN*	*R*	*P*	*D*
0.0	8.868031	0	8.433221	0	8.868031	*DN*	-	*P*	-
0.1	8.497001	0	7.655538	0	8.497001	*DN*	-	*P*	-
0.4	8.390626	0	7.551563	0	8.390626	*DN*	-	*P*	-
0.5	6.817920	0	7.513618	0	7.513618		*R*-	*P*	-
0.6	6.611384	0	7.492803	0	7.492803	-	*R*	*P*	-
0.7	6.579161	0	7.409354	0	7.409354	-	R	*P*	-
0.8	6.574326	9.556136	7.403906	5.030051	7.036201	0.44	0.56	0.85	0.15
1.1	6.477364	8.933314	7.263204	4.719848	6.863415	0.51	0.49	0.84	0.16
1.2	6.183837	8.869649	7.231623	4.685983	6.721767	0.49	0.51	0.80	0.20
1.5	6.110053	8.474689	7.122899	4.237344	6.566229	0.55	0.45	0.81	0.19
1.6	6.081463	8.392972	7.038725	4.196486	6.510805	0.55	0.45	0.81	0.19
1.7	5.964562	6.831037	6.983073	4.175824	6.204785	0.76	0.24	0.72	0.28
1.8	5.938792	6.614719	6.963334	4.167040	6.138237	0.81	0.19	0.70	0.30
2.0	5.897602	6.576650	6.942564	4.115207	6.099969	0.81	0.19	0.70	0.30

Table 7 The result for $i = 3$ (down), $c = 0.4$, $\alpha = 0.8$

s	Value				GV	*I*		*II*	
	DN vs P	*DN vs D*	*R vs P*	*R vs D*	v	*DN*	*R*	*P*	*D*
0.0	7.679899	0	8.433221	0	8.433221	-	R	*P*	-
0.1	7.647301	0	7.655538	0	7.655538	-	*R*	*P*	-
0.4	7.551563	0	7.551563	0	7.551563	-	*R*	*P*	-
0.5	5.415381	0	7.513618	0	7.513618	-	*R*	*P*	-
0.6	5.154424	0	7.492803	0	7.492803	-	*R*	*P*	-
0.7	5.148572	0	7.409353	0	7.409353	-	R	*P*	-
0.8	5.144705	4.690239	7.403906	5.030051	5.030051	-	R	-	*D*
1.1	5.105788	4.297863	7.263204	4.719848	4.719848	-	*R*	-	*D*
1.2	4.717435	4.267385	7.231623	4.685983	4.685983	-	*R*	-	*D*
1.5	4.667655	4.237344	7.122899	4.237344	4.237344	-	*R*	-	*D*
1.6	6.872290	4.196486	7.038725	4.196486	4.196486	-	*R*	-	*D*
1.7	6.743334	3.017556	6.983073	4.175824	4.175824	-	*R*	-	*D*
1.8	6.713779	2.863917	6.963334	4.167040	4.167040	-	*R*	-	*D*
2.0	6.680185	2.859057	6.942564	4.115207	4.115207	-	*R*	-	*D*

prevents the sleep index getting too large, Player I's mixed strategy does not tend to playing "do nothing" almost all the time but goes to a strategy where one does nothing 80 % of the time.

5 Conclusion

These models investigate the Maintenance and Repair policy for a standby system where the environment of when it is needed is controlled by an opponent. The most obvious context for this problem is the military one either in conventional or peace keeping roles. It could also apply to emergency services that need to respond to terrorist threats. We have shown that if there is no limit on resources available to the "enemy", then the problem reduces to one with a single decision maker dealing with a constantly high risk environment. If more realistically the enemy cannot always be ready to act, but needs time to recuperate, resupply and plan, the situation is much more complex, both in the situation where the restful periods have a long-term effect and when this effect is discounted over time.

One interesting feature is that the optimal policies can be mixed so each period there is a certain probability one should perform maintenance, and a certain probability one does nothing. Clearly if there are a number of such standby units, the mixed policy can translate into what proportion should be given preventive maintenance at that time. If the difference between the benign and the dangerous environment (b_1, b_2) is small, one tends only to perform maintenance when the equipment is close to failure, but in other situations one will maintain the equipment when it is in a good state because one feels the environment is likely soon to be dangerous (especially if the sleep index is high). One always repairs a failed unit, but the "enemy" will seek to take advantage of the failure by making the environment as dangerous as it can in those circumstances.

The models introduced in this chapter are the first to address the question of maintenance in an environment where failure can be catastrophic and where there is an enemy seeking such catastrophes. Clearly, more sophisticated models can be developed but we believe this chapter has indicated that one can get useful insights by addressing the problem as a stochastic game. Moreover, the game theory approach may be used to model Maintenance and Repair policies for equipment which are routinely used to deal with threats such as airport passenger and luggage screening devices.

Acknowledgments It is a privilege for us, particularly for LT to submit an article to this book in honour of Prof Osaki. LT did his doctorate in quantum probability and so was only learning about Operational Research in his post at the University of Manchester in the UK, when Professor Osaki came to the Department of Decision Theory as a visiting professor for a year in 1976. Through him, he learnt so much about using stochastic processes to model problems of replacement, maintenance, inventory and other aspects of logistics. This visit led to two joint papers, but also encouraged LT to work on real problems where there are decisions and uncertainty involved. At the time of his visit, Professor Osaki was publishing papers on the optimal Maintenance and Repair policies for standby systems [15, 16]; so it is appropriate that this paper looks again at the problem of repair and maintenance of standby systems. However, it is an unfortunate telling comment on current times that the paper concentrates on what happens when malevolent people seek to cause havoc by manipulating the operating environment.

References

1. Barlow RE, Proschan F (1965) Mathematical theory of reliability. John Wiley, New York
2. Çinlar, E., 1984. Markov and semimarkov models of deterioration. In Reliability Theory and Models(M. Abdel-Hamid, E. Çinlar, J. Quinn, eds), Academic Press, New York, 3–41
3. Çinlar E, Özekici S (1987) Reliability of complex devices in random environments. Prob Engrg Inform Sci 3:97–115
4. Çinlar E, Shaked M, Shanthikumar JG (1989) On lifetimes influenced by a common environment. Stoch Proc Appl 33:347–359
5. Dekker R (1996) Applications of maintenance optimization models: A review and analysis. Reliab Eng Syst Saf 51(3):229–240
6. Kiessler PC, Klutke G-A, Yang Y (2002) Availability of periodically inspected systems subject to Markovian degradation. J Appl Prob 39:700–711
7. Kim YH, Thomas LC (2006) Repair strategies in an uncertain environment: Markov Decision Processes Approach. J Oper Res Soc 57:957–964
8. Klutke G-A, Wortman M, Ayhan H (1996) The availability of inspected systems subject to random deterioration. Prob Engrg Inform Sci 10:109–118
9. Milhaud X, Lefèvre C (1990) On the association of the lifelengths of components subjected to a stochastic environment. Adv App Prob 22:961–964
10. McCall JJ (1965) Maintenance policies for stochastically failing equipment: A survey. Manage Sci 11(5):493–524
11. Mertens JF, Neyman A (1981) Stochastic Games. Int J Game Theory 10:53–66
12. Monahan GE (1982) A survey of partially observable Markov decision processes: theory, models and algorithms. Manage Sci 28:1–16
13. Murthy DNP, Yeung V (1995) Modelling and Analysis of Maintenance Service Contracts. Mathl Comput Modelling 22(10–12):219–225
14. Murthy DNP, Asgharizadeh E (1999) Optimal decision making in a maintenance service operation. European Journal of Operational Research 116:259–273
15. Nakagawa T, Osaki S (1975) Stochastic behaviour of a two unit priority standby redundant system with repair. Microelectronics Reliability 14:309–313
16. Okumoto K, Osaki S (1976) Optimum policies for a standby system with preventive maintenance, J. Operational Research Society 25:415–423
17. Özekici S (1995) Optimal maintenance policies in random environments. Eur. J. Oper. res. 82:283–294
18. Özekici, S., 1996. Complex systems in random environments. In Reliability and Maintenance of Complex Systems(S. Özekici, ed), Springer Verlag, New York, 137–157
19. Pierskalla WP, Voelker JA (1976) A survey of maintenance models: the control and surveillance of deteriorating systems. Naval Research Logistics Quarterly 23:353–388
20. Shaked M, Shanthikumar JG (1989) Some replacement policies in a random environment. Prob. Engrg. Inform. Sci. 3:117–134
21. Sherif YS, Smith ML (1981) Optimal maintenance models for systems subject to failure-A review. Naval Research Logistics Quarterly 28(1):47–74
22. Thomas LC (1986) A survey of maintenance and replacement models for maintainability and reliability of multi-item systems. Reliability Engineering and System Safety 16(4):297–309
23. Thomas, L.C., Jacobs, P.A., Gaver, D.P., 1987. Optimal Inspection Policies for Standby Systems. Commun Stat-Stoch Models 3(2), 259–273
24. Valdez-Flores C, Feldman RM (1989) A survey of preventive maintenance models for stochastically deteriorating single-unit systems. Naval Research Logistics 36:419–446
25. Wang H (2002) A survey of maintenance policies of deteriorating systems. European Journal of Operational Research 139:469–489
26. Wartman M, Klutke G-A (1994) On maintained systems operating in a random environment. J. Appl. Prob. 31:589–594
27. Yang Y, Klutke G-A (2000a) Improved inspection schemes for deteriorating equipment. Prob. Engrg. Inform. Sci. 14:445–460

28. Yang Y, Klutke G-A (2000b) Lifetime characteristics and inspection schemes for Lèvy degradation processes. IEEE Trans. on Rel. 49:337–382
29. Yeh L (1995) An optimal inspection-repair-replacement policy for standby systems. J. Appl. Prob. 32:212–223

Maintenance Modeling and Policies

Yaping Wang and Hoang Pham

Abstract The systems used in production, transportation services, and communication services constitute the majority part of not only industrial activities but also our daily life. Most of them have many units or components with various structure that will degrade with time or usage, and even suffer from a sudden failure due to the random shocks. For some systems, such as military systems, aircrafts, and nuclear power plants, they are of great importance and cannot afford to any failure. A machine in industrial plant failing to work properly will interpret the whole production assembly line and cost a large amount of capital and labor, while the failure of the aircraft will endanger the life of all the passengers. Therefore, maintenance on these systems is necessary due to the two aspects: (1) prolong the service life of the products; (2) improve the system reliability to avoid unnecessary failure.

The development of maintenance theory is experiencing a long history:

A. At first, most studies about maintenance are focussed on the repair and replacement. All of the actions are perfect to renew the system as good as new.
B. Then, people think that the system should not be maintained only when the system fails. As a result, the preventive maintenance come to our attention to consider the proactive maintenance actions before the system failure.
C. Also, imperfect maintenance including imperfect repair, imperfect monitoring, and imperfect preventive maintenance becomes a hot issue in maintenance issue.
D. Because system will suffer from more than one failure modes, maintenance model for deterioration systems, complex systems subject to competing risks, multi-component systems, or systems with redundancy have been extensively investigated in the past several decades.
E. Furthermore, the maintenance model only minimizing the cost rate or maximizing the system availability cannot satisfy the requirements of more and more complex

Y. Wang (✉) · H. Pham
Rutgers University, Piscataway, NJ, USA
e-mail: hopham@rci.rutgers.edu

T. Dohi and T. Nakagawa (eds.), *Stochastic Reliability and Maintenance Modeling*, Springer Series in Reliability Engineering 9, DOI: 10.1007/978-1-4471-4971-2_7,

maintenance problems in our real life. We need to pay attention to multiattribute of maintenance issues, including cost, availability, reliability, profit and so on.

F. Many of them also care about the maintenance polices under the warranty.

This chapter aims to provide a literature review on maintenance aspects for various maintenance models and policies including repair maintenance, age replacement, block replacement, periodic replacement, imperfect maintenance, preventive maintenance, inspection policy, optimal maintenance model for complex systems, multiobjective maintenance, maintenance policy under warranty, and other related results on this subject.

1 Repair Maintenance

Delia and Rafael [1] analyze the maintenance policies with two types of repair modes, including preventive and corrective repairs, and phase-type distributed repair times for a cold standby system subject to multistage degradation. Delia and Rafael [2] examine the replacement policy for a Markovian degraded system submitted to internal or external failures with holding time on various system levels, external repair time and internal repair time, all of which follow the phase-type distribution. A later work of Delia and Rafael [3] considers a maintenance model with failure and inspection following arrival processes and two types of repair modes, minimal and perfect, distributed as different phase-type distributions for a deteriorating system suffering from both internal and external failures. Tang and Lam [4] study a δ-shock maintenance model for a deteriorating system with shocks occurring according to a renewal process, where the interarrival time of shocks follows a Weibull or gamma distribution. Because the system is deteriorating, the deadlock threshold for the δ-shock is geometrically nondecreasing after each repair, and the repair time is modulated by an increasing geometric process.

Finkelstein [5] introduces a generalized Strehler-Mildvan model to estimate the first passage time of the survival function for the system subject to cumulative damage due to biological aging and sudden killing event. The asymptotic aging properties for the repairable system are discussed. Vaurio [6] develops the advanced models to the general recursive equations for the availability and mission-failure probability of the standby structure system by considering different durations for testing and repairs, as well as various failure types, including start-up, standby, and during mission failures, and two additional human errors. Ram and Singh [7] study the availability, mean time to failure (MTTF), transition state probability, and cost analysis for the complex system consisting of two repairable subsystems with "1-out-of-2: F" and "1-out-of-n: F" structure under the "preemptive-resume repair discipline" by using Gumbel–Hougaard copula in repair.

2 Age Replacement

Qian et al. [8] focus on the analysis of maintenance policies for an extended cumulative shock model with shocks occurring at a nonhomogeneous Poisson process. The system will be maintained when cumulative shock do not exceeds the failure threshold, repaired when cumulative shock exceeds the threshold, and replaced at failure N or time T. Jiang and Ji [9] study a multi-attribute value maintenance model with age replacement policy to consider cost, availability, reliability, and lifetime as objectives by using weighted average mean method. Pandey et al. [10] considered an age replacement policy for the gamma deterioration model where the component is replaced when the system fails or reaches a specific age, whichever occurs first. Sheu et al. [11] propose an age-dependent replacement policy with two types of failure, type I failure removed by minimal repairs, and type II failure removed by replacement. The replacement will be performed under three conditions that the number of type I failure reaches N or at age T or first type II failure happens, whichever occurs first. Ozekici [12] studies the optimal age repair and replacement policies in the presence of random environments characterized as the semi-Markov chain, in which the measurement of the device age in the maintenance analysis is based on the intrinsic clock but not the real age.

Dayanik and Gurler [13] utilize two adaptive Bayesian decision model with various levels of information to analyze the optimal age replacement policy for the maintained system which will suffer from a critical failure with a given probability. Dekker and Plasmeijer [14] develops an opportunity age replacement policy with two parameters based on the marginal cost approach by modeling as the regenerative process: (1) control limit, which indicates the age when replacement is done preventively; (2) planned replacement time, which indicates the time when replacement should be done if it has not happened yet.

3 Block Replacement

Nguyen and Murthy [15] put forward a replacement policy combining the repair limit and block replacement to determine the replacement interval T and the replacement threshold cost x. Murthy [16] extends the block replacement by considering the penalty cost for the inventory shortage. Marathe and Nair [17] compare two multistage planned replacement strategies of block replacement and age replacement with the one stage replacement strategy to consider failure cost and economic gain. Nakagawa and Mizutani [18] extend three usual maintenance models of simple replacement, block replacement, and periodic replacement from an infinite time span to a finite time span and also consider the optimal solutions for periodic time and sequential actions of PM. Sheu [19] derives the expressions of expected long-run replacement cost rate and total α-discounted cost for a system subject to nonhomogeneous Poisson process shock with two failure types, minor failure removed by minimal repair or catastrophic failure restored by unscheduled replacement, under both age replacement and block replacement policies. Satow et al. [20] focus on a replacement policy for a unit that suffers from cumulative damage due to aging process and shocks in order to obtain the optimal replacement level k^* which minimizes the expected cost rate.

Sheu and Griffith [21] present an extended block replacement policy with two types of failures, where the maintenance action is decided based on the shock number upon last replacement. Anisimov [22] studies the asymptotic results of block replacement policies with periodic inspection for a multi-component system under fast Markov switches and provides the simplified result for the system with exponential failures. Scarf and Cavalcante [23] develop a novel hybrid replacement polices combing the block replacement and age-based inspection maintenance for the series system consisting of heterogeneous nature components with successive replacement.

4 Periodic Replacement

Nakagawa and Kijima [24] propose a periodic replacement policy with minimal repair at failure for the cumulative shock model in order to obtain the optimal solution for the time T^*, shock N^*, and damage Z^* at which time point the replacement is done. Chaudhuri and Suresh [25] put forward an algorithm based on fuzzy set theory to develop a periodic replacement maintenance policy with three maintenance action types of minor, medium, and major to determine the optimal replacement schedule. Sim and Endrenyi [26] consider a Markov process to model the maintenance policy with periodically minimal maintenance and major maintenance after a number of minimal maintenances for a continuously operating system subject to degradation and Poisson failures. The optimal solution to minimize either cost rate or unavailability is derived. Because in real world, not only the inspection, repair, and replacement costs and times, but also the system operating cost, will increase with the system aging, Chiang and Yuan [27] present a state-dependent maintenance policy $R_{i,j}(T, N, \alpha)$ for a continuously deteriorating system subject to degradation and fatal shocks using a continuous-time Markov process, where T is the system inspection interval, N is the system boundary for replacement, and α is the probability that repair will restore the system to a better state.

5 Imperfect Maintenance

Sheu and Chang [28] put forward a generalized extended periodic preventive maintenance in the presence of imperfect maintenance characterized as improvement factor by effective age and hazard rate distribution for the system with age-dependent failure type of minor failure and catastrophic failure. Liu and Huang [29] apply the non-homogeneous continuous time Markov model (NHCTMM) to model the optimal replacement policy for the multi-state system with the imperfect maintenance that utilizes the quasi-renewal process to describe the stochastic behavior of the multi-state aging element after each imperfect repair. Cassady et al. [30] explore the imperfect repair based on the Kijima's first virtue age model by validating the simulation results using 2^3 factorial experiment and converting reliability and maintainability parameters into coefficients of availability model using metamodels to determine the optimal replacement interval according to the system average cost. Satow and Kawai [31] present an imperfect inspection with upper and lower inspection threshold for a bivariate failure distribution. Flammini et al. [32] take into account the imperfect maintenance for computer systems with N-modular redundant using the multi-formalism modeling which allows for two independent models: a maintenance model based on state-based formulism and a failure model based on Bayesian networks. You et al. [33] put forward two types of control-limit PM polices of reliability limit PM (RLPM) policy and hazard rate limit PM policy based on extended proportional hazards model (EPHM) with the joint effects from the imperfect PM and time-varying operation conditions.

Doyen and Gaudoin [34] introduce two kinds of imperfect repair treatment methods: one is that upon each repair, the failure intensity is reduced; the other is upon each repair, the virtual age is reduced. Park et al. [35] present a periodic maintenance policy for a repairable system, where instead of restoring the system level, the imperfect PM will slow down the system degradation rate while upon each PM the system hazard rate still increases monotonically. Kallen [36] puts forward an imperfect maintenance model based on the superposition of renewal processes and approximate the distribution of inter-repair time. In the model, all of the failure cells are repaired only when a fraction of cells are damaged. Kahle [37] considers optimal maintenances with incomplete repair using the treatment in Kijima [38, 39], which assumes that the imperfect repair will impact the system failure intensity by reducing the system virtual age. Tsai et al. [40] focus on the optimal preventive maintenance model combining the minimal repairs for each failures and imperfect maintenance at some prespecific time for the repairable products and derive the maintenance time to minimize the total expected cost.

Soro et al. [41] apply the continuous-time Markov chain to model the preventive maintenance policy for the multi-state system in the presence of imperfect maintenance and minimal repair and evaluate the three performance indicators of reliability function, production rate, and system availability in the maintenance problem. Wang et al. [42] combine the imperfect maintenance into the delay time model (DTM) to model the optimal inspection-maintenance policies of maximizing the long-run availability by Genetic algorithm (GA). Makis and Jardine [43] consider the optimal replacement policy for a general model incorporating imperfect repair which will restore the system to a functioning state just prior to system failure, or "as good as new" state, or lead to a scrap required for an addition replacement.

6 Preventive Maintenance

6.1 Condition-Based Maintenance

Compared with the time-based replacement policy, Huynh et al. [44] introduce a condition-based maintenance model for the degradation-threshold-shock (DTS) model to take the dependence between degradation process and shock process into account. Wang et al. [45] consider a novel maintenance model combing the condition-based replacement, periodical inspections, and (S, s) type provisioning policy, noted as (T, S, s, L_p) policy, where T denotes the inspection interval, S the maximum inventory level, s the reorder point, and L_p the replacement threshold. Furthermore, a simulation model is established to modulate the uncertain deterioration process and finally the maintenance scheduling is optimized to minimize the cost rate using genetic algorithm. The study of Camci [46] aims to design a comprehensive maintenance tool to combine the corrective maintenance, preventive maintenance, and condition-based maintenance with regard of prognostic information to balance the two objectives of cost and availability by using Genetic algorithm. Grall et al. [47] propose a condition-based maintenance model including both the inspection and replacement policies based on a multilevel control-limit rule for a stochastically deteriorating system for the purpose of obtaining the optimal replacement threshold and inspection scheduling to minimize the long run expected cost.

Van Noortwijk and Frangopol [48] describe two maintenance models of condition-based maintenance and reliability-based maintenance for the deteriorating civil infrastructures for the purpose of minimizing the life cycle cost under the constraint of adequate reliability level. Deloux et al. [49] propose a maintenance policy that combines the statistical process control (SPC) and condition-based maintenance (CBM) for a continuously deteriorating system with two kinds of failure mechanisms, deteriorating and random shocks. The system failure is governed by deteriorating process as a function of the deterioration level and the system time but an associated failure acceleration factor due to stress is taken into account when the stress intensity exceeds some critical level. Van der Weide et al. [50] derive the reliability estimation and the optimal solution for calculating the discounted cost based on both condition-based and age-based policy for a maintained system that deteriorates due to both transient shocks and cumulative degradation process governed by a stochastic point process. Van der Weide and Pandey [51] present a periodic inspection condition-based maintenance policy for the systems subject to shock and cumulative stochastic damage in the presence of hidden or latent failures. Instead of using renewal process, nonhomogeneous poisson process (NHPP) is applied to model the nonlinear damage increments.

6.2 Other PM Models

Curcuru et al. [52] put forward a predictive maintenance policy in the presence of imperfect monitoring, in which the stochastic degradation process is characterized by a first-order autoregressive model with drift and the a priori information of monitoring system is provided by Bayesian approach. Wu and Clements-Croome [53] consider three optimal maintenance policies for the system whose failure processes can be modulated by Geometric process (GP) with different costs for the up- and down-time : (1) only corrective maintenance (CM); (2) imperfect PM and CM; and (3) periodic PM and CM. Amari [54] discusses the bounds and approximation methods for the mean time between failures (MTBF) for repairable systems with preventive maintenance at periodic intervals under different scenarios of various failure rate distribution.

Yeh et al. [55] present the two periodical PM policies with reduction age for the second-hand products to determine the optimal PM number and each PM degree by minimizing the expected maintenance cost, where the initial age is known and the prespecific length of usage is Weibull distributed. Casto et al. [56] suggest an age-based PM policy for the gradually degraded critical structures characterized by gamma process to determine the optimal replacement time and derive the analytical optimal solutions to minimize the expected maintenance cost rate. Zhao [57] presents a preventive maintenance policy for a deteriorating system with a critical reliability level to satisfy the preference of field managers where the imperfect PM effect is modulated by a parameter of degradation ratio.

Sun et al. [58] introduce a preventive maintenance policy for the systems with failure interaction by using a new technique of extended Split System Approach. Panagiotidou and Tagaras [59] investigate the optimal preventive maintenance policy for production equipments submitted to deterioration in presence of two quality states of in-control state and out-of-control state with different income. In the model, both the failure distribution and shift distribution of the system states are dependent on the equipment age and system actual state. Roux et al. [60] model the optimal preventive maintenance policy for the multi-component systems in the industrial production problems by means of combination techniques of PDEVs and timed Petri nets and derive the optimal simulated solutions via Nelder–Mead method. Nguyen and Murthey [61] study the two types of optimal preventive maintenance policies for repairable systems whose failure rate will increase with the repair number to minimize the cost rate for an infinite time span.

7 Inspection Policy

Apeland and Scarf [62] discuss an inspection maintenance modeling with time-delay by using fully Bayesian approach, which is different from the classic probabilistic approach in the probability distribution assumption and uncertainty treatment. Hariga [63] develops an inspection scheduling for one single machine with random failure and also extends the heuristic procedure to find the optimal inspection intervals of maximizing the expected profit for the exponential distributed failure. Wortman et al. [64] examine the maintenance strategy with the inspection time modulated by a renewal process for a nonself-announcing failure system subject to deterioration governed by random shocks. Chelbi and Ait-Kadi [65] develop the expression of the time-stationary availability for a hidden failure system subject to the transient shocks with a predetermined inspection time in order to generate an optimal solution for the target availability level with limit resources. Kiessler et al. [66] examine the limiting average availability of a hidden-failure deterioration system with the periodic inspections where deterioration rate is governed by a Markov model. Yang and Klutke [67] characterize the properties of the lifetime distribution for the Levy degradation process and then illustrate the implement of the results to inspection scheduling for the maintained system with nonself-announcing failures.

A generalized Petri Net is proposed by Hosseini et al. [68] to formulate a new condition-based maintenance model for a system subject to deterioration failures and Poisson failures. In order to maximize the system throughout, an optimal inspection policy based on minimal maintenance, major maintenance, and major repairs is obtained. Kharoufeh et al. [69] utilize the Laplace-Stieltjes transform to explicitly derive the lifetime distribution as well as the limiting availability for a periodically inspected single-unit system with hidden failure, which is subject to the degradation wear due to its random environment characterized by a continuous Markov chain, and random shocks modulated by a homogeneous Poisson process. Klutke and Yang [70] present a maintenance policy for the periodically inspected systems with nonself-announcing failure, submitted to cumulative damage due to both graceful degradation and random shocks for the purpose of optimizing the system performance from the limiting average availability point of view. Wang [71] determines an optimal inspection interval time in terms of an cost function for a three-stage failure system of normal, minor, and severe detective stages with two model options: one is upon inspection when we found the minor detective, we will replace or repair immediately; the other is instead of instant maintenance, we will shorten the inspection interval to half.

Chan and Wu [72] apply the cumulative count of conforming chart (CCC chart) to develop the inspection and maintenance policies for the production systems with fractions of conforming and nonconforming products under different scenarios of minor/major inspection and minor/major maintenance. Kurt and Kharoufeh [73] model the optimal periodic inspection-maintenance policies by the infinite-horizon Markov decision process to minimize the expected discounted cost rate due to operation, failure, inspection, repair, and maintenance. In the model, upon each inspection, three maintenance actions should be chosen from: (1) leave it as it were till the next inspection time, (2) replace it with a new one, and (3) repair the system.

8 Optimal Maintenance Models for Complex Systems

8.1 Multi-Degraded Systems

Gurler and Kaya [74] approximate the explicit expression of the long-run average cost rate in a maintenance control policy for a multi-component system each with several stages, which can be further divided as good, doubtful, PM, and down, by using the multidimensional Markov process. Flores-Colen and Brito [75] discuss a symmetric maintenance schedule of preventive and predictive maintenance for the building facades based on different simulated maintenance scenarios for the performance-degradation models by using life cycle cost analysis. Huang and Yuan [76] propose a two-stage PM policy with imperfect maintenance under periodic inspection for the multi-state Markova deterioration system, where the transition probabilities and risks of imperfect maintenance are updated upon completing of each PM.

Saassouh et al. [77] propose a two-mode stochastically deteriorating model with a sudden change point in the degradation path, where the increments of deterioration follow a gamma law when the system is in the first mode, and the mean deterioration rate increases when it flips into the second mode. Based on the definition of the model, the decision rules for an online maintenance policy are determined to optimize the system performance from the angle of asymptotic unavailability. Ponchet et al. [78] compare two condition-based maintenance (CBM) models for a gradually deteriorating system submitted to random change in its degradation rate. In the first CBM model, the decision is based only on the degradation level, while in the second model, the maintenance decision depends on both the degradation level and degradation rate. Chen et al. [79] utilize the Geometric process (GP) to model the maintenance problem of the repairable deteriorating systems and derive the Bayesian inference of parameters in GP using the combination method of Gibbs sampler and Metropolis algorithm. Zhao et al. [80] discuss the condition-based maintenance policy with inspection and replacement for a deteriorating system characterized as a nonmonotone stochastic process with environmental covariates modeled by a finite state Markov chain in order to minimize the expected maintenance cost rate. Van der Wang [81] compares the various maintenance polices for the deteriorating systems with both single unit and multiple units.

8.2 Competing Risk Systems

Frostig and Kenzin [82] derive the limiting average availability in a maintenance model for a hidden-failure system that suffers from the wear out and cumulative shock damage with a Poisson process. Two models are discussed: Model 1 assumes the wear out process and shock will not receive any impact from the external environment; In Model II, the shock magnitude, the rate of shock and wear out process depend on the external environment modulated by a Markov process. Lam and Zhang

[83] study a replacement policy for two systems embedded in the δ-shock model: One is deteriorating system with a nondecreasing threshold after repair times and geometrically increasing repair times; the other is improving system with decreasing threshold after repair and geometrically deceasing repair times. Chen and Li [84] analyze a deteriorating system subject to extreme shock. The deterioration process is governed by both the external shocks and internal loading from the point of views: (1) the magnitude of the random shock the system can bear will be decreasing with the numbers of repairs; (2) the repair time will be increasing after each repair. Finally, an optimal replacement policy N^*, at which failure number the system will be repaired, is determined by minimizing the long-run average cost.

Zequeria and Berenguer [85] study a maintenance policy, considering three types of actions: minimal repairs, preventive maintenance, and replacement, for a system with two dependent competing failure modes of maintenance and nonmaintenance by minimizing the system cost rate during an infinite time. In the model, the improvement factor for the failure rate upon preventive maintenance actions depend on the time when the actions are performed. Zhu et al. [86] examine the maintenance model for a competing risk of degradation and sudden failure, in which the unit is renewed when it reaches a predetermined degradation level or comes to a sudden failure within the limit of certain degradation threshold. Also a preventive maintenance (PM) is done at the scheduled time. The maintenance scheduling variables of degradation threshold and scheduled time to preventive maintenance are determined by maximizing the system availability with the constraint of repair cost.

Li and Pham [87] focus on the condition-based maintenance model for a generalized multi-state degradation system subject to multiple competing failure processes, consisting of two degradation processes and cumulative random shock to minimize the average long-run cost rate function by Nelder–Mead downhill simplex method. Wu et al. [88] propose two types of multi-state system maintenance policies of preventive replacements and corrective replacements based on the expected discounted maintenance cost rate to determine two threshold variables, consisting of threshold level on current system state and threshold level on residual life, under the case of finite life cycle by using Laplace transform and inversion approach as well as approximation method.

8.3 Multi-Unit Systems

Barros et al. [89] concern the problem of imperfect monitoring in a maintenance policy, in which each unit have a given probability to be detected failure, for a two-unit parallel system with stochastic dependency. A new delay time model for a multi-component system with multiple failure modes is proposed by Wang et al. [90]. In the model, each component is modeled separately based on its failure mode and then a common inspection schedule is made for the large subsystem according to individual component analysis. Zhang and Wang [91] consider the replacement policy for a cold standby system with two components both following a geometric repair. In the model, component 1 is given a use priority and the replacement is performed when the failure number of the component 1 reaches N. Marseguerra et al. [92] propose a condition-based maintenance model for a continuously monitored multi-component system subject to stress-dependent degradation process by using the coupled methods of Genetic Algorithm and Monte Carlo Simulation to derive the optimal solution for the multi-objective maintenance problem considering the interest, availability and profit as the objectives.

Tian and Liao [93] focus on the study of condition-based maintenance model based on proportional hazards model for the multi-component systems with economic dependency. Taghipour and Banjevic [94] propose a maintenance optimization model for a multi-component repairable system submitted to hidden failure in order to determine the optimal periodical inspection intervals. In the work by Tian et al. [95], a condition-based maintenance model with two failure probability threshold is introduced for the wind power generating systems consisting of multiple components with economic dependency to minimize the total operational and maintenance cost by using the Artificial neural network (ANN). Bouvard et al. [96] present a dynamic method to develop the condition-based maintenance model for the commercial heavy vehicles, where there exist multiple components with grouped maintenance policies and each maintenance action upon inspection is determined upon the individual degradation level for each component. In Wang and Lin [97], an improved particle swarm optimization (IPSO) is introduced to minimize the total maintenance cost with preventive maintenance and replacement for the series-parallel systems. Mahmoud and Moshref [98] derive the explicit expressions for the mean time to failure (MTTF), steady state availability, buy periods, and system profit gain for a stochastic model of a two-unit cold standby system subject to human error failures, hardware failures, and preventive maintenance (PM).

Laggoune et al. [99] consider an opportunistic preventive maintenance policy for a multi-component system, where the age-based policy is used to analyze the maintenance model for individual components and then multi-grouping approach is applied to derive the cost function for the whole system. The optimal solutions are derived by Monte Carlo simulation combining with the informative search method. Cepin [100] determines the optimal scheduling to improve the safety of equipment outages in nuclear power plants by minimizing the mean value of the selected time-dependent risk measure. The large uncertainty in the safety assessment is considered.

9 Multi-Objective Maintenance

Martorell et al. [101] propose a new integrated Multi-Criteria Decision-making (IMCDM) method to determine the parameters in the technical specifications and maintenance (TSM) of Safety-related Equipment using multi-objective GA based on the reliability, availability, and maintenance (RAM) criterion. An example of emergency diesel generator system illustrates the application and viability of the proposed method. Martorell et al. [102] addressed the multi-objective problem of surveillance requirements at Nuclear Power Plants with dependable variables of Testing Intervals (TI) and Testing Planning using a novel double-loop multiple objective evolutionary algorithm.

In Quan et al. [103] a new approach, which combines the preference with evolutionary algorithm by using utility theory to search the Pareto frontier rather than conducting a dominated Pareto search, was developed to find the optimal solutions for a multi-objective preventive maintenance scheduling. Sanchez et al. [104] put forward a genetic algorithm based approach using distribution free tolerance intervals to address a multi-objective optimization of unavailability and cost model embedded within the uncertainty of the imperfect maintenance. Okasha and Frangopol [105] considered two strategies of selecting maintenance actions, maintenance scheduling, and maintenance structural components for optimization programs to design and construct structural systems in terms of system reliability, redundancy, and life cycle cost as criterion by multi-objective GA. Two numerical examples are used to illustrate these two strategies. Marseguerra and Zio [106] introduce a multi-objective optimization approach to determine the optimal surveillance Test Interval (STI) based on genetic algorithm search toward solutions of optimal performance with high assurance. Wang and Pham [107] studied a multi-objective maintenance optimization embedded within the imperfect PM and replacement for one single-unit system subject to the dependent competing risk of degradation wear and random shocks by simultaneously maximizing the system asymptotic availability and minimizing the system cost rate using the fast elitist nondominated Sorting Genetic Algorithm (NSGA-II).

Nosoohi and Hejazi [108] propose a novel multi-objective maintenance optimization model simultaneous considering the four different objectives of cost, corrective failure number per cycle, residual lifetime, and investment cost in order to determine preventive replacement times and the numbers of spare parts using ε-constraint method. Bocchini and Frangopol [109] utilize the Genetic Algorithm (GA) to achieve the multi-objective optimization of minimizing the total maintenance cost and maximizing the maintenance performance indicators to determine the schedule of the PM implement. Papakostas et al. [110] describe a multiple criteria analysis, consisting of cost, operational risk, remaining lifetime, and flight delays, for a set of aircraft maintenance planning alternatives with economic and operational constraints.

10 Maintenance Policy Under Warranty

Chen and Popova [111] consider a new maintenance policy combining the minimal repair and replacement under the two-dimension warranty of maximum warranty usage limits and maximum warranty period time by using optimization approach based on Monte Carlo simulation. Pan and Thomas [112] extend the research by Zuo et al. [113] by considering a larger state space with time-dependent parameters for the multistage deteriorating products with a free repair warranty (FRW) policy by using the continuous time Markov Chain. Wu et al. [114] study a periodic PM policy considering two parts: (1) maintenance cost model include the preventive maintenance and minimal repair, in which the first PM is triggered at the time chosen by the buyer till the end of the cycle; (2) the value of the maintenance is reflected as the system aging losses by PM actions.

Vahdani et al. [115] develop the optimal replacement-repair policy under the renewal free replacement warranty (RFRW) for the multi-state deteriorating systems to minimize the warranty service cost. Yeh and Lo [116] derive the optimal preventive maintenance schedule under warranty policy, including the PM numbers and maintenance degree, for the repairable products to minimize the total warranty cost.

Sahin and Polatoglu [117] minimize the warranty costs under both the renewing and nonrenewing polices for the unit with increasing failure rate (IFR) when two types of replacements policies are considered. Jung and Park [118] propose the periodic PM policies to minimize the expected long-run maintenance cost rate following the post-warranty period when two warranty policies are considered: renewing warranty and nonrenewing warranty. Jung et al. [119] compare the expected maintenance cost based on the various expected life cycles under the product user's view from two post-warranty policies: (1) replacement model proposed by Sahin and Polatoglu [117], and (2) optimal PM model proposed by Jung and Park [118].

References

1. Delia MC, Rafael PO (2006) A deteriorating two-system with two repair modes and sojourn times phase-type distributed. Reliab Eng Syst Saf 91:1–9
2. Delia MC, Rafael PO (2006) Replacement times and costs in a degrading system with several types of failure: the case of phase-type holding times. Eur J Oper Res 175:1193–1209
3. Delia MC, Rafael PO (2008) A maintenance model with failures and inspection following Markovian arrival processes and two repair modes. Eur J Oper Res 186:694–707
4. Tang Y, Lam Y (2006) A δ-shock maintenance model for a deteriorating system. Eur J Oper Res 168:541–556
5. Finkelstein M (2009) On damage accumulation and biological aging. J Stat Plann Infer 139(5):1643–1648
6. Vaurio JK (2001) Unavailability analysis of periodically tested standby components. IEEE Trans Reliab 44(3):512–522
7. Ram M, Singh SB (2008) Availability and cost analysis of a parallel redundant complex system with two types of failure under preemptive-resume repair discipline using Gumbel-Hougaard family copula in repair. Int J Reliab Qual Saf Eng 4:341–365
8. Qian C, Nakamura S, Nakagawa T (2003) Replacement and minimal repair polices for a cumulative damage model with maintenance. Comput Math Appl 46:1111–1118
9. Jiang R, Ji P (2002) Age replacement policy: a multi-attribute value model. Reliab Eng Syst Saf 76:311–318
10. Pandey MD, Yuan XX, van Noortwijk JM (2005) Gamma process model for reliability analysis and replacement of aging structural components, safety and reliability of engineering systems and structures. In: Proceedings of the 9th international conference on structural safety and reliability (ICOSSAR). Rome, Italy, pp 2439–2444
11. Sheu SH, Griffith WS, Nakagawa T (1995) Extended optimal replacement model with random minimal repair costs. Eur J Oper Res 85:636–649
12. Ozekici S (1995) Optimal maintenance policies in random environments. Eur J Oper Res 82:283–294

13. Dayanik S, Gurler U (2002) An adaptive Bayesian replacement policy with minimal repair. Oper Res 50(3):552–558
14. Dekker R, Plasmeijer RP (2001) Multi-parameter maintenance optimization via the marginal cost approach. J Oper Res Soc 52:188–197
15. Nguyen DG, Murthy DNP (1984) A combined block and repair limit replacement policy. J Oper Res Soc 35(7):653–658
16. Murthy DNP (1982) A note on block replacement policy. J Oper Res Soc 33(5):481–483
17. Marathe VP, Nair KPK (1966) Multistage planned replacement strategies. Oper Res 14(5):874–887
18. Nakagawa T, Mizutani S (2009) A summary of maintenance policies for a finite interval. Reliab Eng Syst Saf 94:89–96
19. Sheu SH (1998) A generalized age and block replacement of a system subject to shocks. Eur J Oper Res 108:345–362
20. Satow T, Teramoto K, Nakagawa T (2000) Optimal replacement policy for a cumulative damage model with time deterioration. Math Comput Model 31:313–319
21. Sheu SH, Griffith WS (2002) Extended block replacement policy with shock models and used items. Eur J Oper Res 140(1):50–60
22. Anisimov VV (2005) Asymptotic analysis of stochastic block replacement policies for multi-component systems in a Markov environment. Oper Res Lett 33(1):26–34
23. Scarf PA, Cavalcante CAV (2010) Hybrid block replacement and inspection policies for a multi-component system with heterogeneous component lives. Eur J Oper Res 206(2):384–394
24. Nakagawa T, Kijima M (1989) Replacement policies for a cumulative damage model with minimal repair at failure. IEEE Trans Reliab 28:581–584
25. Chaudhuri D, Suresh PV (1995) An algorithm for maintenance and replacement policy using fuzzy set theory. Reliab Eng Syst Saf 50:79–86
26. Sim SH, Endrenyi J (1993) A failure-repair model with minimal and major maintenance. IEEE Trans Reliab 42(1):134–140
27. Chiang JH, Yuan J (2001) Optimal maintenance policy for a Markovian system under periodic inspection. Reliab Eng Syst Saf 71:165–172
28. Sheu SH, Chang CC (2009) An extended periodic imperfect preventive maintenance model with age-dependent failure type. IEEE Trans Reliab 58(2):397–405
29. Huang Chun-Chen, Yuan John (2010) A two-stage preventive maintenance policy for a multi-state deterioration system. Reliab Eng Syst Saf 95:1255–1260
30. Cassady CR, Iyoob IM, Schneider K, Pohl EA (2005) A geometric model of equipment availability under imperfect maintenance. IEEE Trans Reliab 54(4):564–571
31. Satow T, Kawai H (2010) An inspection threshold of bivariate failure. In: Proceedings of the 16th ISSAT international conference on reliability and quality in design, Washington, pp 230–234
32. Flammini F, Marrone S, Mazzocca N, Vittorini V (2009) A new modeling approach to the safety evaluation of N-modular redundant computer systems in presence of imperfect maintenance. Reliab Eng Syst Saf 94:1422–1432
33. You MY, Li HG, Meng G (2011) Control-limit preventive maintenance policies for components subject to imperfect preventive maintenance and variable operational conditions. Reliab Eng Syst Saf 96:590–598
34. Doyen L, Gaufoin O (2004) Classes of imperfect repair models based on reduction of failure intensity or virtual age. Reliab Eng Syst Saf 84:45–56
35. Park DH, Jung GM, Yun JK (2000) Cost minimization for periodic maintenance policy of a system subject to slow degradation. Reliab Eng Syst Saf 68:105–112
36. Kallen MJ (2011), Modeling imperfect maintenance and the reliability of complex systems using superposed renewal process. Reliability Engineering and System Safety (in press)
37. Kahle W (2007) Optimal maintenance polices in incomplete repair modes. Reliab Eng Syst Saf 92:563–565

38. Kijima M, Morimura H, Suzuki Y (1988) Periodic replacement problem without assuming minimal repair. Eur J Oper Res 37:194–203
39. Kijima M (1998) Some results for repairable systems with general repair. J Appl Probab 26:89–102
40. Tsai TR, Liu PH, Lio YL (2011) Optimal maintenance time for imperfect maintenance actions on repairable product. Comput Ind Eng 60:744–749
41. Soro IW, Nourelfath M, Ait-Kadi D (2010) Performance evaluation of multi-state degraded systems with minimal repairs and imperfect preventive maintenance. Reliab Eng Syst Saf 95:65–69
42. Wang L, Hu HJ, Wang YQ, Wu W, He PF (2011) The availability model and parameters estimation method for the delay time model with imperfect maintenance at inspection. Appl Math Model 35:2855–2863
43. Makis V, Jardine AKS (1992) Optimal replacement policy for a general model with imperfect repair. J Oper Res Soc 43(2):111–120
44. Huynh KT, Barro A, Berenguer C, Castro IT (2011) A periodic inspection and replacement policy for systems subject to competing failure modes due to degradation and traumatic events. Reliab Eng Syst Saf 96:497–508
45. Wang L, Chu J, Mao WJ (2009) A condition-based replacement and spare provisioning policy for deteriorating systems with uncertain deterioration to failure. Eur J Oper Res 194:184–205
46. Camci F (2009) System maintenance scheduling with prognostics information using genetic algorithm. IEEE Trans Reliab 58(3):539–552
47. Grall A, Berenguer C, Dieulle L (2002) A condition-based maintenance policy for stochastically deteriorating systems. Reliab Eng Syst Saf 76:167–180
48. Van Noortwijk JM, Frangopol DM (2004) Two probabilistic life-cycle maintenance models for deteriorating civil infrastructures. Probab Eng Mech 19:345–359
49. Deloux E, Castanier B, Berenguer C (2009) Predictive maintenance policy for a gradually deteriorating system subject to stress. Reliab Eng Syst Saf 94(2):418–431
50. Van der Weide JAM, Pandey MD, van Noortwijk JM (2010) Discounted cost model for condition-based maintenance optimization. Reliab Eng Syst Saf 95:236–246
51. van der Weide JAM, Pandey MD (2011) Stochastic analysis of shock process and modeling of condition-based maintenance. Reliability Engineering and System Safety (in press)
52. Curcuru G, Galante G, Lombardo A (2010) A predictive maintenance policy with imperfect monitoring. Reliab Eng Syst Saf 95:989–997
53. Wu SM, Clements-Croome D (2005) Optimal maintenance policies under different operational schedules. IEEE Trans Reliab 54(2):338–346
54. Amari SV (2006) Bounds on MTBF of systems subjected to periodic maintenance. IEEE Trans Reliab 55(3):469–474
55. Yeh RH, Lo HC, Yu RY (2011) A study of maintenance policies for second-hand products. Comput Ind Eng 60:438–444
56. Casto IT, Barros A, Grall A (2011) Age-based preventive maintenance for passive components submitted to stress corrosion cracking. Mathematical and Computer Modeling (in press)
57. Zhao XJ, Fouladirad M, Berenguer C, Bordes L (2010) Condition-based inspection/replacement policies for non-monotone deteriorating systems with environmental covariates. Reliab Eng Syst Saf 95:921–934
58. Sun Y, Ma L, Mathew J (2009) Failure analysis of engineering systems with preventive maintenance and failure interactions. Comput Ind Eng 57:539–549
59. Panagiotidou S, Tagaras G (2007) Optimal preventive maintenance for equipment with two quality states and general failure time distributions. Eur J Oper Res 180:329–353
60. Roux O, Duvivier D, Quesnel G, Ramat E (2011) Optimization of preventive maintenance through a combined maintenance-production simulation model. International Journal of Production Economics (in press)
61. Nguyen DG, Murthy DNP (1981) Optimal preventive maintenance policies for repairable systems. Oper Res 29(6):1181–1194

62. Apeland S, Scarf PA (2003) A fully subjective approach to modeling inspection maintenance. Eur J Oper Res 148:410–425
63. Hariga MA (1996) A maintenance inspection model for a single machine with general failure distribution. Microelectron Reliab 36(3):353–358
64. Wortman MA, Klutke GA, Ayhan H (1994) A maintenance strategy for systems subjected to deterioration governed by random shocks. IEEE Trans Reliab 43(3):439–445
65. Chelbi A, Ait-Kadi D (2000) Generalized inspection strategy for randomly failing systems subjected to random shocks. Int J Prod Econ 64:379–384
66. Kiessler PC, Klutke GA, Yang Y (2002) Availability of periodically inspected systems subject to Markovian degradation. J Appl Probab 39:700–711
67. Yang Y, Klutke GA (2000) Lifetime-characteristics and inspection-schemes for Levy degradation process. IEEE Trans Reliab 49(4):377–382
68. Hosseini MM, Kerr RM, Randall RB (2000) An inspection model with minimal and major maintenance for a system with deterioration and Poisson failures. IEEE Trans Reliab 49(1):88–987
69. Kharoufeh JP (2003) Explicit results for wear processes in a Markovian environment. Oper Res Lett 31:237–244
70. Klutke GA, Yang Y (2002) The availability of Inspected systems subject to shocks and graceful degradation. IEEE Trans Reliab 51(3):371–374
71. Wang WB (2011) An inspection model based on a three-stage failure process. Reliability Engineering and System Safety (in press)
72. Chan LY, Wu SM (2009) Optimal design for inspection and maintenance policy based on the CCC chart. Comput Ind Eng 57:667–676
73. Kurt M, Kharoufeh JP (2010) Optimally maintaining a Markovian deteriorating system with limited imperfect repairs. Eur J Oper Res 205:368–380
74. Gurler U, Kaya A (2002) A maintenance policy for a system with multi-state components: an approximate solution. Reliab Eng Syst Saf 76:117–127
75. Flores-Colen I, Brito JD (2010) A systematic approach for maintenance budgeting of buildings facades based on predictive and preventive strategies. Constr Building Mater 24:1718–1729
76. Huang CC, Yuan J (2010) A two-stage preventive maintenance policy for a multi-state deterioration system. Reliab Eng Syst Saf 95:1255–1260
77. Saassouh B, Dieulle L, Grall A (2007) Online maintenance policy for a depredating system with random change of mode. Reliab Eng Syst Saf 92:1677–1685
78. Ponchet A, Fouladirad M, Grall A (2010) Assessment of a maintenance model for a multi-deteriorating mode system. Reliab Eng Syst Saf 95:1244–1254
79. Chen JW, Li KH, Lam Y (2010) Bayesian computation for geometric process in maintenance problems. Math Comput Simul 81:771–781
80. Zhao XJ, Fouladirad M, Berenguer C, Bordes L (2010) Condition-based inspection/replacement policies for non-monotone deteriorating systems with environmental covariates. Reliab Eng Syst Saf 95:921–934
81. Wang HZ (2002) A survey of maintenance policies of deteriorating systems. Eur J Oper Res 139:469–489
82. Frostig E, Kenzin M (2009) Availability of inspected systems subject to shocks-A matrix algorithmic approach. Eur J Oper Res 193:168–183
83. Lam Y, Zhang YL (2004) A shock model for the maintenance problem of repairable system. Comput Oper Res 31:1807–1820
84. Chen JY, Li ZH (2008) An extended extreme shock maintenance model for a deteriorating system. Reliab Eng Syst Saf 93:1123–1129
85. Zequeria RI, Berenguer C (2006) Periodic imperfect preventive maintenance with two categories of competing failure modes. Reliab Eng Syst Saf 91:460–468
86. Zhu Y, Elsayed EA, Liao H, Chan LY (2010) Availability optimization of systems subject to competing risk. Eur J Oper Res 202(3):781–788
87. Li WJ, Pham H (2005) An inspection-maintenance model for systems with multiple competing processes. IEEE Trans Reliab 54(2):318–327

88. Wu J, Adam Ng TS, Xie M, Huang HZ (2010) Analysis of maintenance policies for finite life-cycle multi-state system. Comput Ind Eng 59:638–646
89. Barros A, Berenguer C, Grall A (2006) A maintenance policy for two-unit parallel systems based on imperfect monitoring information. Reliab Eng Syst Saf 91:131–136
90. Wang WB, Banjevic D, Pecht M (2010) A multi-component and multi-failure mode inspection model based on the delay time concept. Reliab Eng Syst Saf 95:912–920
91. Zhang YL, Wang GJ (2011) An optimal repair-replacement policy for a cold standby system with use priority. Appl Math Model 35:1222–1230
92. Marseguerra M, Zio E, Podofillini L (2002) Condition-based maintenance optimization by means of genetic algorithms and Monte Carlo simulation. Reliab Eng Syst Saf 77:151–166
93. Tian ZG, Liao HT (2011) Condition based maintenance optimization for multi-component systems using proportional hazards model. Reliab Eng Syst Saf 96:581–589
94. Taghipour S, Banjevic D (2011) Period inspection optimization models for a repairable system subject to hidden failures. IEEE Trans Reliab 60(1):275–285
95. Tian ZG, Jin TD, Wu BR, Ding FF (2011) Condition based maintenance optimization for wind power generation systems under continuous monitoring. Renew Energy 36:1502–1509
96. Bouvard K, Artus S, Berenguer C, Cocquempot V (2011) Condition-based dynamic maintenance operations planning and grouping. Application to commercial heavy vehicles, Reliability Engineering and System Safety (in press)
97. Wang CH, Lin TW (2011) Improved particle swarm optimization to minimize periodic preventive maintenance cost for series-parallel systems. Expert Syst Appl 38:8963–8969
98. Mahmoud MAW, Moshref ME (2010) On a two-unit cold standby system considering hardware, human error failures and preventive maintenance. Math Comput Model 51:736–745
99. Laggoune R, Chateauneuf A, Aissani D (2009) Opportunistic policy for optimal preventive maintenance of a multi-component system in continuous operating units. Comput Chem Eng 33:1499–1510
100. Cepin M (2002) Optimization of safety equipment outages improves safety. Reliab Eng Syst Saf 77:71–80
101. Martorell S, Villanueva JF, Carlos S, Nebot Y, Sanchez A, Pitarch JL, Serradell V (2005) RAMS+C informed decision-making with application to multi-objective optimization of technical specifications and maintenance using genetic algorithms. Reliab Eng Syst Saf 87:65–75
102. Martorell S, Carlos S, Villanueva JF, Sanchez AI, Gavlan B, Salazar D, Cepin M (2006) Use of multiple objective evolutionary algorithms in optimizing surveillance requirements. Reliab Eng Syst Eng 91:1027–1038
103. Quan G, Greenwood GW, Liu DL, Hu SR (2007) Searching for multi-objective preventive maintenance schedules: combining preferences with evolutionary algorithms. Eur J Oper Res 177:1969–1984
104. Sanchez A, Carlos S, Martorell S, Villanueva JF (2009) Addressing imperfect maintenance modeling uncertainty in unavailability and cost based optimization. Reliab Eng Syst Saf 94:22–32
105. Okasha NM, Frangopol DM (2009) Lifetime-oriented multi-objective optimization of structural maintenance considering system reliability, redundancy and life-cycle cost using GA. Struct Saf 31:460–474
106. Marseguerra M, Zio E, Podofillini L (2005) Optimal reliability/availability of uncertain systems via multi-objective optimization of technical specifications and maintenance using genetic algorithms. Reliab Eng Syst Saf 87(1):65–75
107. Wang YP, Pham H (2011) Multi-objective optimization of imperfect preventive maintenance policy for dependent competing risk system with hidden failure. IEEE Trans Reliab 60:3
108. Nosoohi I, Hejazi SR (2011) A multi-objective approach to simultaneous determination of spare part numbers and preventive replacement times. Appl Math Model 35:1157–1166
109. Bocchini P, Frangopol DM (2011) A probabilistic computational framework for bridge network optimal maintenance scheduling. Reliab Eng Syst Saf 96:332–349
110. Papakostas N, Papachatzakis P, Xanthakis V, Mourtzis D, Chryssolouris G (2010) An approach to operational aircraft maintenance planning. Decis Support Syst 48:604–612

111. Chen Tom, Popova Elmira (2002) Maintenance policies with two-dimensional warranty. Reliab Eng Syst Saf 77:61–69
112. Pan Y, Thomas MU (2010) Repair and replacement decisions for warranted products under Markov deterioration. IEEE Trans Reliab 59(2):368–373
113. Zuo MJ, Liu B, Murthy DN (2000) Replacement-repair policy for multi-state deterioration products under warranty. Eur J Oper Res 123:519–530
114. Wu J, Xie M, Adam Ng TS (2011) On a general periodic preventive maintenance policy incorporating warranty contracts and system aging losses. Int J Prod Econ 129:102–110
115. Vahdani H, Chukova S, Mahlooji H (2011) On optimal replacement-repair policy for multi-state deteriorating products under renewing free replacement warranty. Comput Math Appl 61:840–850
116. Yeh RH, Lo HC (2001) Optimal preventive-maintenance warranty policy for repairable products. Eur J Oper Res 134:59–69
117. Sahin I, Polatoglu H (1996) Maintenance strategies following the expiration of warranty. IEEE Trans Reliab 45(2):220–228
118. Jung GM, Park DH (2003) Optimal maintenance policies during the post-warranty period. Reliab Eng Syst Saf 82:173–185
119. Jung KM, Park MJ, Park DH (2010) System maintenance cost dependent on life cycle under renewing warranty policy. Reliab Eng Syst Saf 95:816–821

Reliability of Systems Subjected to Imperfect Fault Coverage

G. Levitin, S. H. Ng, R. Peng and M. Xie

Abstract Due to imperfect fault coverage, the reliability of redundant systems cannot be enhanced unlimitedly with the increase of redundancy. Many works have been done on the reliability modeling and optimization of systems subjected to imperfect fault coverage. The methodologies adopted mainly include combinatorial approach, ordered binary decision diagram and universal generating function. Depending on the type of fault tolerant techniques used, there are mainly three kinds of fault coverage models: (1) element level coverage (ELC). (2) fault level coverage (FLC). and (3) performance-dependent coverage (PDC). This chapter reviews the literatures on the reliability of systems subjected to imperfect fault coverage and shows an extended work.

R. Peng (✉)
Dongling School of Economics and Management, University of Science and Technology Beijing, Beijing, China
e-mail: pengrui1988@gmail.com

S. H. Ng · M. Xie
Department of Industrial and Systems Engineering, National University of Singapore, Singapore, Singapore
e-mail: isensh@nus.edu.sg

G. Levitin
The Israel Electric Corporation Ltd, Haifa, Israel
e-mail: levitin@iec.co.il

M. Xie
Department of Systems Engineering and Engineering Management, City University of Hong Kong, Kowloon, Hong Kong
e-mail: minxie@cityu.edu.hk

T. Dohi and T. Nakagawa (eds.), *Stochastic Reliability and Maintenance Modeling*, Springer Series in Reliability Engineering 9, DOI: 10.1007/978-1-4471-4971-2_8,

1 Introduction

Redundancy is widely used to enhance system reliability, especially for systems with stringent reliability requirements, such as nuclear power controllers and flight control systems [24, 41, 46]. Usually, the fault tolerance is implemented by providing sufficient redundancy and using automatic fault and error handling mechanisms (detection, location, and isolation of faults/failures). However, as the fault and error handling mechanisms themselves can fail, some failures can remain undetected or uncovered, which can lead to the total failure of the entire system or its subsystems [4, 11, 51]. Examples of this effect of uncovered faults can be found in computing systems, electrical power distribution networks, pipelines carrying dangerous materials etc [8, 14]. The probability of successfully covering a fault (avoiding fault propagation) given that the fault has occurred is known as the coverage factor [11]. The models that consider the effects of imperfect fault coverage are known as imperfect fault coverage models or simply fault coverage models or coverage models [5].

Many works have been done on the reliability modeling and optimization of systems subjected to imperfect fault coverage. The methodologies adopted mainly include combinatorial approach, binary decision diagram, and universal generating function (UGF). Combinatorial modeling techniques, such as graph theoretic technique, digraphs, reliability logic diagrams, and particularly fault trees, have long been used for reliability analysis because of their concise representation of system failure combinations [1, 2, 45]. Before the paper [17], combinatorial models were thought to be inadequate to capture the dynamic system behavior associated with fault and error recovery. For this reason, Markov chains were used for reliability assessment of fault-tolerant systems. Markov chains are extremely flexible and can capture the fault coverage mechanisms quite well [11, 19]. However, Markov chains also have some disadvantages. In addition to the computational complexity, it is also difficult to determine the correct Markov model for a given system, because the operational configuration of the system must be specified explicitly and the rate at which the system state changes must be determined. The relative advantages of fault tree and Markov models have been exploited by using behavioral decomposition [47], and converting the fault tree to a Markov chain automatically [9, 48]. The Simple and Efficient Algorithm presented in [6] enables reliability engineers to use their favorite software package for computing system reliability which includes the consideration of fault coverage. Though the inclusion-exclusion method and the sum of disjoint products method have been used by many researchers for evaluating the system availability (reliability), ordered binary decision diagram (OBDD) is shown to be very efficient in terms of computational time and accuracy [22, 23, 53]. For this reason, OBDD has been used to model imperfect fault coverage in some recent papers [37, 50, 52]. The UGF introduced in [49] has also been used in some recent papers to model imperfect fault coverage and it is shown to be very flexible [27, 31].

Depending on the type of fault tolerant techniques used, there are mainly three kinds of fault coverage models: (1) Element level coverage (ELC). A particular coverage factor value is associated with each element. This value is independent of the status of other elements. (2) Fault level coverage (FLC). The coverage factor value depends on the number of good elements that belong to a specific group (i.e., the status of other elements). (3) Performance-dependent coverage (PDC). The coverage factor value depends on the cumulative performance of the available group elements at the moment when the failure occurs. The ELC model is appropriate when the selection among the redundant elements is made on the basis of a self-diagnostic capability of the individual elements. Such systems typically contain a built-in test capability. The FLC model is appropriate for modeling systems in which the selection among redundant elements varies between initial and subsequent failures. In the HARP (Hybrid Automated Reliability Predictor) terminology [10], ELC models are known as single-fault models, whereas FLC models are known as multi-fault models. Multi-fault models have the ability to model a wide range of fault-tolerant mechanisms. An example is a majority voting system among the currently known working elements, see Myers and Rauzy [38]. The performance-dependent coverage considered in Levitin and Amari [29] takes place when the fault detection and recovery functions are performed by system elements in parallel with their main functions. The proposed model is suitable for systems that cannot change the states during task execution, such as alarm systems and data processing systems performing short tasks. The systems usually remain in idle mode, thus fault detection and coverage can be performed only during task execution. When the task arrives, the system can be in one of various states, depending on availability of its elements. Therefore, the coverage probability depends only on the performance available at the moment of task arrival and does not depend on the history of failures.

This chapter reviews the literatures on reliability modeling and optimization of systems subjected to imperfect fault coverage and presents an extended work. Section 2 focuses on the works which employ the combinatorial approach to study imperfect fault coverage. Section 3 focuses on the study of imperfect fault coverage with ordered binary decision diagram. Section 4 focuses on the study on imperfect fault coverage with UGF. Section 5 shows the extended work.

2 Combinatorial Approach

Dugan [18] presented a Dugan fault tree solution (DFTS) algorithm which computes the exact unreliability of systems with incorporation of imperfect coverage using only a fault tree model of the system. The DFTS algorithm determines the unreliability of the system during the enumeration of the operational states that correspond to the fault tree. As the state space is generated, the fault handling behavior is automatically incorporated, and the leakage from each state to the failure state is calculated. The general coverage model used in Dugan [18] is incorporated into the system state transition as shown in Fig. 1

The general coverage model describes the behavior of the system in response to a fault. From a particular operational state $m1$, suppose that the combined failure rate of transient, intermittent, and permanent faults is λ. If the failure of a component

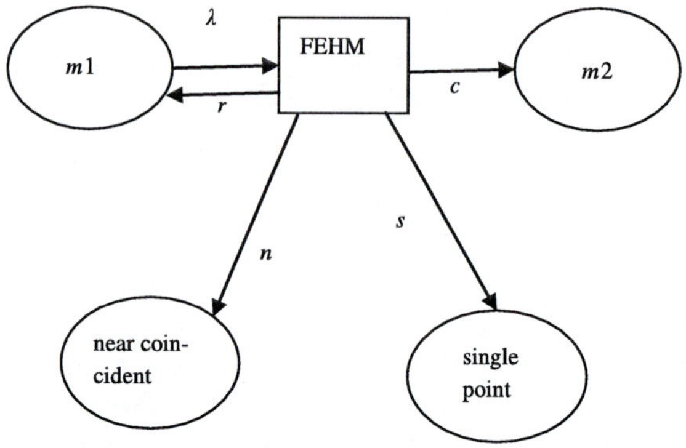

Fig. 1 Incorporating coverage model into state transition

(at constant rate λ) leads to an operational state $m2$, a fault/error handling model (FEHM) is inserted on the corresponding arc. The transient restoration exit of the coverage model leads back to the original state $m1$. The permanent coverage exit of the coverage model leads to the target state $m2$ in which the system functions with at least one fewer component. The two other exits lead to the states single point failure and near-coincident faults. A major advantage of DFTS algorithm is that the generation/solution of the model can be halted as soon as the desired accuracy is achieved. However, the equations needed to solve the system are unwieldy, which hampers their understanding and implementation.

Pham and Pham [44] discussed the problem of determining the optimum number of spares in the dynamic redundant systems. The optimum value of spares that maximizes the reliability of the system with imperfect coverage has been obtained. Pham [42] addressed the design issue for the optimal number of spare units in a triple-modular-redundancy system with spare units, including fault coverage and common-cause failure. Pham [43] examined a model of a high voltage system consisting of a power supply and two transmitters with imperfect coverage in which the failure rate of the fault coverage is constant. This work was extended in Moustafa [33] to a *k*-out-of-*n* system with imperfect coverage. Akhtar [3] studied the reliability of *k*-out-of-*n* systems with imperfect fault coverage. It is assumed that failures not covered can lead to an absorbing failure state with no transition possible to a function state. Recursive expressions for mean time between failures and mean time to failures are obtained for repairable systems. Newton [39] commented on the method proposed in Akhtar [3]. An alternative probability argument is used to obtain the mean time between failures and mean time to failures for *k*-out-of-*n*: G systems. This has the advantage that higher moments of such failure times can also be determined.

Since the DFTS is rather complicated for implementation, a simple alternative solution that uses cut-set solution methods was proposed in Doyel et al. [17]. The Doyel, Dugan, Patterson (DDP) algorithm presented in Doyel et al. [17] combines aspects of behavioral decomposition, sum-of-disjoint products, and multi-state solution methods. Different from Dugan, only three exits are considered: the transient restoration exit, the permanent coverage exit, and the single-point failure exit. The DDP algorithm first obtains all the minimal cut-sets and then produces a sum of disjoint products from the set of cut-sets to determine the system unreliability. For example, one considers a redundant system consisting of 3 units A, B, and C that is operational as long as 1 unit is up, provided that no uncovered failures occur. The minimal cut-sets are $C_1 = \{\underline{A}\}$, $C_2 = \{\underline{B}\}$, $C_3 = \{\underline{C}\}$, and $C_4 = \{\overline{A}, \overline{B}, \overline{C}\}$, where $\underline{X}(\overline{X})$ labels the basic event that component X fails and the failure is uncovered (covered). The system unreliability can be formulated as

$$\begin{aligned} \text{Unreliability} = \Pr\{\bigcup_{i=1}^{4} C_i\} &= \Pr\{C_1\} + \Pr\{\neg C_1 C_2\} \\ &+ \Pr\{\neg C_1 \neg C_2 C_3\} + \Pr\{\neg C_1 \neg C_2 \neg C_3 C_4\} \end{aligned} \tag{1}$$

where $\neg C_i$ denotes the logical negation of event C_i.

The DDP algorithm was further developed and generalized in Amari et al. [6], which presented a SEA (Simple and Efficient Algorithm) to find the unreliability of systems subjected to imperfect fault coverage. The SEA separates the modeling of fault coverage failures into two terms that are multiplied to compute the system reliability. The first term, a simple product, represents the probability that no uncovered fault occurs. The second term comes from a combinatorial model which includes the covered faults that can lead to system failure. The major contribution of SEA is that reliability engineers can use their favorite software package for computing reliability, and can adjust the input and output of that program slightly to produce a result which includes the consideration of fault coverage. Further, SEA is conceptually simpler and more efficient than earlier approaches. Amari et al. [7] studied the optimal reliability of systems subjected to imperfect fault coverage. It is shown that the reliability of systems subjected to imperfect fault coverage decreases with an increase in redundancy after a particular limit. Using the SEA, Amari et al. [7] also computed the reliability expressions of some common systems subjected to imperfect fault coverage. The systems considered include parallel, parallel-series, series-parallel, k-out-of-n, and k-out-of-$(2k-1)$ systems. Amari et al. [8] studied the optimal design of k-out-of-n: G subsystems subjected to imperfect fault coverage. It is assumed that there exists a k-out-of-n: G subsystem in a nonseries-parallel system and, except for this subsystem, the redundancy configurations of all other subsystems are fixed. The overall system reliability is evaluated using the SEA algorithm. Procedures are proposed to solve the optimal cost-effective design policies for k-out-of-n: G subsystems and the optimal design policies which maximize the overall system reliability.

All the above works assume that each system element has a specific coverage, that is, ELC is assumed. The ELC is most appropriate when the selection among redundant elements is made on the basis of a self-diagnostic capability of the individual elements. However, if the redundancy management implementation results in coverage being a function of the fault sequence within the redundant set, the FLC is more appropriate. Myers [34] studied the reliability of k-out-of-n: G systems considering four different coverage models: perfect fault coverage, element level coverage, fault level coverage, and one-on-one level coverage. The one-on-one level coverage is actually a special case of FLC in which faults prior to the one-on-one fault are considered to have perfect coverage. Techniques are presented for both combinatorial and recursive function calculation of k-out-of-n: G system reliability considering imperfect fault coverage. Based on the algorithm presented in Myers [34, 35] studied the achievable limits on the reliability of k-out-of-n: G systems subjected to imperfect fault coverage, with consideration of both ELC and FLC. The system is assumed to consist of n identical and independent elements. The reliability of the system in the case of ELC and FLC is calculated with the combinatorial approach as

$$R\mathrm{iid}_{\mathrm{ELC}}(k, p, c) = \sum_{i=k}^{n} \binom{n}{k} p^i (q \cdot c)^{n-i} \tag{2}$$

$$R\mathrm{iid}_{\mathrm{FLC}}(k, p, \mathbf{c}) = \sum_{i=k}^{n} \binom{n}{k} p^i q^{(n-i)} \prod_{i=1}^{n-k} c_i \tag{3}$$

where p is the reliability of each system element, q equals to $1-p$, c is the coverage probability of each element in the case of ELC, c_i is the coverage probability of i-th failure in the case of FLC, and $\mathbf{c}$ denotes the vector $\{c_1,\ldots c_n\}$. It is shown, over a wide range of realistic coverage values and relative high component reliabilities, that the optimal redundancy level is 2 for ELC systems and 4 for FLC systems. Over this same range of system characteristics, optimal FLC systems outperform ELC systems, in terms of failure probability, by several orders of magnitude. Myers [36] studied the probability of survival for redundant systems utilizing a mission abort policy. Systems having both perfect and imperfect fault coverage are addressed.

3 Ordered Binary Decision Diagram Approach

The binary decision diagram was initially developed as a tool for validating VLSI circuitry design by Bryant [12]. A binary decision diagram is a rooted, directed, and acyclic graph used to represent a Boolean function. It consists of decision nodes and two terminal nodes called 0-terminal and 1-terminal. Each decision node is labeled by a Boolean variable and has two child nodes called low child and high child. As the path descends to a low child (high child) from a node, then that node's variable is assigned to 0 (1). A path from the root node to the 1-terminal represents a (possibly partial) variable assignment for which the represented Boolean function is true. Such a binary decision diagram is called 'ordered' if different variables appear in the same order on all paths from the root.

Xing and Dugan [50] analyzed the reliability of a generalized phased-mission system with consideration of combinatorial phase requirements and imperfect fault coverage. The SEA is used to incorporate the effect of imperfect fault coverage into a binary decision diagram based algorithm to compute the system reliability. The coverage factor of each component is assumed to be constant in each phase, that is, the ELC model is used. This work was extended in Xing [51], which studied the reliability of a general phased-mission system with consideration of both imperfect fault coverage and common-cause failures.

Chang et al. [13] presented an OBDD based algorithm for the calculation of the time-specific as well as the steady-state failure frequency of a repairable system. The algorithm is also extended to incorporate imperfect fault coverage into the system availability evaluation. Markov chains are used to model the state of the components and the conditional probabilities from SEA are used to incorporate imperfect fault coverage. The coverage factor of each system component is assumed to be constant or time dependent. Chang et al. [14] proposed a model for multi-state systems with imperfect fault coverage. An OBDD based approach for the evaluation of multi-state system reliability and the Griffith's importance measure has also been proposed.

Xing [52] proposed an efficient approach for fully incorporating both imperfect fault coverage and common-cause failures into network reliability and sensitivity analysis. The consideration of imperfect coverage and common-cause failures is separated from the combinatorics of the solution based on reduced OBDD. It is shown that the reduced OBDD-based algorithm requires less memory than other traditional methods (such as inclusion–exclusion and sum of disjoint products) and is more efficient in reliability evaluation.

Myers and Rauzy [37] derived a table-based algorithm to compute the unreliability of k-out-of-n: F systems with imperfect fault coverage, from a principle proposed by Dutuit and Rauzy [20]. The encoding procedures of ELC and FLC models by means of classical fault tree gates are shown, followed by the binary decision diagram representation based on the Shannon decomposition. A digital flight control system test case is shown to illustrate the importance of the effect of imperfect fault coverage and the efficiency of the proposed algorithm. Myers and Rauzy [38] proposed a more efficient algorithm than the algorithm in Myers and Rauzy [37]. It also explained in detail the difference between ELC and FLC. In order to show the binary decision diagram representation of systems subjected to imperfect fault coverage and the difference between ELC and FLC, we consider a 1-out-of-3: G system for perfect coverage, ELC, and FLC models. If the reliabilities of the three components are denoted as p_1, p_2, and p_3, the binary decision diagram for the perfect coverage model is as shown in Fig. 2 The 0-terminal node is omitted in order to make the binary decision diagram more compact.

The reliability of this system is the sum of all paths from the terminal node labeled 1 to the top node, p_1, with the components subsequent to a connection by a dashed line being complemented. The reliability of the 1-out-of-3: G system shown in Fig. 2 is then

$$R_{perfect} = p_1 + p_2(1 - p_1) + p_3(1 - p_1)(1 - p_2) \tag{4}$$

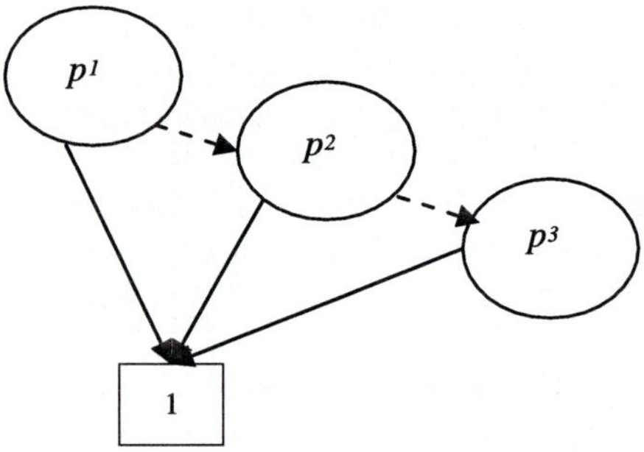

Fig. 2 Binary decision diagram for the perfect coverage model

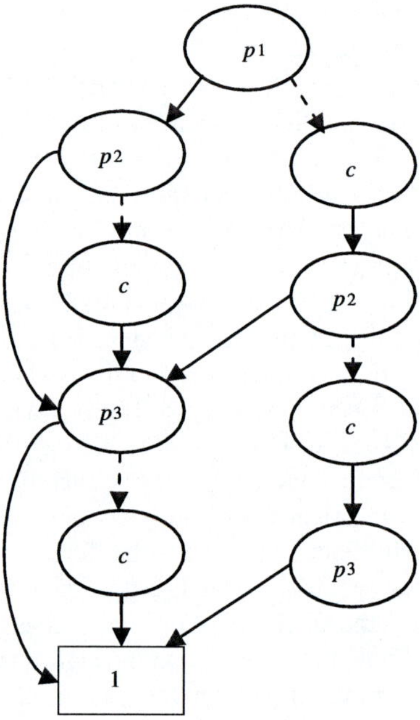

Fig. 3 Binary decision diagram for the ELC model

The binary decision diagram for the ELC model is as shown in Fig. 3, where c is the fault coverage factor. The reliability of the 1-out-of-3: G system shown in Fig. 3 is then

$$\begin{aligned} R_{ELC} = {} & p_1[p_2 + (1 - p_2)c][p_3 + (1 - p_3)c] \\ & + (1 - p_1)c\{p_2[p_3 + (1 - p_3)c] + p_3[p_2 + (1 - p_2)c]\} \end{aligned} \tag{5}$$

The binary decision diagram for the FLC model is as shown in Fig. 4, where $c_i (i = 1, 2)$ is the fault coverage factor for the i-th failure. The reliability of the 1-out-of-3: G system shown in Fig. 4 is then

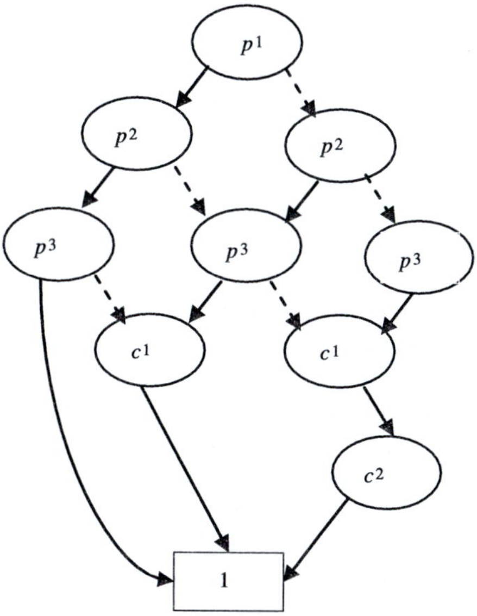

Fig. 4 Binary decision diagram for the FLC model

$$R_{FLC} = p_1[p_2(p_3 + (1 - p_3)c_1) + (1 - p_2)c_1(p_3 + (1 - p_3)c_2)] \\ + (1 - p_1)c_1[p_2p_3 + (1 - p_2)c_2p_3 + (1 - p_3)c_2p_2] \tag{6}$$

4 Universal Generating Function Approach

The UGF was introduced in Ushakov [49] and proved to be extremely effective in evaluating reliability of complex multi-state systems. Much research has been done on incorporating UGF into reliability analysis of various k-out-of-n systems, series-parallel systems, weighted voting systems, acyclic information networks, and manufacturing systems [16, 26, 32, 54, 55]. The UGF of a discrete random value X is defined as a polynomial

$$u_X(z) = \sum_{h=0}^{H} \varepsilon_h z^{x_h}, \tag{7}$$

where the variable X has H+1 possible values and $\varepsilon_h = \Pr\{X = x_h\}$.

To obtain the UGF representing the pmf of a function of two independent random variables $\varphi(X, Y)$, the following composition operator is used:

$$U_{\varphi(X,Y)}(z) = u_X(z) \underset{\varphi}{\otimes} u_Y(z)$$
$$= \left(\sum_{h=0}^{H} \varepsilon_h z^{x_h}\right) \underset{\varphi}{\otimes} \left(\sum_{d=0}^{D} \varepsilon_d z^{y_d}\right) = \sum_{h=0}^{H}\sum_{d=0}^{D} \varepsilon_h \varepsilon_d z^{\varphi(x_h, y_d)} \tag{8}$$

Levitin [27] presented an efficient UGF-based approach for reliability analysis of complex multi-state systems taking into account imperfect fault coverage. In Levitin and Xing [31], the approach was extended to the case when each element has specific subset of other elements affected by the uncovered propagated failure. In both works, the uncovered failure is incorporated by assuming that state 0 of each system component corresponds to uncovered failure. A subsystem fails as long as one single component in this subsystem is in state 0. This approach allows obtaining the performance distribution of complex multi-state systems using a generalized reliability block diagram method (recursive aggregating multi-state elements and replacing them by single equivalent ones). Since the probability for each component to be in state 0 is assumed to be constant and independent from the states of other elements, this approach is only suitable for modeling ELC.

A more general approach proposed in Levitin and Amari [30] is able to evaluate the reliability of multi-state systems with FLC. In the case of FLC, one needs to incorporate the coverage probabilities depending on the number of failed elements into the performance distribution of each group affected by FLC. Thus one has to know not only entire group performance but also the total number of failed elements in each state of this group (combination of states of its elements). To obtain both these indices, the performance distribution for system elements is described by a modified UGF as

$$\tilde{u}_j(z) = \sum_{h=0}^{k_j} p_{jh} z^{s_{jh}, g_{jh}}, \tag{9}$$

where s_{jh} represents the realization of the random number of failed elements in state h. The UGF of the entire system is done by recursively aggregating the UGF of system elements and replacing them by single equivalent ones. The computational complexity of the proposed algorithm for solving multi-state systems with FLC is the same as the computational complexity of an equivalent multi-state system without consideration of imperfect fault coverage.

Levitin and Amari [29] suggested the PDC model for the case when the effectiveness of recovery mechanisms in a subsystem depends on the entire performance level of this subsystem. The UGF of each PDC group is obtained by first obtaining the UGF representing the conditional group performance distribution given that all the faults are covered and then incorporating the performance-dependent fault coverage factors to get the UGF representing the unconditional group performance distribution. For example, if the conditional performance distribution of PDC group k in a system given that all the faults are covered is represented by

$$U_{w_k}(z) = \sum_{h=0}^{n_k} p_{kh} z^{g_{kh}}, \tag{10}$$

the UGF representing the unconditional performance distribution of PDC group k can be obtained as

$$\tilde{U}_{w_k}(z) = \sum_{h=0}^{n_k} p_{kh} c_k(g_{kh}) z^{g_{kh}} + [1 - \sum_{h=0}^{n_k} p_{kh} c_k(g_{kh})] z^{g_{k0}} \tag{11}$$

where $ck(\cdot)$ is the fault coverage probability function depending on the group performance.

Levitin [28] presented a model of series-parallel multi-state systems with two types of task parallelization: parallel task execution with work sharing, and redundant task execution. It is assumed that the elements in each subsystem can be distributed into different work sharing groups in order to achieve both performance and reliability requirement. A framework to solve the optimal balance of the two kinds of parallelization which maximizes the system reliability is proposed based on the assumption that the ELC applies in each work sharing group. It is shown that the greatest system reliability can be achieved by proper balance between two kinds of parallelization. Considering the different types of fault handling mechanisms in practice, the ELC model alone cannot adapt to all the cases. The optimal system structure problem can be extended to consider different kinds of fault coverage models. Section 5 shows an extension of the optimal structure problem to the case of FLC presented in [40].

5 Optimal System Structure in the Case of FLC

5.1 Model Description and Problem Formulation

The fault tolerant structure is very common in task processing and data-transmission systems. An example is a multichannel data transmission system in which data packages are divided into subpackages transmitted through different channels. If some channels fail, the automatic data exchange management is able to distribute the transmission task among the available channels. In this case, the system remains operating, though with a lower performance. However, when a failure of any channel is undetected (uncovered failure), the management system cannot make proper reconfiguration and still assigns some subpackages to the unavailable channel. In this case, some information is lost and the entire data transmission task fails.

Consider a system consisting of M subsystems connected in series. Each subsystem m contains E_m different elements connected in parallel. The performance rate G_j of element j at any time instant is a random variable that takes its values from $\mathbf{g}_j = \{g_{j0}, g_{j1}, \ldots, g_{jk_j}\}$. The probability associated with different states of any system element j can be represented by the set $\mathbf{p}_j = \{p_{j0}, p_{j1}, \ldots, p_{jk_j}\}$, where $p_{jh} = \Pr\{G_j = g_{jh}\}$. The state 0 corresponds to the total element failure, and other k_j states correspond to the working states with full or partial performance.

We assume that the states of multi-state system elements are mutually independent. The elements belonging to the same subsystem can be separated into independent work sharing groups. The available elements belonging to a work sharing group share their work in an optimal way that maximizes the performance of the entire group. In the case of detected failures of some elements, the task is able to be redistributed among the available elements. An undetected failure of any element belonging to a work sharing group cannot be covered within this group, and causes the failure of the entire group. Different work sharing groups belonging to the same subsystem perform the same task in parallel providing the task execution redundancy.

Assume that the entire system has $K+1$ different states, and that v_i is the entire system performance rate in state i. The multi-state system performance rate is a random variable V that takes values from the set $\{v_0, \ldots, v_K\}$. The system structure function $V = \phi(G_1, \ldots, G_n)$, which maps the spaces of the elements' performance rates into the space of the system's performance rates, is determined by the system structure. The elements' distribution among work sharing groups in each subsystem m can be represented by the vector $\boldsymbol{\alpha}_m = \{\alpha_{mj}, 1 \le j \le E_m\}$, where α_{mj} is the index of the subset to which element j belongs. Concatenation of vectors $\boldsymbol{\alpha} = \{\boldsymbol{\alpha}_1, \ldots, \boldsymbol{\alpha}_M\}$ determines the distribution of elements among the work sharing group for the entire system. For any given $\boldsymbol{\alpha}$, and given pmf of the system elements, one can obtain the pmf of the entire system performance V in the form

$$Q_i, v_i, 0 \le i \le K, \quad \text{where } Q_i = \Pr\{V = v_i\}. \tag{12}$$

The multi-state system reliability is defined as the probability that the multi-state system satisfies the demand Levitin [26]. For example, in applications where the system performance is defined as its productivity/capacity, and θ^* is the minimum allowed capacity, the multi-state system reliability takes the form

$$R(\theta^*) = \sum_{i=1}^{K} Q_i 1(v_i > \theta^*) \tag{13}$$

The multi-state system structure optimization problem is formulated as follows. Find vector $\boldsymbol{\alpha} = \{\boldsymbol{\alpha}_1, \ldots, \boldsymbol{\alpha}_M\}$, which maximizes multi-state system reliability $R(\theta^*)$ for a given demand θ^*,

$$\boldsymbol{\alpha} = \arg\max\{R(\boldsymbol{\alpha}, \theta^*)\}. \tag{14}$$

5.2 System Reliability Evaluation and Structure Optimization

The performance distribution for system elements is described by a modified UGF as

$$\tilde{u}_j(z) = \sum_{h=0}^{k_j} p_{jh} z^{s_{jh}, g_{jh}}, \tag{15}$$

where s_{jh} represents the realization of the random number of failed elements in state h. The UGF of an individual element takes the form

$$\tilde{u}_j(z) = p_{j0} z^{1, g_{j0}} + \sum_{h=1}^{k_j} p_{jh_j} z^{0, g_{jh}}, \tag{16}$$

where g_{j0} corresponds to the case of failure of the element (1 failure), $g_{jh}(1 \le j \le k_j)$ corresponds to the h-th working state of element j (0 failure). Applying the operator

$$\tilde{U}_{\{i,j\}}(z) = \tilde{u}_i(z) \underset{\omega}{\otimes} \tilde{u}_j(z) = \sum_{h=0}^{k_i} \sum_{d=0}^{k_j} p_{ih} p_{jd} z^{s_{ih}+s_{jd}, \omega(g_{ih}, g_{jd})} \tag{17}$$

recursively one can obtain the UGF of the entire work sharing group i in subsystem m in the form.

$$\tilde{U}_{mi}(z) = \sum_{h=0}^{n_{mi}} P_{mih} z^{s_{mih}, g_{mih}} \tag{18}$$

that represents the distribution of the number of failed elements and the corresponding performance of the work sharing group. Here ω is the performance composition function for elements connected in parallel with work sharing, P_{mih} is the probability that work sharing group i in subsystem m contains exactly s_{mih} failed elements and functions at the performance level g_{mih}, given all the failures are covered (g_{mi0} correspond to the failures of all the elements in the group).

A. *The UGF of a work sharing group in the case of FLC*

In the case of FLC, the coverage probability of a failure is determined by the total number of elements in the work sharing group and the number of failed elements in this group (which affects the load on the monitoring system). Let $c_m(|\Phi_{mi}|, j)$ be the fault coverage probability in the case of jth failure in work sharing group i in subsystem m (when $j-1$ elements are already unavailable), and $r_{mi}(k)$ be the probability that the group does not fail after k failures have consecutively occurred. It can be seen, that

$$r_{mi}(k) = \prod_{j=0}^{k} c_m(|\Phi_{mi}|, j) \tag{19}$$

By definition$r_{mi}(0) = c_m(|\Phi_{mi}|, 0) = 1$ and $c_m(|\Phi_{mi}|, |\Phi_{mi}|) = 0$.

The uncovered failures can be incorporated into the UGF by applying the following operator ε:

$$\begin{aligned} U_{mi}(z) = \varepsilon(\underset{mi}{\tilde{U}}(z)) &= \varepsilon\left(\sum_{h=0}^{n_{mi}} P_{mih} z^{s_{mih}, g_{mih}}\right) \\ &= \sum_{h=0}^{n_{mi}} P_{mih} r(s_{mih}) z^{g_{mih}} + \left[1 - \sum_{h=0}^{n_{mi}} P_{mih} r(s_{mih})\right] z^{g_{mi0}} \end{aligned} \tag{20}$$

This UGF represents the unconditional distribution of performance of entire work sharing group i in subsystem m.

B. *The UGF of the entire system*

Applying $U_{\{mi,mj\}}(z) = U_{mi}(z) \underset{\varpi}{\otimes}{}_{mj}(z)$ recursively one can obtain the UGF of subsystem m in the form $U_m(z) = \sum_{h=0}^{n_m} P_{mh} z^{g_{mh}}$. Here ϖ is the performance composition function for elements connected in parallel without work sharing, P_{mh} is the probability that the performance of subsystem m equals to g_{mh}. Applying $U_{\{m,l\}}(z) = U_m(z) \underset{\pi}{\otimes} U_l(z)$ recursively one can obtain the UGF of the entire system in the form $U_s(z) = \sum_{h=0}^{n_s} P_h z^{g_h}$. Here π is the performance composition function for elements connected in series, P_h is the probability that the performance of the entire system equals to g_h. From the UGF $U_s(z)$, representing the pmf of the entire multi-state system performance (12), the system reliability can be obtained using (13).

C. *Performance composition functions*

The choice of functions φ depends on the type of connection between the elements, and on the type of the system. Consider, for example, a data transmission system with performance defined as transmission capacity (bandwidth). Assume that each element j has a random data transmission capacity G_j. If two elements i and j transmit the same data, providing data transmission redundancy, the transmission capacity of the pair of elements is determined by $\varpi(G_i, G_j) = \max(G_i, G_j)$. If the parallel elements share their work, then the entire capacity that they provide is given by $\omega(G_i, G_j) = G_i + G_j$. If data flow is transmitted by two consecutive elements, the performance of the two elements is determined by $\pi(G_i, G_j) = \min(G_i, G_j)$.

D. *System structure optimization*

The optimal system structure determination problem formulated by (14) is an NP complete set partitioning problem. An exhaustive examination of all possible solutions is not realistic, considering reasonable time limitations. The genetic algorithm has proven to be an effective optimization tool for a large number of complicated problems in reliability engineering [15, 21, 25]. In our genetic algorithm, solutions are represented by integer strings $S = \{s_1, s_2, \ldots s_n\}$, where each s_i belongs to the range $(1, \max_{m=1}^{M} E_m)$.

The following procedure determines the fitness value for an arbitrary solution defined by integer string $S = \{s_1, s_2, \ldots s_n\}$.

1. For each subsystem m=1,...,M:

1.1. Determine the number of WSG for each element of the mth component:

$$\alpha_{mj} = 1 + \mathrm{mod}_{E_m}(s_{x+j}), 1 \le j \le E_m, \tag{21}$$

where $x = \sum_{k=1}^{m-1} E_k$.

1.2. For each WSG $i(1 \le i \le E_m)$, create set Φ_{mi} using the recursive procedure

$$\Phi_{mi} = \emptyset, \quad \text{for } i = 1, \ldots, E_m :$$
$$\text{if } \alpha_{mj} = i, \ \Phi_{mi} = \Phi_{mi} \cup \{x + j\}.$$

2. Determine the UGF of the entire system and calculate system reliability using (13). Assign the obtained system reliability to the solution fitness.

5.3 Illustrative Examples

Consider a data transmission system consisting of two consecutive multichannel communication lines. Each channel can have failure state with zero transmission capacity and two working states with full and reduced transmission capacity. The system is able to work properly if the system capacity is greater than the minimal allowed capacity C^*. The distributions of the performances (transmission capacities) of channels are presented in Table 1

As an illustration, we assume that the coverage probability of the jth failure in work sharing group i in any subsystem m decreases with $|\Phi_{mi}|$ and increases with j as given in Table 2

Table 3 contains the optimal system configurations for $C^* = 20$ Kb/sec, $C^* = 30$ Kb/sec, and $C^* = 40$ Kb/sec obtained using the GA and characteristics of the corresponding transmission systems.

Table 1 Performance distributions of data transmission channels

Sub-system	Element	Performance levels					
		Probability p_{j0}	Capacity g_{j0}	Probability p_{j1}	Capacity g_{j1}	Probability p_{j2}	Capacity g_{j2}
1	1	0.15	0	0.7	10	0.15	20
	2	0.15	0	0.65	12	0.20	20
	3	0.20	0	0.60	15	0.20	25
	4	0.15	0	0.60	18	0.25	25
	5	0.15	0	0.70	14	0.15	20
	6	0.10	0	0.80	11	0.10	24
	7	0.20	0	0.50	20	0.30	30
2	8	0.20	0	0.60	12	0.20	25
	9	0.20	0	0.60	14	0.20	24
	10	0.20	0	0.70	15	0.10	25
	11	0.15	0	0.65	20	0.20	30
	12	0.15	0	0.70	12	0.15	20
	13	0.10	0	0.80	18	0.10	30
	14	0.25	0	0.65	10	0.10	20

Table 2 Coverage probability of the j-th failure in a work sharing group with $|\Phi_{mi}|$ elements

$\|\Phi_{mi}\|$	1	2	3	4	5	6	7
$j = 1$	0	0.99	0.63	0.36	0.22	0.15	0.08
$j = 2$	–	0	0.99	0.63	0.36	0.22	0.15
$j = 3$	–	–	0	0.99	0.63	0.36	0.22
$j = 4$	–	–	–	0	0.99	0.63	0.36
$j = 5$	–	–	–	–	0	0.99	0.63
$j = 6$	–	–	–	–	–	0	0.99
$j = 7$	–	–	–	–	–	–	0

Table 3 Parameters of solutions

	No sharing	$C^* = 20$	$C^* = 30$	$C^* = 40$	No redundancy
Max capacity	30	60	65	74	164
$R(0)$	≈1.0	0.9998	0.9998	0.9894	0.0952
$R(20)$	0.3640	0.9772	0.9733	0.9375	0.0952
$R(30)$	0.0	0.8461	0.9405	0.8656	0.0952
$R(40)$	0.0	0.2393	0.2650	0.7526	0.0952
Subsystem 1	(1),(2), (3),(4), (5),(6), (7)	(1,2,5), (3,7), (4,6),	(1,2,3), (4,5), (6,7)	(1,4,5), (2,6,7), (3)	(1,2,3,4, 5,6,7)
Structure Subsystem 2	(8),(9), (10),(11), (12),(13), (14)	(8),(9), (10,12,14), (11,13)	(8,9,14), (11,12), (10,13)	(8,9,13) (10,11,12), (14)	(8,9,10,11, 12,13,14)

6 Conclusions

Computer systems that are used in life-critical applications and designed with sufficient redundancy are vulnerable to uncovered failures, which foil the automatic recovery mechanism. Many works have been developed to model the reliability of systems subjected to imperfect fault coverage. The DFTS algorithm is able to determine the unreliability of a system subjected to imperfect fault coverage during enumeration of the operational states that correspond to the fault tree. A simple alternative, the DDP algorithm, uses existing cut-set solutions methods and combines aspects of behavioral decomposition, sum-of-disjoint products, and multi-state solution method. The DDP algorithm was further developed into SEA, which can be used with any combinatorial solution technique. Recently, some OBDD-based algorithms have been proposed to study the reliability of systems subjected to imperfect fault coverage. The OBDD technique is used because of its high efficiency compared to other techniques, such as inclusion-exclusion and sum of disjoint products. Some recent papers have also proposed UGF-based algorithms to study imperfect fault coverage. The UGF is shown to be very flexible in modeling different kinds of fault coverage. This chapter also presents an extension of the optimal system structure problem formulated in Levitin [28] to the case of FLC.

References

1. Abraham JA (1979) An improved algorithm for network reliability. IEEE Trans Reliab 28(1):58–61
2. Aggarwal KK, Misra KB, Gupta JS (1975) A fast algorithm for reliability evaluation. IEEE Trans Reliab 24(1):83–85
3. Akhtar S (1994) Reliability of k-out-of- n: G systems with imperfect fault-coverage. IEEE Trans Reliab 43(1):101–106
4. Arnold TF (1973) The concept of coverage and its effect on the reliability model of a repairable system. IEEE Trans Comput 22(3):325–339
5. Amari S (1997) Reliability, risk and fault-tolerance of complex systems. PhD Thesis, Indian Institute of Technology, Kharagpur
6. Amari SV, Dugan JB, Misra RB (1999a) A separable method for incorporating imperfect fault-coverage into combinatorial models. IEEE Trans Reliab 48(3):267–274
7. Amari SV, Dugan JB, Misra RB (1999b) Optimal reliability of systems subject to imperfect fault-coverage. IEEE Trans Reliab 48(3):275–284
8. Amari S, Pham H, Dill G (2004) Optimal design of k-out-of- n: G subsystems subjected to imperfect fault-coverage. IEEE Trans Reliab 53(4):567–575
9. Bavuso SJ, Dugan JB, Trivedi KS, Rothmann EM, Smith WE (1987) Analysis of typical fault-tolerant architectures using HARP. IEEE Transactions on Reliability 36(2):176–185
10. Bavuso SJ et al. (1994) HiRel: hybrid automated reliability predictor (HARP) integrated reliability tool system (Version 7.0), 4 vols, NASA TP 3452
11. Bouricius WG, Carter V, Schneider PR (1969) Reliability modeling techniques for self-repairing computer systems. In: Proceedings of the 24th national conference, ACM, pp 295–309
12. Bryant R (1986) Graph based algorithms for Boolean function manipulation. IEEE Trans Comput 35(8):677–691

13. Chang YR, Suprasad VA, Kuo SY (2004) Computing system failure frequencies and reliability importance measures using OBDD. IEEE Trans Comput 53(1):2004
14. Chang YR, Amari SV, Kuo SY (2005) OBDD-based evaluation of reliability and importance measures for multistate systems subject to imperfect fault coverage. IEEE Trans Dependable Secure Comput 2(4):336–347
15. Coit D, Smith A (1996) Reliability optimization of series-parallel systems using genetic algorithm. IEEE Trans Reliab 45(2):254–266
16. Ding Y, Zuo MJ, Lisnianski A, Li W (2010) A framework for reliability approximation of multi-state weighted k-out-of- n systems. IEEE Trans Reliab 59(2):297–308
17. Doyel SA, Dugan JB, Patterson-Hine FA (1995) A combinatorial approach to modeling imperfect coverage. IEEE Trans Reliab 44(1):87–94
18. Dugan JB (1989) Fault trees and imperfect coverage. IEEE Transactions on Reliability 38(2):177–185
19. Dugan JB, Trivedi KS (1989) Coverage modeling for dependability analysis of fault-tolerant systems. IEEE Transactions on Computers 38(6):775–787
20. Dutuit Y, Rauzy A (2001) New insights in the assessment of k-out-of-n and related systems. Reliability Engineering and System Safety 72(3):303–314
21. Huang HZ, Qu J, Zuo MJ (2009) Genetic-algorithm-based optimal apportionment of reliability and redundancy under multiple objectives. IIE Trans 41(4):287–298
22. Kuo SY, Lu SK, Yeh FM (1999) Determining terminal-pair reliability based on edge expansion diagrams using OBDD. IEEE Trans Reliab 48(3):234–246
23. Kuo SY, Yeh FM, Lin HY (2007) Efficient and exact reliability evaluation for networks with imperfect vertices. IEEE Trans Reliab 56(2):288–300
24. Lee YJ, Na MG (2009) Design of delay-tolerant controller for remote control of nuclear reactor power. Nuclear Eng Technol 41(1):71–78
25. Levitin G, Lisnianski A, Beh-Haim H, Elmakis D (1998) Redundancy optimization for series-parallel multi-state systems. IEEE Trans Reliab 47(2):165–172
26. Levitin G (2005) Universal generating function in reliability analysis and optimization. Springer, London
27. Levitin G (2007) Block diagram method for analyzing multi-state systems with uncovered failures. Reliab Eng Syst Saf 92(6):727–734
28. Levitin G (2008) Optimal structure of multi-state systems with uncovered failures. IEEE Trans Reliab 57(1):140–148
29. Levitin G, Amari SV (2008a) Multi-state systems with static performance-dependent fault coverage. Proc Inst Mech Eng, Part O J Risk Reliab 222(2):95–103
30. Levitin G, Amari SV (2008b) Multi-state systems with multi-fault coverage. Reliab Eng Syst Saf 93(11):1730–1739
31. Levitin G, Xing LD (2010) Reliability and performance of multi-state systems with propagated failures having selective effect. Reliab Eng Syst Saf 95(6):655–661
32. Li CY, Chen X, Yi XS, Tao JY (2010) Heterogeneous redundancy optimization for multi-state series-parallel systems subject to common cause failures. Reliab Eng Syst Saf 95(3):202–207
33. Moustafa M (1997) Reliability of K-out-of- N: G systems with dependent failures and imperfect coverage. Reliab Eng Syst Saf 58(1):15–17
34. Myers AF (2007) k-out-of- n: G system reliability with imperfect fault coverage. IEEE Trans Reliab 56(3):464–473
35. Myers A (2008) Achievable limits on the reliability of k-out-of- n: G systems subject to imperfect fault coverage. IEEE Trans Reliab 57(2):349–354
36. Myers A (2009) Probability of loss assessment of critical k-out-of- n: G systems having a mission abort policy. IEEE Trans Reliab 58(4):694–701
37. Myers A, Rauzy A (2008a) Assessment of redundant systems with imperfect coverage by means of binary decision diagrams. Reliab Eng Syst Saf 93(7):1025–1035
38. Myers A, Rauzy A (2008b) Efficient reliability assessment of redundant system subject to imperfect fault coverage using binary decision diagrams. IEEE Trans Reliab 57(2):336–348

39. Newton J (1995) Comment on: Reliability of k-out-of- n: G systems with imperfect fault-coverage. IEEE Trans Reliab 44(1):137–138
40. Peng R, Levitin G, Xie M, Ng SH (2011) Reliability modeling and optimization of multi-state systems with multi-fault coverage. submitted to the Seventh International Conference on mathematical methods in reliability-theory. Methods. Applications
41. Perhinschi MG, Napolitano MR, Campa G, Seanor B, Burken J, Larson R (2006) Design of safety monitor schemes for a fault tolerant flight control system. IEEE Trans Aerospace Electron Syst 42(2):562–571
42. Pham H (1992a) Optimal cost-effective design of triple-modular-redundancy-with-spares systems. IEEE Transactions on Reliability 42(3):369–374
43. Pham H (1992b) Reliability analysis of a high voltage system with dependent failures and imperfect coverage. Reliab Eng Syst Saf 37(1):25–28
44. Pham H, Pham M (1992) Reliability analysis of dynamic redundant systems with imperfect coverage. Reliab Eng Syst Saf 35(2):173–176
45. Schneeweiss W (1987) Approximate fault-tree analysis with prescribed accuracy. IEEE Transactions on Reliability 36(2):250–254
46. Tian ZG, Zuo MJ, Huang HZ (2008) Reliability-redundancy allocation for multi-state series-parallel systems. IEEE Trans Reliab 57(2):303–310
47. Trivedi KS, Geist R (1983) Decomposition in reliability analysis of fault-tolerant systems. IEEE Trans Reliab 32(5):463–468
48. Trivedi KS, Dugan JB, Geist R, Smotherman M (1984) Hybrid reliability modeling of fault-tolerant computer-systems. Comput Electr Eng 11(2–3):87–108
49. Ushakov I (1987) Optimal standby problems and a universal generating function. Soviet J Comput Syst Sci 25(4):79–82
50. Xing LD (2002) Analysis of generalized phased-mission system reliability, performance, and sensitivity. IEEE Trans Reliab 51(2):199–211
51. Xing LD (2007) Reliability evaluation of phased-mission systems with imperfect fault coverage and common-cause failures. IEEE Trans Reliab 56(1):58–68
52. Xing LD (2008) An efficient binary-decision-diagram-based approach for network reliability and sensitivity analysis. IEEE Trans Syst Man Cybern Part A Syst Humans 38(1):105–115
53. Yeh FM, Lu SK, Kuo SY (2002) OBDD-Based evaluation of k-terminal network reliability. IEEE Trans Reliab 51(4):443–451
54. Yeh WC (2009) A convolution universal generating function method for evaluating the symbolic one-to-all-target-subset reliability function of acyclic multi-state information networks. IEEE Trans Reliab 58(3):476–484
55. Youssef AMA, ElMaraghy MA (2008) Performance analysis of manufacturing systems composed of modular machines using the universal generating function. J Manuf Syst 27(2):55–69

Replacement and Maintenance Policies of Devices: A Review

Mohamed Abdel-Hameed

Abstract In this chapter, we discuss the contributions of the author to replacement and maintenance of devices. Some related works are also discussed.

1 Introduction

During the past decades, the research on maintenance of engineering structures and devices has increased continuously. A characteristic feature of the maintenance and replacement policies is that decisions often must be made under uncertainty (such as in deterioration and cost). In maintenance management, the most important uncertainty is generally the uncertainty in the time to failure (lifetime), and/or the rate of deterioration. Up to the early 1990s, most mathematical maintenance models were based on describing the uncertainty in aging using a lifetime distribution. A disadvantage of a lifetime distribution, however, is that it only quantifies whether a component is functioning or not. In order to represent aging on the basis of lifetime distributions, the celebrated failure rate function can be applied. The failure rate, however, is only useful for making inferences for a large population of components rather than for a single component.

For engineering structures and infrastructures it is generally more attractive to base a failure model on the physics of failure and the characteristics of the operating environment. Therefore, it is recommended to model deterioration in terms of a time-dependent stochastic process. Markov processes properly model the temporal variability of deterioration. Such processes include stochastic processes with independent increments like the Brownian motion with drift, the compound Poisson

M. Abdel-Hameed (✉)
Department of Statistics, College of Business and Economics,
United Arab Emirates University, Al Ain, United Arab Emirates
e-mail: abdelham@hotmail.com

T. Dohi and T. Nakagawa (eds.), *Stochastic Reliability and Maintenance Modeling*,
Springer Series in Reliability Engineering 9, DOI: 10.1007/978-1-4471-4971-2_9,

process, and the gamma process. For the stochastic modeling of monotonic and gradual deterioration, the gamma process is most appropriate.

In this chapter, we review some of the work done by the author on maintenance models, using lifetime distribution as well as observing the amount of degradation the device is subject to over time.

Section 2 deals with maintenance models based on the life distribution. Section 3 deals with maintenance models based on observing the amount of degradation the device is subjected to over time.

2 Periodic Maintenance Policies and Periodic Maintenance Policies with Imperfect Repairs

Model 1

Abdel-Hameed [6] considers a periodic replacement policy for a device. The device is replaced every T units of time (planned replacement), each at cost c_0 . At failure, the device is either restored to its condition prior to failure (minimal repair) or replaced (unplanned replacement). If the device failed at age t, it is replaced with probability $p(t)$ or minimally repaired with probability $q(t) = 1- p(t)$. The cost of the ith minimal repair of a failed device at age t is $c_i(t)$, and the cost of each unplanned replacement is c_∞. This procedure repeats itself after each replacement, planned or unplanned.

We define F to be the distribution function of the failure time, $\overline{F} = 1 - F$ to be the survival function, and $R = -ln\overline{F}$ be the cumulative hazard function. Let $S = (S_n, n = 1, 2, \ldots)$ be the process describing the times of successive failures. For $n \geq 1$, let S_n^* be the age of the device that replaced the nth failed device. Furthermore, we let $(N(t), t > 0)$ be the stochastic process denoting the number of unplanned replacements; when $T = \infty$. Assuming τ to be the time of first jump of the process N, throughout, we will denote the renewal function of the process N by M. It is seen that τ has a distribution function G with survival probability

$$\overline{G}(t) = exp[-R_p(t)]$$

where

$$R_p(t) = \int_0^t p(y)R(dy)$$

Let $\hat{N} = (\hat{N}(t), t > 0)$ be the process describing the number of minimal repairs. If $A(T)$ denotes the long-run average cost per unit of time, from standard renewal argument it follows that,

$$A(T) = T^{-1}[c_0 + c_\infty M(T) + E \sum_{i=1}^{\hat{N}(T)} c_n(S_n^*)].$$

Define $f(t)$ to be equal to the expected cost of minimal repairs in $[0, t)$, and let $v = min(\tau, T)$. Let

$$V(T) = E \sum_{i=1}^{\hat{N}(v)} c_n(S_n^*).$$

We have the following

Theorem 2.1 *The expected cost of minimal repairs in* $[0, T)$, $f(T)$, *satisfies the following integral equation:*

$$f(T) = V(T) + M * V(T)$$

where $M * V$ *is the convolution of* M *and* V.

The following theorem establishes the optimal value of the periodic replacement time, and gives conditions for its uniqueness.

Theorem 2.2 *(a) Let* f *be as given in Theorem 2.1,* m *be the renewal density of the renewal function* M, *and* f' *be the derivative of* f.

Then the optimal periodic replacement time is the unique solution of the equation

$$c_\infty[Tm(T) - M(T)] + [Tf'(T) - f(T)] = c_0.$$

(b) The solution in (a) is unique if both $M(t)$ *and* $f(t)$ *are convex functions, in their respective arguments.*
For proofs of the above theorems and more detailed investigations the reader is referred to the above-mentioned reference.

Model 2 (see Ref. [4])

A system is subject to shocks, the normal cost of running the system per unit of time is denoted by $a > 0$, and each shock to the system increases the running cost by an amount $c > 0$, per unit of time. The cost of completely replacing the system is c_o. The system is to be completely replaced at times $T, 2T, \ldots$ at a cost $c_o > 0$. The value T is known as the period of the policy. In practice, reliability analysts are often asked to find the optimal value of the period, that is to say, the value of T that minimizes some functional of the cost. Such functional is normally taken to be the long-run average cost per unit of time or the discounted total cost.

Let $N = \{N(t), t \geq 0\}$ be the process describing the number of shocks that the system is subject to during the interval $[0, t)$. Throughout we assume N as a counting process whose jumps are of one unit magnitude. For $t \geq 0$, we define $M(t)$ as the expected number of shocks in $[0, t)$. From Fubini's theorem it follows that the expected total cost of running the system per period is given by

$$aT + c\int_0^T M(t)\mathrm{d}t + c_o$$

From standard renewal theory argument it follows that the long-run average cost per unit of time is given by

$$A(T) = [aT + c\int_0^T M(t)\mathrm{d}t + c_o]/T.$$

The following theorem gives the form of the optimal periodic replacement time.

Theorem 2.3 *The optimal value of the periodic replacement time always exists and is equal to the unique solution of the integral equation*

$$\int_0^T [M(T) - M(t)]\mathrm{d}t = c_0/c.$$

Moreover, it is finite if and only if

$$\lim_{T\longrightarrow\infty}\int_0^T [M(T) - M(t)]\mathrm{d}t > c_0/c.$$

Now we discuss the case when the maintenance and replacement costs are time dependent. Let $(\tau_n, n = 1, 2, \ldots)$ *be the sequence describing the successive jump times of the shock process* N*. The normal cost of running the system per unit of time is* $a > 0$ *and the cost of completely replacing the system is* c_o*. For* $t \in [\tau_n, \tau_{n+1})$*,* $c_n(t)$ *is the additional cost of operating the system per unit of time. Let*

$$h(t) = E(c_{N((t)}(t)).$$

The following is the theorem analogous to the one above in this general case.

Theorem 2.4 *If* h *is continuous, increasing, then the optimal value of the periodic replacement time exists and is the unique solution of the integral equation*

$$\int_0^T [h(T) - h(t)]\mathrm{d}t = c_o.$$

Moreover, it is finite if and only if

$$\lim_{T\longrightarrow\infty}\int_0^T [M(T) - M(t)]\mathrm{d}t > c_0.$$

For a more detailed examination of the above model and the proofs of the above theorem the reader is referred to the above-mentioned reference. In the mentioned reference, the special cases when the shock process is a non-homogeneous birth and death process, as well as a renewal process are discussed.

3 Replacement and Maintenance Policies of Devices Subject to Degradation

In 1975, Abdel-Hameed [1] proposed to use the gamma process as a model for degradation occurring randomly in time. During the past 3 decades, gamma processes were satisfactorily fitted to data on creep of concrete, fatigue crack growth, corroded steel gates, thinning due to corrosion, and chloride ingress into concrete. On the basis of the gamma degradation processes, case studies have been performed to determine optimal maintenance decisions for steel coatings, and optimal inspection intervals for high-speed railway. (see Refs. [9, 10] and [11] for more detailed investigations of these topics).

In 1977 [2] the author discussed the case where a device is subject to a non-homogeneous gamma process. He considers the cases where the degradation is monitored continuously or monitored periodically. The device is replaced at failure [corrective maintenance (CM)] or when the deterioration level exceeds a predetermined level [preventive maintenance (PM)]. The cost of CM is fixed, while the cost of PM depends on the deterioration level at the time when the maintenance is performed. He obtains an explicit formula for the long-run average cost per unit of time. van Noortwijk [11] applies this result to maintenance of a cylinder on a swing bridge.

In 1984 [3] the author extended the above results to the case where the degradation is an increasing Levy process.

Abdel-Hameed [5] studied condition-based maintenance of a system subject to stochastic degradation, where the degradation process is assumed to be a non-decreasing jump process, denoted by X. The system has a threshold Y and it fails once the deterioration exceeds or equals the threshold. We assume G to be the distribution function of Y and we define $\overline{G} = 1 - G$. Examples of pure jump processes are: (1) compound Poisson processes with positive jumps, (2) gamma processes, (3) pure-birth processes, (4) stable processes, as well as Levy processes. The degradation level is monitored periodically at times $k\tau, k = 1, 2, \ldots, \tau > 0$. The two decision variables are the inspection interval and the PM level. In the operations research literature, such a policy is called a "control-limit policy" with the PM level called the "control limit". A failure is defined as the event in which the degradation exceeds a failure random threshold (failure level). A failure is detected only by inspection. The system is renewed when an inspection reveals either that the PM level is crossed while no failure has occurred (preventive replacement) or that the failure level Y is crossed (corrective replacement). If a replacement occurs before failure, when the degradation level is x, the device is replaced by a new and identical one at a cost of

$c_l(x)$. A failure is discovered only by inspection; upon detection of failure the device must be replaced by a new and identical one at a cost of $k = c_1(\infty)$. We assume that $c_1()$ is an increasing function in its argument, and $c_1(0) = 0$. The regular cost of operating the system per unit of time, when the deterioration level is x, is denoted by $c_2(x)$, and is assumed to be an increasing function in its argument. A renewal brings the system back to its "as good as new" condition. The cost of preventive replacement is a function of the degradation and the cost of corrective replacement is fixed, where the former is less than or equal to the latter. Also, the cost of inspection per unit time is a function of the inspection interval and the cost of system operation per unit time is a function of the deterioration. Inspection does not degrade the system and is perfect in the sense that deterioration will be observed with certainty and $g(\tau)$ is the cost of inspection per unit of time; we assume that g is a decreasing function in its argument. Both inspection and replacement take negligible time. The optimal maintenance decision is determined by minimizing the long-term average cost per unit time. This cost is computed by applying renewal reward theory. He finds the 'optimal inspection policy', where by optimal inspection policy is meant the policy that minimizes the long-run average cost per unit of time. Moreover, he determines appropriate conditions on the cost functions and the parameters of the deterioration process which ensure that the optimal inspection policy is a control-limit policy. Let ς be the failure time of the device, i.e.,

$$\varsigma = \inf\{t > 0 : X_t \geq Y\}.$$

Furthermore, let $N(\tau)$ be the inspection time index at which a failure is detected, i.e.,

$$N(\tau) = \inf\{n : n\tau \geq \varsigma\}.$$

Define F to be the smallest sigma-algebra generated by $\{X(n\tau), n \leq N(\tau)\}$. An inspection policy is defined as any stopping time with respect to F; we denote the class of such inspection policies by κ. Observe that any inspection policy has $\{\tau, 2\tau, \ldots\}$ as its support.

Below we summarize the main results obtained in this paper, without proofs. Let

$$\begin{aligned} c_1^0(x) &= c_l(x) - k \quad \text{if} \quad x < \infty \\ &= 0 \qquad\qquad\quad \text{if} \quad x = \infty. \end{aligned}$$

Then it follows from standard renewal theory arguments that the long-run average cost of replacement per unit time when using an inspection policy $T \in \kappa$ is

$$\psi(T) \stackrel{\text{def}}{=} [E(T)]^{-1}[E \int_0^T c_2(X_t) + Ec_1^0(X_T) + k] + g(\tau)].$$

He proves that under appropriate assumptions on the parameters of the degradation process and its infinitesimal generator, the optimal maintenance policy is a control-limit policy.

Abdel-Hameed and Nakhi [8] treat the maintenance policy for devices subject to degradation, when the degradation process is an increasing semi-Markov process. Specifically, let the degradation process (Z) be an increasing semi-Markov process with embedded Markov renewal process $(X, T) = (X_n, T_n; n \in N)$, where $X_n = Z(T_n)$. Let $Q = \{Q(x, A, t), x, t \in R_+, A \subset R_+\}$ be the semi-Markov kernel associated with (X, T), that is $Q(x, A, t) = \Pr\{X_{n+1} \in A, T_{n+1} - T_n \leq t | X_n = x\}$, and Markov renewal kernel $R = \{R(x, A, t), x, t \in R_+, A \subset R_+\}$, where $R(x, A, t) = \sum_{n=0}^{\infty} Q^{(n)}(x, A, t)$. The system has a resistance level (denoted by random variable Y), and the device fails once the degradation level crosses the resistance level. The resistance level and the degradation process are assumed to be independent. We denote the failure time by ρ. Let $\hat{Z}$ be the degradation process, obtained by killing the process Z at the failure time, that is, $\hat{Z} = (Z_t, t < \rho)$. Define $\hat{Q}$ and $\hat{R}$ as the corresponding semi-Markov kernel and Markov renewal kernel, respectively. The system can be replaced before or at failure, and is maintained continuously. The maintenance and non-failure costs are state dependent. They determine the optimal maintenance policy, using the total discounted as well as the long-run average cost per unit of time. Let $g : R_+ \to R$ be the function describing the maintenance rate. In the case, where the state space is countable, we define, for degradation levels i, j in the state space,

$$\begin{aligned} q(i, j) &= P\{X_{n+1} = j | X_n = i\}, \\ m(i) &= E_i(T), \\ \hat{q}(i, j) &= q(i, j) \frac{\bar{G}(j)}{\bar{G}(i)}; \\ \hat{h}(i) &= P_i\{T_1 = \rho\}. \end{aligned}$$

Assume that the costs of a preventative (corrective) maintenance are c_1 and c_2, $(c_2 > c_1)$, respectively, and define the matrix $Q = (q(i, j))$. The optimal replacement policy that minimizes the long-run average cost per unit time can be summarized in the following algorithm. For more detailed explanations the reader is referred to the reference above.

Algorithm. Assume that the degradation level at time zero is equal to i, normally taken equal to zero.

Step 1. let $j = i$.

Step 2. Compute the matrix $\hat{R}$, using the well known relationship $\hat{R} = [I - \hat{Q}]^{-1}$, where I is the identity matrix of proper dimensions.

Step 3. For i, j let $\hat{r}(i, j) = \hat{R}(i, j) - \hat{R}(i, j-1)$.

Step 4. Compute

$$b_j(k) = c_2[\frac{m(k)\hat{h}(j)}{m(j)} - \hat{h}(k)] + m(k)(g(j) - g(k))$$

for $k = i, \ldots, j$.

Step 5. Compute

$$F(j) = \sum_{i}^{j} \hat{r}(i, j) b_j(k).$$

Step 6. If $F(j) \geq c_1$, then j is the optimum replacement level, otherwise $j = j+1$ and go to step 2.

Abdel-Hameed [7] considers the optimal maintenance policy for a system subject to degradation. The degradation level is only observed at successive inspection times. It is assumed that the degradation levels at inspections, and the times of successive inspections form a Markov renewal process. Failure is detected only by inspection, at this point in time the system goes through a CM. The system is also maintained when the degradation exceeds a predetermined level (PM). He determines the optimal maintenance policy using both the total discounted as well as long-run average cost criteria. The system has a nominal life Y, with right tail probability $\overline{G}$ and once the degradation exceeds Y, the system fails. The states of the system are only observed by inspection, also failures are only detected by inspection.

Let $\hat{T} = \{\hat{T}_n, n = 1, 2, \ldots\}$ be the times of successive inspections. Define the process, $\hat{X} = \{\hat{X}_n, n = 1, 2, \ldots\}$ as the process describing the degradation levels at the successive inspection times. The system is replaced when a failure is detected (CM), or once the observed degradation exceeds level M (PM). If at inspection the system did not fail and the degradation is below level M, the system is left alone. Each PM costs c_1 and each CM costs $c_1 + d$; $d \geq 0$. The system is as good as new after each maintenance (Corrective or Preventive), and maintenance is instantaneous. If at inspection the degradation level is x, an inspection cost $c(x)$ occurs. The process $(\hat{X}, \hat{T}) = \{(\hat{X}_n, \hat{T}_n), n = 1, 2, \ldots\}$ is assumed to be a Markov renewal process with state space $[0, \infty)$. Let $\hat{Q}$ and $\hat{R}$ be the semi-Markov kernel and Markov renewal function corresponding to $(\hat{X}, \hat{T})$. Define

$$L_1 = \inf\{n : \hat{X} > Y\} \quad \text{and}$$

$$L_2 = \inf\{n : \hat{X} > M\}.$$

Then $\hat{T}_{L_1}$ and $\hat{T}_{L_2}$ are the times of first corrective and PMs, respectively. We denote these times by ζ, T_M, respectively, and we define $L = L_1 \wedge L_2$. Let $(\hat{X}^{(1)}, \hat{T}^{(1)}) = \{(\hat{X}_n, \hat{T}_n), n < L_1\}$. It follows that $(\hat{X}^{(1)}, \hat{T}^{(1)})$ is a Markov renewal process with state space $[0, M)$. Furthermore, for $x < M, y < M$ and $t \in R_+$, their corresponding semi-Markov and Markov renewal functions (denoted by $\hat{Q}^{(1)}$ and $\hat{R}^{(1)}$) are given by

$$\hat{Q}^{(1)}(x, dy, t) = \hat{Q}(x, dy, t)\frac{\bar{G}(y)}{\bar{G}(x)} \quad \text{for } y \geq x$$
$$= 0 \quad \text{for } y < x,$$

$$\hat{R}^{(1)}(x, dy, t) = \hat{R}(x, dy, t)\frac{\bar{G}(y)}{\bar{G}(x)} \quad \text{for } y \geq x$$
$$= 0 \quad \text{for } y < x.$$

Let S be the time of first maintenance. For $k \geq 2$, let $\hat{X}^{(k)}, \hat{T}^{(k)}, S^{(k)}$ be independent copies of $\hat{X}^{(1)}, \hat{T}^{(1)}, S$, respectively; let $S^{(1)} = S$, $\tau_0 = 0$, $L_0 = 0$, and for $k \geq 1$, define

$$\tau_k = \sum_{j=1}^{k} S^{(j)},$$

$$N_k = \{n : \hat{T}_n^{(k)} = \tau_k - \tau_{k-1}\},$$

$$L_k = \sum_{j=1}^{k} N_j.$$

Then for $k \geq 1$, τ_k and N_k are the time of the kth maintenance and the index at which such maintenance is performed, respectively.

We define the processes $(X, T) = ((X_n, T_n), n = 1, 2, \ldots\}$ as follows:

$$X_n = \sum_{k=n-L_{k-1}}^{\infty} \hat{X}_{n-L_{k-1}}^{(k)} I(L_{k-1} \leq n < L_k), \quad X_{L_k} = 0 \text{ for } k \geq 1,$$

$$T_n = \sum_{k=1}^{\infty} \hat{T}_{n-L_{k-1}}^{(k)} + \tau_k I(L_{k-1} < n \leq L_k), \quad T_{L_k} = \tau_k \text{ for } k \geq 1.$$

We note that the process (X, T) describes the degradation levels and the inspection times (over the infinite horizon), when maintenances are done at respective minuteness times $(\tau_1, \tau_2, \ldots)$. Furthermore, this process is regenerative with the maintenance times as the successive regeneration points; this process has state space $[0, M)$.

In [7] the total discounted as well as the long-run average costs are computed. Here we will summarize the basic formulas for the long-run average cost only, for the formulas using the total discounted cost the reader is referred to this reference. Since the process (X, T) is a regenerative process, it follows that the long-run average cost of running the system is given by (denoted by $C(M)$)

$$C(M) = \frac{E_0(c(X_n), n < L) + c_1 P_0\{S = T_M\} + (c_1 + d) P_0(S = \zeta)}{E_0(S)}$$
$$= \frac{E_0(c(X_n), n < L) + d P_0(S = \zeta) + c_1}{E_0(S)}$$
$$= \frac{\int_0^M \hat{R}^{(1)}(0, dy, \infty) c(y) + d P_0(S = \zeta) + c_1}{E_0(S)}.$$

Let $m(y) = \hat{E}_y(T_1)$, from Theorem 3 and Corollary 1 of [7] it follows that

$$P_0(S = \zeta) = \int_0^M \hat{R}^{(1)}(0, dy, \infty)[1 - \hat{Q}^{(1)}(y, [0, \infty), \infty)]$$

and

$$E_0(S) = \int_0^M \hat{R}^{(1)}(0, dy, \infty) m(y).$$

4 Conclusion and Perspectives

We discussed above several replacement, maintenance, and degradation models. It is worth noting that the gamma degradation model and maintenance policies for devices subject to such degradation process have attracted the attention of maintenance engineers more than any other models. Perhaps the reason behind this is that the gamma degradation process is the easiest to understand from a mathematical point of view. One would hope that other degradation processes and maintenance policies of devices subject to such degradation processes will be explored by safety and maintenance practitioners. For example, the inverse Gaussian process can be used to model degradation, the results obtained for the gamma degradation process can be easily extended to this case. Furthermore, maintenance policies based on semi-Markov degradation processes, given in Ref. [8], can provide a more accurate model for degradation and maintenance.

Acknowledgments The author thanks Professor Shunji Osaki for his valuable contributions to the fields of reliability theory, applied probability, and operations research. He has published numerous papers in professional journals, organized many international conferences, and has written several books. He served as Editor, Associate Editor, and on the Editorial board of many international journals. His dedication and contribution to the above-mentioned fields are certainly appreciated. The author thanks the anonymous referee for his suggestions on an earlier version of this chapter.

References

1. Abdel-Hameed MS (1975) A gamma wear process. IEEE Trans Reliab 24:152–153
2. Abdel-Hameed MS (1977) Optimal replacement policies for devices subject to a gamma wear process. In: Tsokos CP, Shimi IN (eds) The theory and applications of reliability; with emphasis on Bayesian and nonparametric methods. Academic Press, New York, pp 397–412
3. Abdel-Hameed MS (1984) Life distribution of devices subject to a Levy wear process. Math Oper Res 9:606–614
4. Abdel-Hameed MS (1986) Optimum replacement of systems subject to shocks. J Appl Probab 23:107–114
5. Abdel-Hameed MS (1987) Inspection and maintenance policies for devices subject to deterioration. Adv Appl Probab 19:917–931
6. Abdel-Hameed MS (1987) An imperfect maintenance model with block replacements. Appl Stoch Models 3:63–72
7. Abdel-Hameed MS (2004) Optimal predictive maintenance policies for a deteriorating system: the total discounted and long-run average cost cases. Commun Stat Theory Methods 33:735–745
8. Abdel-Hameed MS, Nakhi Y (1991) Optimal replacement and maintenance of systems subject to semi-Markov damage. Stoch Process Appl 37:141–160
9. Durham SD, Padgett WJ (1977) A cumulative damage model for system failure with applications to carbon and fiber composites. Technometrics 39:34–44
10. Park C, Padgett WJ (2005) Accelerated degradation models for failure based on geometric Brownian motion and gamma processes. Lifetime Data Analysis 11:511–527
11. van Noortwijk JM (1998) Optimal replacement decisions for structures under stochastic deterioration. In: Nowak AS (ed) Proceedings of the eighth IFIP WG. 7.5 working conference on reliability and optimization of structural systems, University of Michigan, pp 273–280.

Dynamical Systems with Semi-Markovian Perturbations and Their Use in Structural Reliability

Julien Chiquet and Nikolaos Limnios

Abstract The aim of this chapter is to present dynamical systems evolving in continuous-time and perturbed by semi-Markov processes (SMP). We investigate both probabilistic modeling and statistical estimation of such models. This work was initially developed in order to study cracking problems for the confinement device in nuclear power plants, where a jump Markov process was used as the perturbing process. The new key element here is the use of SMPs instead of Markov ones for the randomization of the system. Several numerical illustrations in reliability are investigated, accompanied with guidelines for a practical numerical implementation.

1 Introduction

In many industrial applications, structures may suffer degradations induced by the corresponding operating conditions. Degradations may be induced by thermal cyclic loadings, mechanical loadings, seismic activity, neutron irradiation, thermal sever transients, etc., which may lead to the failure of the structure.

Mechanisms that cause failures are complex due to their interdependencies and their different physical time-scales. Besides, these degradation mechanisms cannot always be described through deterministic models. Thus, a stochastic approach is often required. As a motivating example, we rely on the widely studied engineering issue referred to as the "crack-growth" problem [22, 27]: in structural mechanics, the main degradation process that leads to fatigue aging is due to the propagation of small defects into cracks in structures subject to small yet cyclic loadings. Many

J. Chiquet (✉)
UMR CNRS 8071–USC INRA, Université d'Évry Val-d'Essonne, Évry, France
e-mail: julien.chiquet@genopole.cnrs.fr

N. Limnios
Université de Technologie de Compiègne, Compiègne, France
e-mail: nikolaos.limnios@utc.fr

T. Dohi and T. Nakagawa (eds.), *Stochastic Reliability and Maintenance Modeling*, Springer Series in Reliability Engineering 9, DOI: 10.1007/978-1-4471-4971-2_10,

industrial fields are concerned, such as aeronautics, nuclear plants, automobile or bridge building among others. Qualitatively, the process remains the same whatever the material considered (e.g., aluminum in aeronautics, steel for confinement devices or pressure vessels, and concrete for bridges). The modeling methodology could be tackled with the mathematical tools provided in this work.

Even in well-controlled lab experiments supervised with cutting edge technology [20], crack-growth remains a very unstable phenomenon: deterministic models have been provided from structural mechanics, e.g., through computationally intensive finite-elements analysis. Yet, it is now acknowledged that probabilistic modeling are required to handle such degradation processes. Beyond the uncertainty propagation approaches offered by the probabilistic mechanics point of view, many authors rather suggested to completely randomize the modeling through a description relying on stochastic processes and dynamical systems (the pioneers in that domain being, to our knowledge, [19, 26]). The present chapter clearly enters this framework, drawing inspiration from [1, 9, 13, 15, 21] among many others.

As such, the stochastic models developed here do not necessarily aim to provide an exact physical representation of the phenomenon. We rather suggest to describe the evolution of an observable variable that characterizes the degradation process well. Hence, a structure is said to "fail" when its level of degradation exceeds a given threshold. The time evolution of the observable degradation process is described by a positive-valued stochastic process $Z = (Z_t, t \geq 0)$ governed by a first order stochastic differential system:

$$\dot{Z}_t = C(Z_t, X_t), \quad Z_0 = z, \tag{1}$$

where $\dot{Z}_t \doteq \mathrm{d}Z_t/\mathrm{d}t$ stands for the first order derivative of Z_t, C is a positive function, and $z > 0$ is the starting point of Z. The process $X = (X_t, t \geq 0)$ is a pure jump process with a countable state space. This model reflects the following physical point of view: the level of degradation Z increases on continuous sample paths; yet, its evolution shifts at discrete instants of time due to random shocks with random intensities induced by the operating conditions. These changes are modeled by the jump process X.

In the case where X is a jump Markov process, the coupled process $(Z, X) = (Z_t, X_t, t \geq 0)$ with state space $\mathbb{R}_+ \times E$ owns a well-characterized infinitesimal generator. Such a modeling belongs to the wider family of stochastic processes referred to as *Piecewise Deterministic Markov Processes*. These hybrid processes are an alternative to diffusion processes [7, 8, 12]. They virtually give a representation of many stochastic process being the mixture of deterministic motions and random jumps. A schematic view of three sample paths of the system defined in (1) are given in Fig. 1, when (Z, X) is observed from the starting point $t_0 = 0$ up to the random time τ when Z reaches an absorbing point Δ.

The purpose of this work is to model the perturbing process X by a semi-Markov process (SMP) and to derive the basic analysis for the associated dynamical system. We insist on the opportunity of considering X to be a SMP rather than a Markov

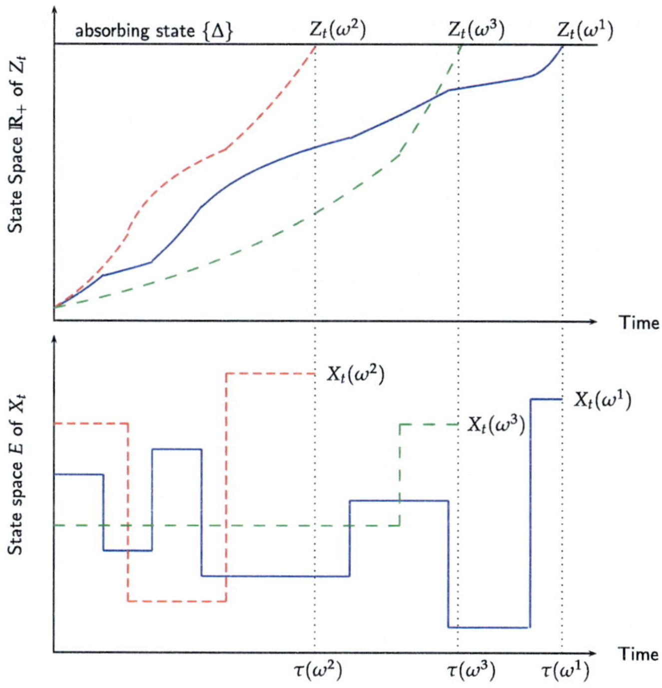

Fig. 1 Modeling degradation paths

process: the more flexible is the randomizing process X, the broader is the model and the wider is the range of its application.

The main contributions described in this chapter are of two kinds:

(1) First, we investigate the probabilistic characterization of the dynamical system (1) when X is semi-Markovian through Markov renewal theory, which allows to calculate the reliability function understood in the following sense: if a threshold Δ define an absorbing state of the system or, equivalently, a failure boundary for the degradation process, the failure time τ is

$$\tau = \inf \{t \geq 0 : Z_t \geq \Delta\},$$

and the associated reliability function turns to

$$R(t) = \mathbb{P}(Z_t < \Delta).$$

Interpreting (Z, X) as an extended SMP, we build a solvable Markov Renewal Equation (MRE) for the associated transition function, then deriving a closed-form. Still, this Markov renewal formulation required numerical resolution:

we propose a detailed guidelines to compute the reliability and give numerical example. This issue is addressed in Sect. 3.

(2) Second, we study the statistical inference of the system, that is, the estimation of the deterministic parameters of function C as well as the estimation of the SMP X. The degradation process being the only process whose paths can be collected during laboratory measurements, we only dispose of some sample paths of Z, observed before the system fails, and defined on the random time interval $[0, \tau]$. From these paths, we develop (1) a method to estimate the parameters of the function C, through an asymptotic analysis of the system (1) followed by a classical regression analysis; (2) a method to estimate the paths of X (as well as its state space E), since samples of X are not directly observed; (3) the construction of the likelihood function associated with the semi-Markov kernel of X and an approached maximum likelihood estimator for the kernel. This is developed in Sect. 4.

Meanwhile, let us start by an introductory section devoted to Markov renewal processes (MRP) theory.

2 Semi-Markov Processes: Background

This section recalls a few basics on SMPs. A larger view can be found for instance in [6, 11, 12, 17, 21, 24], yet the material provided here should hopefully be sufficient for the understanding of the main results developed throughout this chapter.

2.1 Notations and Settings

Consider an infinite countable set, say E, and an E-valued pure jump stochastic process $X = (X_t)_{t \in \mathbb{R}_+}$. Let $0 = S_0 \leq S_1 \leq \ldots \leq S_n \leq S_{n+1} \leq \ldots$ be the jump times of X, and $J_0, J_1, J_2, \ldots$ the successively visited states of X. Note that S_0 may also take positive values. Let $\mathbb{N}$ be the set of non-negative integers. Then, X is connected to (J_n, S_n) through

$$X_t = J_n, \quad if \quad S_n \leq t < S_{n+1}, \quad t \geq 0 \qquad \text{and} \quad J_n = X_{S_n}, \quad n \geq 0.$$

Definition 2.1. The stochastic process $(J_n, S_n)_{n \in \mathbb{N}}$ is said to be a Markov renewal process (MRP), with state space E, if it satisfies, a.s., the following equality

$$\begin{aligned} &\mathbb{P}(J_{n+1} = j, S_{n+1} - S_n \leq t \mid J_0, \ldots, J_n; S_1, \ldots, S_n) \\ &\quad = \mathbb{P}(J_{n+1} = j, S_{n+1} - S_n \leq t \mid J_n) \end{aligned}$$

for all $j \in E$, all $t \geq 0$, and all $n \in \mathbb{N}$. In this case, X is called a SMP.

Remark 2.1. We assume that the above probability is independent of n and S_n, and in this case the MRP is called *time homogeneous*. Only time-homogeneous MRP are considered in the sequel.

The MRP $(J_n, S_n)_{n\in\mathbb{N}}$ is determined by the *initial distribution* α, with $\alpha(i) = \mathbb{P}(J_0 = i), i \in E$ and by the transition kernel

$$Q_{ij}(t) := \mathbb{P}(J_{n+1} = j, S_{n+1} - S_n \leq t \mid J_n = i),$$

called the *semi-Markov kernel* of X. The process (J_n) is a Markov chain with state space E and transition probabilities $p_{ij} := Q_{ij}(\infty) := \lim_{t\to\infty} Q_{ij}(t)$, called the embedded Markov chain (EMC) of X. It is worth noticing that here $Q_{ii}(t) \equiv 0$, for all $i \in E$, but in general we can consider semi-Markov kernels by dropping this hypothesis.

An important point is the following decomposition of the semi-Markov kernel

$$Q_{ij}(t) := \mathbb{P}(J_{n+1} = j, S_{n+1} - S_n \leq t \mid J_n = i) = p_{ij}F_{ij}(t), \quad t \geq 0, \quad i, j \in E,$$

where p_{ij} is the transition kernel of the EMC (J_n), and $F_{ij}(t) := \mathbb{P}(S_{n+1} - S_n \leq t \mid J_n = i, J_{n+1} = j)$ is the conditional distribution function of the sojourn time in the state i given that the next visited state is j, (with $j \neq i$). Let us also, define the distribution function $H_i(t) := \sum_{j\in E} Q_{ij}(t)$ and its mean value m_i, which is the mean sojourn time of X in state i. In general, Q_{ij} is a subdistribution, i.e., $Q_{ij}(\infty) \leq 1$, hence H_i is a distribution function, $H_i(\infty) = 1$, and $Q_{ij}(0-) = H_i(0-) = 0$.

Remark 2.2. A special case of semi-Markov processes is the one where $F_{ij}(\cdot)$ does not depend on j, i.e., $F_{ij}(t) \equiv F_i(t) \equiv H_i(t)$, and

$$Q_{ij}(t) = p_{ij}F_i(t).$$

Any general semi-Markov process can be transformed into one of this kind (see, e.g., [17]).

Example 2.1. A Markov process with state space $E = \mathbb{N}$ and generating matrix $\mathbf{A} = (a_{ij})_{i,j\in E}$ is a special semi-Markov process with semi-Markov kernel

$$Q_{ij}(t) = \frac{a_{ij}}{a_i}(1 - e^{-a_i t}), \quad i \neq j, \quad a_i \neq 0,$$

where $a_i := -a_{ii}$, $i \in E$, and $Q_{ij}(t) = 0$, if $i = j$ or $a_i = 0$. In this case, the transition function of the EMC is $p_{ij} = a_{ij}/a_i$ and we recover an exponential distribution for the conditional distribution function of the sojourn time such as $F_i(t) = 1 - \exp(-a_i t)$, with $t \geq 0$.

A usual restriction that fits practical applications is to assume a *regularity* condition for the SMP of interest. To specify this condition, we introduce the counting process $(N(t), t \geq 0)$ which counts the number of jumps of X in the time interval

$(0, t]$, by $N(t) := \sup\{n \geq 0 : S_n \leq t\}$. Also, define $N_i(t)$ to be the number of visits of X to state $i \in E$ in the time interval $(0, t]$. That is to say,

$$N_i(t) := \sum_{n=0}^{N(t)} \mathbf{1}_{\{J_n = i\}} = \sum_{n=0}^{\infty} \mathbf{1}_{\{J_n = i, S_n \leq t\}}.$$

If we consider the (eventually delayed) renewal process $(S_n^i)_{n \geq 0}$ of successive times of visits to state i, then $N_i(t)$ is the counting process of renewals. Now, a SMP X is said to be regular if

$$\mathbb{P}_i(N(t) < \infty) = 1,$$

for any $t \geq 0$ and any $i \in E$.

For regular SMPs we have $S_n < S_{n+1}$, for any $n \in \mathbb{N}$, and $S_n \to \infty$. In the sequel, we are concerned with regular SMPs.

Let us also have a brief discussion about the nature of the different states of an MRP. An MRP is irreducible, if, and only if, its EMC (J_n) is irreducible. A state i is recurrent (transient) in the MRP, if, and only if, it is recurrent (transient) in the EMC. For an irreducible finite MRP, a state i is positive recurrent in the MRP, if, and only if, it is recurrent in the EMC and if for all $j \in E$, $m_j < \infty$. If the EMC of an MRP is irreducible and recurrent, then all the states are positive-recurrent, if, and only if, $m := \nu m := \sum_i \nu_i m_i < \infty$, and null-recurrent, if, and only if, $m = \infty$ [where ν is the stationary probability of EMC (J_n)]. A state i is said to be periodic with period $a > 0$ if $G_{ii}(\cdot)$ (the distribution function of the random variable $S_2^i - S_1^i$) is discrete concentrated on $\{ka : k \in \mathbb{N}\}$. Such a distribution is also said to be periodic. In the opposite case it is called aperiodic. Note that the term *period* has a completely different meaning from the corresponding one of the classical Markov chain theory.

2.2 Markov Renewal Equation

An essential tool in semi-Markov theory is the MRE which can be solved using the so-called *Markov renewal function*. To unveil this function, we first need to introduce the convolution in the Stieljes-sense.

For $\phi(i, t)$, $i \in E, t \geq 0$ a real-valued measurable function, the convolution of ϕ by Q is defined by

$$Q * \phi(i, t) := \sum_{k \in E} \int_0^t Q_{ik}(ds)\phi(k, t - s).$$

Now, consider the n-fold convolution of Q by itself. For any $i, j \in E$,

$$Q_{ij}^{(n)}(t) = \begin{cases} \sum_{k \in E} \int_0^t Q_{ik}(ds) Q_{kj}^{(n-1)}(t-s) & n \geq 2, \\ Q_{ij}(t) & n = 1, \\ \delta_{ij} \mathbf{1}_{\{t \geq 0\}} & n = 0, \end{cases}$$

where δ_{ij} is the Kronecker delta, that is to say, $\delta_{ij} = 1$ if $i = j$, 0 otherwise.

It is easy to prove (e.g., by induction) the following fundamental equality

$$Q_{ij}^{(n)}(t) = \mathbb{P}_i(J_n = j, S_n \leq t),$$

where, as usual, $\mathbb{P}_i(\cdot)$ means $\mathbb{P}(\cdot \mid J_0 = i)$, and $\mathbb{E}_i$ is the corresponding expectation. The Markov renewal function $\psi_{ij}(t),\ i, j \in E, t \geq 0$ is defined by

$$\begin{aligned} \psi_{ij}(t) &:= \mathbb{E}_i[N_j(t)] = \mathbb{E}_i \sum_{n=0}^{\infty} \mathbf{1}_{\{J_n = j, S_n \leq t\}} \\ &= \sum_{n=0}^{\infty} \mathbb{P}_i(J_n = j, S_n \leq t) = \sum_{n=0}^{\infty} Q_{ij}^{(n)}(t). \end{aligned}$$

In matrix form, this writes

$$\psi(t) = (I(t) - Q(t))^{(-1)} = \sum_{n=0}^{\infty} Q^{(n)}(t).$$

This can also be written as

$$\psi(t) = I(t) + Q * \psi(t), \tag{2}$$

where $I(t) = I$ (the identity matrix), if $t \geq 0$ and $I(t) = 0$, if $t < 0$.

Equation (2) is a special case of what is called a MRE. A general MRE is one of the following form:

$$\Theta(t) = g(t) + Q * \Theta(t), \tag{3}$$

where $\Theta(t) = (\Theta_{ij}(t))_{i,j \in E}$, $g(t) = (g_{ij}(t))_{i,j \in E}$ are matrix-valued measurable functions, with $\Theta_{ij}(t) = L_{ij}(t) = 0$ for $t < 0$. The function $g(t)$ is a given while $\Theta(t)$ is unknown.

The following Theorem bring some results about existence and unicity of a solution to MRE as (3).

Theorem 2.1. *(Markov Renewal Theorem [25]) Let* $\mathbf{B}$ *be the space of all locally bounded, on* $\mathbb{R}_+$*, matrix functions* $\Theta(t)$*, i.e.,* $\|\Theta(t)\| = \sup_{i,j} |\Theta_{i,j}(t)|$ *is bounded on sets* $[0, \xi]$*, for every* $\xi \in \mathbb{R}_+$*. Also, denote by* $\overline{H}_i(t) := 1 - H_i(t)$*. Let the following conditions be fulfilled:*

(1) The EMC (J_n) is ergodic, i.e., irreducible and positive-recurrent, with stationary probability $\nu = (\nu_i, i \in E)$.
(2) The mean sojourn time in every state is finite, i.e., for every $i \in E$,

$$m_i := \int_0^\infty \overline{H}_i(t)dt < \infty, \quad \text{and} \quad m := \sum_{i \in E} \nu_i m_i > 0.$$

(3) The distribution functions $H_i(t)$, $i \in E$, are nonperiodic.
(4) The functions $L_{ij}(t)$, $t \geq 0$, are direct Riemann integrable, i.e., they satisfy the following two conditions, for any $i, j \in E$:

$$\sum_{n \geq 0} \sup_{n \leq t \leq n+1} |L_{ij}(t)| < \infty,$$

and

$$\lim_{\Delta \downarrow 0} \left\{ \Delta \sum_{n \geq 0} \left[\sup_{n\Delta \leq t \leq (n+1)\Delta} L_{ij}(t) - \inf_{n\Delta \leq t \leq (n+1)\Delta} L_{ij}(t) \right] \right\} = 0.$$

Then Eq. (3) has a unique solution $\Theta = \psi * L(t)$ belonging to **B**, *and*

$$\lim_{t \to \infty} \Theta_{ij}(t) = \frac{1}{m} \sum_{\ell \in E} \nu_\ell \int_0^\infty L_{\ell j}(t)dt. \tag{4}$$

Finally, we unveil another very important function to characterize the process, namely, the semi-Markov transition function

$$P_{ij}(t) := \mathbb{P}(X_t = j \mid X_0 = i), \quad i, j \in E, t \geq 0,$$

which is the conditional marginal law of the process. It can be shown that P verifies a particular MRE, which will be essential in the development of our probability assessments in the next section.

Proposition 2.1. *The transition function $P(t) = (P_{ij}(t))$ satisfies the following MRE*

$$P(t) = I(t) - H(t) + Q * P(t),$$

which, under Conditions (1–3) of Theorem 2.1, has the unique solution

$$P(t) = \psi * (I(t) - H(t)),$$

and, for any $i, j \in E$,

$$\lim_{t \to \infty} P_{ji}(t) = \nu_i m_i / m =: \pi_i.$$

Here $H(t) = diag(H_i(t))$ is a diagonal matrix.

It is worth noticing that, in general, the stationary distribution π of the SMP X is not equal to the stationary distribution ν of the EMC (J_n). Nevertheless, we have $\pi = \nu$ when, for example, m_i is independent of $i \in E$.

3 A Dynamical Differential System for Structural Reliability Study

We now turn back to the main motivation of the chapter, that is, investigating the following differential system:

$$\dot{Z}_t = C(Z_t, X_t), \qquad Z_0 = z. \tag{5}$$

To ensure that (5) owns a unique solution, we set the usual regularity assumption for C, that is, $C : \mathbb{R}_+ \times E \longrightarrow \mathbb{R}_+$ is measurable and Lipschitz w.r.t. the first argument, uniformly on the second.

We also set some restrictions for the reliability study of (5). Looking toward the description of what is understood here as the degradation process, the following assumptions naturally rise from physical considerations:

- the level of degradation is positive and increases across time;
- the failure domain is defined by a threshold $\Delta \in \mathbb{R}_+^* = (0, \infty)$.

These assumptions require that the function $C : (x, i) \rightarrow C(x, i)$ is strictly positive for all $x \in \mathbb{R}_+$, $i \in E$. Moreover, we set $\Delta > z > 0$ to ensure that the system does not starts in a failure state.

Now, to be specific with the reliability analysis of (5), we define $U = [z, \Delta)$ the set of working states with $0 < z < \Delta$ and $D = [\Delta, \infty)$ the set of down states. Assuming a nonreparable system and thanks to the continuous, increasing evolution of Z, failure occurs as soon as point Δ is reached: this point is an absorbing state of the system. The failure time can thus be written as a function of the coupled process (Z, X):

$$\tau = \inf\{t \geq 0 : Z_t \in D\} \equiv \inf\{t \geq 0 : (Z_t, X_t) \in D \times E\}. \tag{6}$$

The reliability and the cumulative distribution function (CDF) of τ turn to

$$R(t) = \mathbb{P}((Z_t, X_t) \in U \times E) = 1 - F_\tau(t). \tag{7}$$

In the remaining of this section, we interpret (Z, X) as an *extended* MRP. We then derive a solvable MRE whose solution is the transition function of the (Z_t, X_t). Then, reliability (7) has a closed-form which can be computed numerically. A numerical illustration is investigated that confirmed our theoretical results and that hopefully bring some insights on the understanding of the semi-Markov kernel associated with (Z, X).

3.1 The Coupled Process as an Extended Markov Renewal Process

In a previous work [4], we considered the system (5) with X a jump Markov process. Here, a more general assumption is made regarding the nature of the perturbing process: we set $X = (X_t, t \geq 0)$ a SMP with finite state space E, which describes random variations in the environment of $Z = (Z_t, t \geq 0)$. The pure jump process X is defined by its semi-Markov kernel

$$Q_{ij}(t) = \mathbb{P}(J_{n+1} = j, S_{n+1} - S_n \leq t | J_n = i), \tag{8}$$

where $i, j \in E$ and $t \geq 0$. As from the previous section, the process $(J_n, S_n, n \in \mathbb{N})$ is the embedded MRP of the SMP X, where $(S_n, n \in \mathbb{N})$ is the random sequence describing the jump times. The random sequence $J_n = X_{S_n}$ is the EMC with transition probabilities $(p_{ij})_{i,j \in E}$, such as $p_{ij} = Q_{ij}(\infty)$. We also put $\alpha_i = \mathbb{P}(X_0 = i)$ the initial distribution of X. Besides, we consider with no loss of generality that the conditional CDF of the sojourn time does not depend on the arrival point j as in Remark 2.2, that is, $F_{ij}(t) \equiv F_i(t)$. The semi-Markov kernel of X thus writes $Q_{ij}(t) = p_{ij} F_i(t)$.

From now, we start to be specific to the couple (Z, X) defined by (5): for any $t < S_1$, we denote by $\varphi_{z,i}(t)$ the deterministic function describing the solution to (5), when $X_0 = i$. Hence, $\varphi_{z,i}(t)$ is the solution before the first jump time of X, conditionally on the starting value $(Z_0, X_0) = (z, i)$. Note that we assume that Z_0 and X_0 are independent.

We are finally ready to associate to (Z, X) the "extended" MRP $(\zeta_n, J_n, S_n, n \in \mathbb{N})$, by extending the "standard" MRP (J_n, S_n) with a third component as follows:

$$\zeta_n = Z_{S_n}, \qquad J_n = X_{S_n}, \qquad n \in \mathbb{N}.$$

As for a usual MRP, we may introduce the appropriate mathematical tools. Thenceforth, consider the semi-Markov kernel associated with the triplet (ζ_n, J_n, S_n): it is denoted by L and defined, for $t > 0$, by

$$L_{ij}(z, B, t) := \mathbb{P}_{z,i}(\zeta_1 \in B, J_1 = j, S_1 - S_0 \leq t), \tag{9}$$

where B is a subset of $\mathscr{B}$, the Borel σ−field of $\mathbb{R}_+$ and $\mathbb{P}_{z,i}(\cdot) := \mathbb{P}(\cdot | Z_0 = z, X_0 = i)$. The Stieltjes-convolution of L with a measurable function ϕ on the space $\mathbb{R}_+ \times E$, denoted by "$*$", is defined by

$$(L * \phi)_{ij}(z, t) = \sum_{k \in E} \int_{\mathbb{R}_+} \int_0^t L_{ik}(z, \mathrm{d}y, \mathrm{d}s) \phi_{kj}(y, t - s),$$

for $i, j \in E$ and $z > 0$. In the same way, the n-fold convolutions of the semi-Markov kernel L are defined recursively. For $n = 0, 1$,

$$L_{ij}^{(0)}(z,B,t) = \mathbb{1}_{\{i=j\}}\mathbb{1}_B(z)\mathbb{1}_{\mathbb{R}_+}(t), \qquad L_{ij}^{(1)}(z,B,t) = L_{ij}(z,B,t),$$

where $\mathbb{1}_B(x)$ is the indicator function, i.e., $\mathbb{1}_B(x) = 1$ if $x \in B$, 0 otherwise. For $n \geq 2$, the n-fold convolution turns to

$$L_{ij}^{(n)}(z,B,t) := (L * L^{(n-1)})_{ij}(z,B,t).$$

The Markov renewal function Ψ of the triplet is

$$\Psi_{ij}(z,B,t) = \sum_{n\geq 0} L_{ij}^{(n)}(z,B,t).$$

In the case at hand, we consider that (ζ_n, J_n, S_n) is a *normal* MRP, that is, $\Psi_{ij}(z,B,t) < \infty$ for any fixed $t > 0, z > 0, B \in \mathscr{B}$ and $i, j \in E$, which implies also that the SMP Z is regular.

For the process (ζ_n, J_n, S_n), a MRE has the following form

$$\Theta_{ij}(z,B,t) = g_{ij}(z,B,t) + (L * \Theta)_{ij}(z,B,t), \tag{10}$$

where $g_{ij}, i, j \in E$ are known functions and $\Theta_{ij}, i, j \in E$ are the unknown functions. The solution to (10), thanks to the results of the previous section, is

$$\Theta_{ij}(z,B,t) = (\Psi * g)_{ij}(z,B,t). \tag{11}$$

3.2 The Transition Function

Consider the transition function P of the couple process (Z, X), defined by

$$P_{ij}(z,B,t) := \mathbb{P}_{z,i}(Z_t \in B, X_t = j), \qquad i, j \in E, B \in \mathscr{B}. \tag{12}$$

We aim at building a MRE suitable for P. For this purpose, we first need a closed-form expression for L. This is achieved in the following Lemma.

Lemma 3.1. *The semi-Markov kernel L of the extended MRP (ζ_n, J_n, S_n) satisfies, for $i \neq j$,*

$$L_{ij}(z,B,\mathrm{d}t) = \delta_{\varphi_{z,i}(t)}(B)Q_{ij}(\mathrm{d}t),$$

where $\delta_x(B)$ is the Dirac distribution, equal to 1 if $x \in B$, 0 otherwise. When $i = j$, we have $L_{ii}(\cdot,\cdot,\cdot) = 0$.

Proof. Conditioning on definition (9), and by definition (8) of Q, we get,

$$L_{ij}(z, B, \mathrm{d}t) = Q_{ij}(\mathrm{d}t) \times \mathbb{P}_{z,i}(\zeta_1 \in B | J_1 = j, S_1 = t).$$

Then, Z_t is fully characterized by $\varphi_{z,i}(t)$ before the first jump time S_1, thus $\mathbb{P}_{z,i}(\zeta_1 \in B | J_1 = j, S_1 = t) = \mathbb{P}_{z,i}(Z_t \in B) = \delta_{\varphi_{z,i}(t)}(B)$, and the result follows. □

Note that, by considering the decomposition $Q_{ij}(t) = p_{ij} F_i(t)$, Lemma 3.1 implies that

$$L_{ij}(z, B, \mathrm{d}t) = \delta_{\varphi_{z,i}(t)}(B) p_{ij} f_i(t) \mathrm{d}t,$$

where $f_i(t) = \mathrm{d}F_i(t)/\mathrm{d}t$ is the conditional probability density function of the sojourn time.

Example 3.1. Consider the special case of X a jump Markov process as defined in the Example 2.1. Then,

$$L_{ij}(z, B, \mathrm{d}t) = a_{ij} e^{-a_i t} \delta_{\varphi_{z,i}(t)}(B) \mathrm{d}t.$$

We may now proceed to the result on the transition function of the coupled process (Z, X).

Proposition 3.1. *The transition function P satisfies the MRE*

$$P_{ij}(z, B, t) = g_{ij}(z, B, t) + (L * P)_{ij}(z, B, t),$$

whose unique solution is $P_{ij}(z, B, t) = (\Psi * g)_{ij}(z, B, t)$, with

$$g_{ij}(t) = [1 - F_i(t)] \mathbb{1}_B(\varphi_{z,i}(t)) \mathbb{1}_{\{i=j\}}. \tag{13}$$

Proof. From (12), it holds that

$$P_{ij}(z, B, t) = \underbrace{\mathbb{P}_{z,i}(Z_t \in B, X_t = j, S_1 > t)}_{P_1} + \underbrace{\mathbb{P}_{z,i}(Z_t \in B, X_t = j, S_1 \leq t)}_{P_2}.$$

Before the first jump, $X_t = X_0$ and Z_t evolves according to $\varphi_{z,i}(t)$. Thus, we easily see that $P_1 = [1 - F_i(t)] \mathbb{1}_B(\varphi_t(z, i)) \mathbb{1}_{\{i=j\}}$. From Total Probability Theorem, P_2 turns to

$$P_2 = \sum_{\substack{k \in E \\ k \neq i}} \int_0^t \mathbb{P}_{z,i}(Z_t \in B, X_t = j | J_1 = k, S_1 = s) \mathbb{P}_{z,i}(J_1 = k, S_1 \in \mathrm{d}s).$$

By definition (8), $\mathbb{P}_{z,i}(J_1 = k, S_1 \in \mathrm{d}s) = Q_{ik}(\mathrm{d}s)$. Noticing that $\mathbb{P}_{z,i}(Z_t \in B, X_t = j | J_1 = k, S_1 = s) = P_{kj}(\varphi_{z,i}(s), B, t - s)$, then P_2 is fully known. Thus, with L given as in Lemma (3.1), expression $P_1 + P_2$ turns to

$$P_{ij}(z, B, t) = [1 - F_i(t)] \mathbb{1}_B(\varphi_{z,i}(t)) \mathbb{1}_{\{i=j\}}$$
$$+ \sum_{k \in E} \int_{\mathbb{R}_+} \int_0^t L_{ik}(z, \mathrm{d}y, \mathrm{d}s) P_{kj}(y, B, t - s).$$

This last equation is of the general form of (10), with g equaling (13). Since the first term into the right-hand side is bounded, its solution is given by (11) and is unique. □

3.3 Application to Reliability Calculus

Enjoying a closed-form for the transition function P, this section intends to show its implication for reliability calculus. Recall that $U = [z, \Delta)$ is the set of working states and $D = [\Delta, \infty)$ is the set of down states. The reliability function is easily expressed as a function of the transition function P of the couple:

$$R(t) = \mathbb{P}((Z_t, X_t) \in U \times E) = \sum_{i,j \in E} \alpha_i P_{ij}(z, U, t).$$

Through Proposition 3.1, P is known. Hence R (as well as F_τ) is fully characterized:

$$R(t) = 1 - F_\tau(t) = \sum_{i,j \in E} \alpha_i \times (\Psi * g)_{ij}(z, U, t).$$

The computation of R thus requires Ψ, determined by summing the n-fold convolutions of the kernel L, which is the essential block of the whole process. We have a closer look to this quantity in the next paragraph.

3.3.1 Insights on the Semi-Markov Kernel

The kernel L can be calculated at a given time point $t > 0$ for the Borel subset U of working state, by integrating the expression given in Lemma 3.1. To this hand, we introduce the quantity

$$t_{z,i}(\Delta) = \inf \left\{ t \geq 0 : \varphi_{z,i}(t) \geq \Delta \right\},$$

which represent the (deterministic) time for the system to enter D when no jump is observed, and when the system starts from $(Z_0, X_0) = (z, i)$. Then, the kernel is easily seen to equals

$$L_{ij}(z, U, t) = p_{ij} \int_0^t f_i(s) \mathbb{1}_U(\varphi_{z,i}(s)) \mathrm{d}s = p_{ij} F_i \left(\min \left\{ t, t_{z,i}(\Delta) \right\} \right).$$

Conversely, the same kind of computation holds when considering the set of down states D, and we have

$$L_{ij}(z, D, t) = p_{ij} \left(F_i(t) - F_i(t_{z,i}(\Delta)) \right) \mathbb{1}_{\{t > t_{z,i}(\Delta)\}}.$$

To illustrate this, let us consider again the special case of a Markov jump process.

Example 3.2. Assume X is a jump Markov process as defined in Example 2.1. Then, we have the following closed-form for the kernel L when considering subset U and D:

$$L_{ij}(z, U, t) = \frac{a_{ij}}{a_i} \left(1 - e^{-a_i \min(t_{z,i}(\Delta), t)} \right),$$

and

$$L_{ij}(z, D, t) = \frac{a_{ij}}{a_i} \left(e^{-a_i t} - e^{-a_i t_{z,i}(\Delta)} \right) \mathbb{1}_{\{t > t_{z,i}(\Delta)\}}.$$

These expressions pave the way for the numerical implementation that leads to the evaluation of the reliability, as detailed in the next paragraph.

3.3.2 Numerical Implementation

The numerical calculation of R successively requires the kernel L, the n-fold convolutions $L^{(n)}$ for each $n \geq 0$, the Markov renewal function Ψ built upon the $L^{(n)}$ and the transition function P, by a convolution between g and Ψ. Since convolution products are time-consuming, any simplification would mean a great time-saving. By Lemma 3.1, the n-fold convolution of L turns to

$$L_{ij}^{(n)}(z, B, t) = \sum_{\substack{k \in E \\ k \neq i}} p_{ik} \int_0^t f_i(s) L_{kj}^{(n-1)}(\varphi_{z,i}(s), B, t-s) \mathrm{d}s, \tag{14}$$

hence removing the integral on $\mathbb{R}_+$, thanks to the Dirac distribution. Since our main interest is the reliability, we compute P just for the subset $B \equiv U$, that is,

$$P_{ij}(z, U, t) = \int_U \int_0^t \Psi_{ij}(z, \mathrm{d}y, \mathrm{d}s) f_j(t-s) \mathbb{1}_U(\varphi_{y,j}(t-s)). \tag{15}$$

Indeed, the sum on E has been removed thanks to the structure of g. Furthermore, the integration on $y \in \mathbb{R}_+$ is limited on U since $\mathbb{1}_U(\varphi_{y,j}(t-s))$ is zero elsewhere.

Now, these functions have to be properly discretized to achieve the numerical computation. In the following, a function with an upper index "#" means its discretized version. This discretization must be operated on both intervals $U = [z, \Delta)$ and $[0, t]$, thus we set two numerical partitions

$$\left\{z = y_0 < y_1 < \cdots < y_\ell < \cdots < y_L = \Delta^-\right\}$$
$$\text{and } \{0 = t_0 < t_1 < \cdots < t_m < \cdots < t_M = t\}.$$

Both L and M, being the respective numbers of discretization steps for $[z, \Delta)$ and $[0, t]$, have to be sufficiently large. When $L, M \to \infty$ each numerical function tends to the associated "true" one. For instance, when $L, M \to \infty$, then $L^\# \to L$ uniformly w.r.t a given matrix norm, for example, $||L|| = \max_{i,j} L_{ij}(z, y, t)$ with t, z, B fixed. Hence, the discrete (numerical) version of (15) is

$$P^\#_{ij}(z, U, t) = \sum_{y_\ell \in [z,\Delta)} \sum_{t_m \in (0,t]} \Delta_{yt} \Psi^\#_{ij}(z, y_\ell, t_m) f_j(t - t_m) \mathbb{1}_{\varphi_{y_\ell, j}(t-t_m)}(U),$$

where $\Delta_{yt} \Psi^\#_{ij}(z, y_\ell, t_m)$ is the only unknown, which stands for the numerical evaluation of $\Psi(z, \mathrm{d}y, \mathrm{d}s)$ in (15). It can be evaluated through

$$\Delta_{yt} \Psi^\#_{ij}(z, y_\ell, t_m) = \sum_{n \geq 0} \Delta_{yt} L^{\#(n)}(z, y_\ell, t_m).$$

The difference $\Delta_{yt} L^{\#(n)}$ is calculated by finite differences on y and t:

$$\begin{aligned}\Delta_{yt} L^{\#(n)}(z, y_\ell, t_m) = {} & [L^{\#(n)}(z, y_\ell, t_m) - L^{\#(n)}(z, y_{\ell-1}, t_m)] \\ & - [L^{\#(n)}(z, y_\ell, t_{m-1}) - L^{\#(n)}(z, y_{\ell-1}, t_{m-1})].\end{aligned}$$

Each element in $L^{\#(n)}$ is obtained by the discretized version of (14):

$$L^{\#(n)}_{ij}(z, y_\ell, t_m) = \sum_{\substack{k \in E \\ k \neq i}} p_{ik} \sum_{t_m \in (0,t]} f_i(t_m) L^{\#(n-1)}_{kj}(\varphi_{z,i}(t_m), y_\ell, t - t_m) \Delta t_m,$$

with $\Delta t_m = t_m - t_{m-1}$, the time-step discretization. Finally, we point out that the sum on the n-fold convolutions of the kernel in the evaluation of $\Psi^\#$ is truncated from the rank n^*, provided that $||L^{\#(n^*)}|| < \varepsilon$. We put ε a small real number, chosen closed to the machine precision. Note that the integer n^* is finite since $L^{\#(n)}_{ij}(z, y, t) \xrightarrow[n \to \infty]{} 0$ for a normal MRP with fixed values of $i, j \in E, t > 0, z > 0$ and $y \in [z, \Delta]$.

3.4 Numerical Illustration

As an illustration to the results and the methodology presented along this section, we suggest to study the process Z governed by

$$\dot{Z}_t = aZ_t \times c(X_t), \qquad Z_0 = z, \tag{16}$$

with $a = 0.01$, $z = 1$, $\Delta = 10$. To fix the idea, we set X a five-state jump *Markov* process with $E = \{1, 2, 3, 4, 5\}$ and a matrix generator given by

$$\mathbf{A} = \begin{pmatrix} -0.2 & 0.16 & 0 & 0.04 & 0 \\ 0.12 & -0.2 & 0.08 & 0 & 0 \\ 0.14 & 0 & -0.2 & 0 & 0.06 \\ 0 & 0.07 & 0 & -0.1 & 0.03 \\ 0 & 0 & 0.05 & 0.05 & -0.1 \end{pmatrix}$$

The initial law is $\alpha = (1/4\ 1/2\ 1/4\ 0\ 0)$.Finally, the function $c : \{1, 2, 3, 4, 5\} \to \{0.5, 1, 1.5, 2, 4\}$ is a one-to-one mapping introduced to "control" the randomizing process X. Note that the multiplicative form of system (16) is reminiscent of stochastic crack-growth modeling and is suitable to describing a wide family of degradation processes.

Before we carry on reliability computations, we suggest to get a better insight into the semi-Markov kernel L of (Z, X) as defined in system (16): X being Markovian, the expressions of L on subsets U and D exactly match the Example 3.2 and can be straightforwardly computed. Rather than plotting L, consider the functions $H_i(z, B, t) = \sum_{j\in E} L_{ij}(z, B, t)$ and the CDF

$$\mathbf{H}_i(t) := \mathbb{P}_i(S_1 \le t) = H_i(z, U, t) + H_i(z, D, t).$$

The function $\mathbf{H}_i$ is the CDF of the sojourn time for the jump process X to be in the state $X_0 = i$. The function $H_i(z, B, \cdot)$ is a sub distribution: when $B \equiv U$, $H_i(z, U, t)$ represents the probability for the system, starting from (z, i), to remain in a safe state when X is jumping for the first time. Similarly, $H_i(z, U, t)$ describes the probability for the system to be in a failure state when the first jump occurs. These remarks are illustrated in Fig. 2, representing $H_i(z, B, \cdot)$ for $i = 1, 5$, respectively on U and D. The function $H_1(z, D, \cdot)$ is approximately zero, meaning that $H_1(z, U, \cdot) \cong \mathbf{H}_1(\cdot)$ is a CDF. Conversely, starting from $(z, 5)$, this probability is strictly greater than zero. As a matter of fact, state 5 for X_t corresponds to a "shock" inducing a strong multiplicative change to the system (16): Z increases a lot faster to the absorbing point Δ. Also remark that we graphically establish that $\mathbf{H}_5(t) = H_5(z, U, \cdot) + H_5(z, D, \cdot)$ is a CDF.

Let us now evaluate the reliability of system described by Eq. (16) through a Markov renewal argument. To do this, the numerical resolution of the MRE is performed with $M = L = 100$ points of discretization.

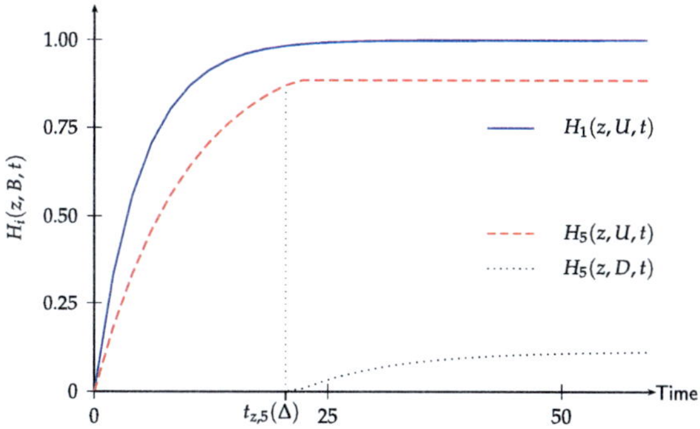

Fig. 2 Function $H_i(z, B, \cdot)$ associated with the semi-Markov kernel

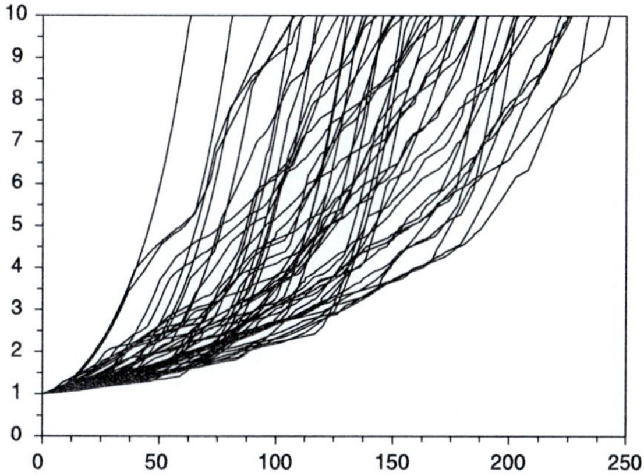

Fig. 3 50 randomly simulated paths of Z_t

As a comparison, we compute the reliability thanks to the usual Monte-Carlo method, which consists in simulating a large number of paths of Z and counting when the state $\{\Delta\}$ is reached or not. This principle is illustrated in Fig. 3 for $K = 50$ trajectories. By the way, these trajectories helps to catch the nature of this particular numerical illustration.

We use the empirical estimator computed on $K = 50,000$ paths $(Z_t^k)_{k=1,\ldots,K}$ simulated through Monte-Carlo techniques, that is $\widehat{R}(t) = \frac{1}{K}\sum_{k=1}^{K} \mathbb{1}_{\{Z_t^k < \Delta\}}$. This estimator is compared with the direct calculus of R through the MRE developed here. Results can be found on Fig. 4 where the Monte-Carlo estimator is used as a reference for sanity-check of the validity of both theoretical results and numerical implementation. One can take note of the very good similarity between the reliability curves obtained via the two methods.

Moreover we represent in Fig. 5 an evaluation of f_τ, the probability density function of the failure time. With the very same kind of argument that we devel-

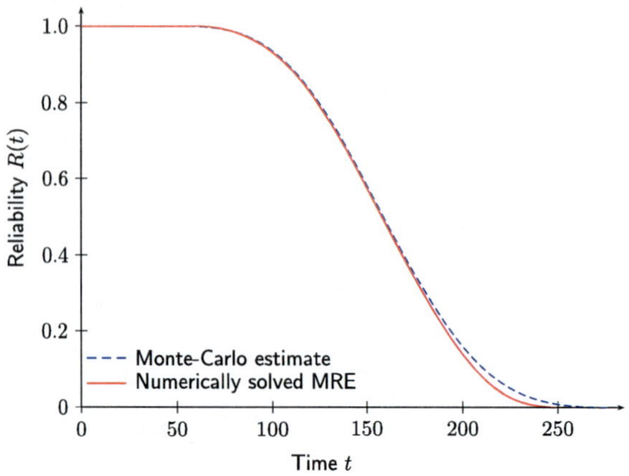

Fig. 4 Comparing the Monte-Carlo estimator to the numerically solved Markov renewal equation for reliability function

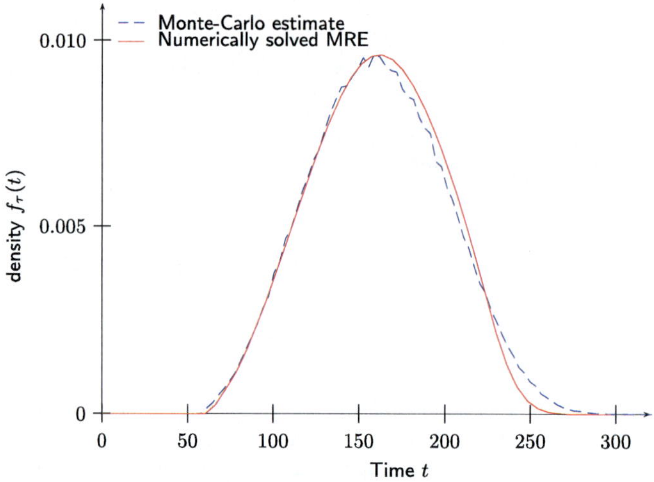

Fig. 5 Comparing the Monte-Carlo estimator to the numerically solved Markov renewal equation for the density function of the failure time

oped, f_τ can be obtained by solving a MRE or via Monte-Carlo simulation. This is done through some routine calculus starting from the basic fact that

$$f_\tau(t) = -\frac{\mathrm{d}}{\mathrm{d}t}R(t).$$

Again, we acknowledge the very good correlation between our proposal and the Monte-Carlo estimate. Moreover, we observe that our proposal is smoother as compared to the Monte-Carlo approach, which would require a huge number of simulation to get a similar result.

4 Statistical Inference

This section addresses the estimation issue related to the system described by Eq. (5) in Sect. 3. It partially follows the exposition given in [5], where we studied the simpler Markov case for X in a real data study related to crack-growth analysis. The SMP estimation developed here consists of new material. We also provide statistical methods which are more robust regarding the estimation of the trajectories of the jump process, derived from the literature related to the segmentation/clustering of piecewise constant signals.

Recall the observation scheme, as plotted in Fig. 1: the only data that can reasonably be made available from experimental feedback are recorded paths of the degradation process Z. Typically, the process is observed from a starting point z which represents the smallest level of degradation that can be characterized, until it reaches the failure threshold Δ at a random time τ. This is exactly how measurements of crack-growth are acquired in [27]. Thus, the sample training data are only composed by some K paths $\left\{Z_t^k, t = 0, \ldots, \tau^k\right\}_{k=1}^K$ where τ^k is the hitting time for the k-th path.

Basically, the most ambitious goal that we would like to aim is to successively estimate (1) the function C and (2) the randomizing SMP X_t, only by considering paths of Z with right censoring. We propose in this section a first methodological effort in that sense, which, requires some additional assumptions to carry out the inference process:

- the observed paths Z are independent and identically distributed;
- the function $C \equiv C_\theta : (z, i) \to C(z, i)$ is a known parametric function, with parameters θ remaining unknown;
- there exists a function G_θ giving X as a function of Z and its first derivative, that is, the function C_θ in the dynamical system (5) may reverse so as

$$X_t = G_\theta\left(Z_t, \dot{Z}_t\right). \tag{17}$$

The first assumption of i.i.d. paths is quite usual in statistics and well motivated in the framework of structural reliability. The second assumption (the parametric modeling of C) clearly eases the inference process. Yet we underline that it has been initially motivated by application purpose. In fact, the modeling of a particular degradation process often owns a physical framework in which scientists have an idea about the general form of C_θ, with θ the parametric adjustment which remains the only unknown. Finally, concerning the third assumption, it is acknowledged that such a function does not always exists, yet it is required to evaluate the paths of the jumping random component. Indeed, when the stochastic process X is a linear additive or a multiplicative term in the function C, we may easily find the corresponding function G, which concerns a broad family of problems. Besides, this is truly the case for most of the stochastic crack-growth formulations that we met and that initially motivated this work [19, 22, 27].

The rest of this section splits into two parts:

(1) First, we describe the estimation of C, which relies on the Bogolyubov's averaging principle [2]. A regression analysis can be performed on the asymptotic, deterministic system so as to estimate the fixed parameter θ in function C.
(2) Second, we address the estimation of the random component, that is, the jump SMP X, which is not directly observed. Once some paths of X and its state space are recovered, we can build the likelihood function, keeping in mind that the paths are defined on randomly censored time intervals. The semi-Markov kernel Q of X are then estimated by maximizing an approached likelihood function.

We finally give a numerical application to illustrate the whole estimation scheme.

4.1 Bogolyubov's Averaging Principle

An approximation of Z is obtained by analyzing the system (5) in a series scheme as in [12], that is by studying the weak convergence, when $\varepsilon \to 0$, of

$$\frac{\mathrm{d}Z_t^\varepsilon}{\mathrm{d}t} = C_\theta(Z_t^\varepsilon, X_{t/\varepsilon}), \qquad Z_0^\varepsilon = z, \tag{18}$$

where X is assumed to be ergodic and θ are the parameters of C. In fact, the change of scale $t \to t/\varepsilon$ is performed for X in order to see the behavior of the dynamical system when the random component X just adds the information it would add after a very long time of observation of (18), since $t/\varepsilon \to \infty$ when $\varepsilon \to 0$. This so-called *averaging approximation* was first introduced by Bogolyubov [2] who showed that (18) converges weakly when $\varepsilon \to 0$ to the following deterministic system

$$\frac{\mathrm{d}\widetilde{z}_t}{\mathrm{d}t} = \overline{C}_\theta(\widetilde{z}_t), \qquad \widetilde{z}_0 = z, \tag{19}$$

with $\widetilde{z}_t$ the limit deterministic process and $\overline{C}_\theta$ a mean function defined by

$$\overline{C}_\theta(z) = \lim_{T\to\infty} \frac{1}{T} \int_0^T C_\theta(z, X_t)\mathrm{d}t, \qquad a.s.$$

An illustration of this principle is provided in Fig. 6, where $\widetilde{z}_t$ is represented among a set of sample paths of Z.

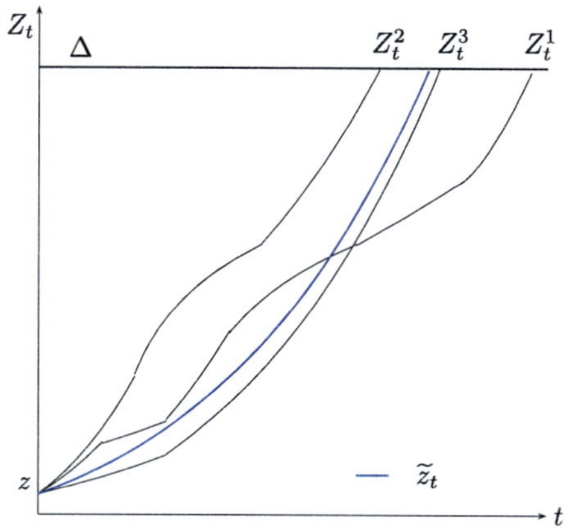

Fig. 6 Application of the Bogolyubov's averaging principle

In the particular case where X is an ergodic SMP with a stationary law π, we have

$$\overline{C}_\theta(z) = \sum_{i\in E} C_\theta(z,i)\pi_i.$$

Through this averaging technique, we have a limit deterministic system (19) associated with stochastic differential system (5). The fixed parameters θ appearing in the function C_θ are the same as the ones appearing in $\overline{C}_\theta$ but in (19) the random part was "eliminated": with the K sample paths Z_t^k, we can perform a classical regression analysis on (19) to estimate the fixed parameters θ.

4.2 The Semi-Markov Process Estimation

The SMP X is fully characterized by its kernel Q and its initial law α. Meanwhile, prior to any estimation of Q or α, some representations of the paths $\{X_t^k, t = 0, \ldots, \tau^k\}_{k=1}^K$ are needed.

4.2.1 Trajectories Estimation

Assume that there exists a function G_θ as defined in (17); hence, we may obtain a first estimation for the X_t^k's through

$$\widetilde{X}_t^k = G_\theta\left(Z_t^k, \widehat{\dot{Z}}_t^k\right), \tag{20}$$

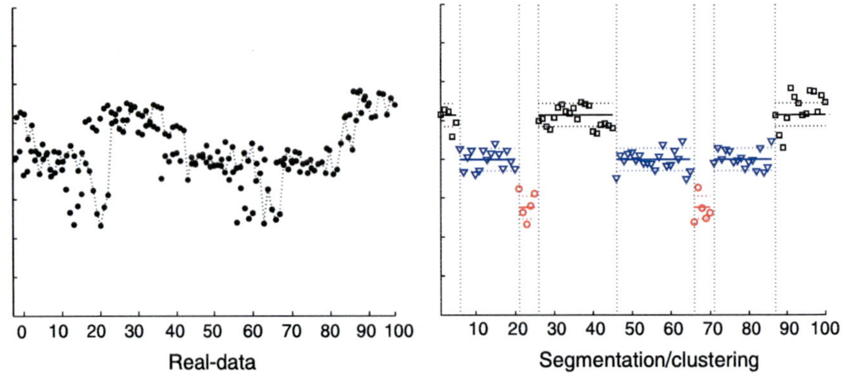

Fig. 7 Illustration of a segmentation/clustering algorithm (*Source* Picard [23])

where the derivative of Z_t can be estimated by various straightforward methods, e.g., the secant method

$$\hat{\dot{Z}}_t^k = \frac{Z_{t+\Delta t}^k - Z_t^k}{\Delta t},$$

with Δt being the time discretization step of the data set.

Note that the parameters θ are required, whose estimation could be performed relying on the averaging principle argument just developed above. Hence, by (20), we basically extract from the trajectories of Z the "random" part that is unexplained by the averaging, deterministic process in (19). By this mean, we obtain some noisy paths taking their values in $\mathbb{R}$, in which values may be quite nearby, as illustrated in Fig. 7. Our model requires a finite state space for the underlying SMP X with some piecewise-constant shape paths, thus it is appropriate to "regroup" the values which are very close from each other to an unique state. This problem can be interpreted as the widely studied segmentation/clustering problem: basically, one wishes to perform (1) the segmentation of a signal assumed to be piecewise constant into, says, q change-points corresponding to the jumps of X and (2) the clustering of the q segments into, says, p clusters corresponding to the states of X.

The segmentation/clustering process is illustrated in Fig. 7 (from Picard [23]): the segmentation is performed on the x-axis while the clustering is performed in the y-axis.

Traditionally, this problem has been studied using hidden Markov models. This is a quite well-studied issue where the segmentation step is usually treated via dynamic programming and the clustering step may be treated through various algorithms such as the popular K-means algorithm. We adopt this naive approach for our problem (see e.g. [5]), yet we want to underline the fact that the segmentation/clustering problem received much attention recently. As a matter of fact, the treatment of huge amount of data with signal lengths up to the million of entries has been required for bioinformatics purpose. Consequently, very powerful variants and implementations of the segmentation/clustering problem have been investigated, regarding the analysis of CGH microarray data. Authors use the traditional HMM modeling (e.g.,[10]) and also mixture modeling (e.g., [14, 23]). They provide very competitive and well integrated model selection approaches to chose both the number of segment p and the number of cluster q. Our problem of paths estimation of X is in very small dimension as compared to the problem of biological data, and can be treated very efficiently. We thus rely on these approaches to process the noisy paths $\{\widetilde{X}_t^k, k = 1\dots, K\}$ from (17), thus leading to K piecewise constant approximated paths $\{\widehat{X}_t^k, k = 1, \dots, K\}$ defined on a finite state space E. The $\widehat{X}^k$s are then used for further estimations linked to the SMP X, namely for estimating its initial distribution α and its kernel Q. This issue is addressed in the following paragraph, based upon an approached maximum likelihood estimator which is equivalent to empirical estimators of the semi-Markov kernel.

4.2.2 K-Histories Empirical Estimators

For clarity purpose, we drop the "hat" on the $\widehat{X}^k$s an related quantities along this section. Note that the writing of the likelihood greatly simplifies when writing a path of X as an ordered sequence:

$$\mathscr{H}_\tau = \big((J_0, W_0), \dots (J_{N(\tau)-1}, W_{N(\tau)-1}), (J_{N(\tau)}, U_\tau)\big),$$

where

- $N(\tau)$ is the number of jumps on $[0, \tau]$,
- $J_n = X_{S_n}, n \in \mathbb{N}$ are the visited states,
- $W_n = S_{n+1} - S_n, n \in \mathbb{N}$ are the sojourn times,
- $U_\tau = \tau - S_{N(\tau)}$.

The density $f_{\mathscr{H}_\tau}$ of $\mathscr{H}_\tau$ is function of $f_\tau(t)$, the density of τ:

$$f_{\mathscr{H}_\tau}(h_t) = f_{\mathscr{H}_t}(h_t) f_\tau(t),$$

where h_t is a realization of $\mathscr{H}_\tau$.

Consider K independent MRPs $(J_n^k, S_n^k, n \geq 0)$, $\quad k = 1, \dots, K$, defined by the same kernel Q and initial distribution α, and K copies $\tau_k, k = 1, \dots, K$ of τ. The same for N^k, U^k. The likelihood for the K histories writes, for t_k a realization of τ_k,

$$\mathscr{L} = \prod_{k=1}^{K} f_{\mathscr{H}_t}(h_{t_k}^k) \cdot \prod_{k=1}^{K} f_\tau(t_k).$$

As an approximation, we assume τ, $\mathscr{H}$ independent. Then, the maximization of the likelihood does not rely on the term

$$\prod_{k=1}^{K} f_\tau(t_k).$$

Hence, the approached likelihood function associated with $(\mathscr{H}^k, 1 \le k \le K)$ writes

$$\tilde{\mathscr{L}}(K) = \prod_{k=1}^{K} \alpha(J_0^k)\left(1 - \sum_{\ell \in E} Q_{J^k_{N^k(t_k)},\ell}(U^k_{t_k})\right) \times \prod_{\ell=1}^{N^k(t_k)} p_{J^k_{\ell-1},J^k_\ell} dF_{J^k_{\ell-1}J^k_\ell}(X^k_\ell),$$

where we remind the decomposition $Q_{ij}(t) = p_{ij}F_{ij}(t)$.

It is clear that the MLE of the initial distribution is $\widehat{\alpha}(i) = n_i/K$, where n_i is the number of trajectories starting from the state i.

The estimator of the kernel which maximized the approached likelihood is easily written by introducing the additional following statistics:

- $N_i(\tau, K)$ the number of visits in state i observed on the K censored paths:

$$N_i(\tau, K) = \sum_{k=1}^{K} \sum_{n=0}^{N^k(\tau_k)-1} \mathbb{1}_{\{J_n^k = i\}} = \sum_{n=0}^{\infty} \mathbb{1}_{\{J_n^k = i, S^k_{n+1} \le \tau_k\}},$$

- $N_{ij}(\tau, K)$ the number of transitions from state i to state j observed on the K censored paths:

$$N_{ij}(\tau, K) = \sum_{k=1}^{K} \sum_{n=0}^{N^k(\tau_k)-1} \mathbb{1}_{\{J_n^k = i, J^k_{n+1} = j\}} = \sum_{n=0}^{\infty} \mathbb{1}_{\{J_n^k = i, J^k_{n+1} = j, S^k_{n+1} \le \tau_k\}},$$

- $M_{ij}(t; \tau, K)$ the number of time the sojourn in i going to j is less than t on the K censored paths:

$$M_{ij}(t; \tau, K) = \sum_{k=1}^{K} \sum_{n=0}^{N^k(\tau_k)-1} \mathbb{1}_{\{J^k_{n+1} = j, J_n^k = i, W_n^k \le t\}}.$$

We finally get the following estimator by straightforward generalization of Moore and Pyke, provided that $F_\tau \ne \delta_0$:

$$\hat{Q}_{ij}(t; \tau, K) = \hat{p}_{ij}(\tau, K)\hat{F}_{ij}(t; \tau, K),$$

with

$$\hat{p}_{ij}(\tau, K) = \frac{N_{ij}(\tau, K)}{N_i(\tau, K)} \text{ and } \hat{F}_{ij}(t; \tau, K) = \frac{M_{ij}(t; \tau, K)}{N_{ij}(\tau, K)}.$$

In fact, the above estimators are the empirical ones.

4.3 Numerical Illustration

We now wish to provide an numerical example which integrates the whole process of estimation, as well as the probabilistic results depicted in the third section. To this aim, we study the following dynamical system, which is in the same vein as system (16):

$$\dot{Z}_t = aZ_t \times c(X_t), \qquad Z_0 = z. \tag{21}$$

As compared to (16), changes are of two kinds: first, the values of the parameters are $a = 0.02$, $z = 5$, $\Delta = 30$. Second, this is the major difference, the randomizing process X is now a three-state space SMP with $E = \{1, 2, 3\}$. The mapping c is such as $c : (1, 2, 3) \to (0.5, 1, 2)$. We also put

$$\alpha = (1/3\, 2/3\, 0)$$

the initial distribution of X. The associated semi-Markov kernel is such as $Q_{ij}(t) = p_{ij}F_{ij}(t)$, with $\mathbf{P} = (p_{ij})_{i,j\in E}$ the transition matrix and $\mathbf{F}(t) = (F_{ij}(t))_{i,j\in E}$ the distribution of sojourn times, given by

$$\mathbf{P} = \begin{bmatrix} 0 & 1 & 0 \\ 0.9 & 0 & 0.1 \\ 1 & 0 & 0 \end{bmatrix}, \qquad \mathbf{F}(t) = \begin{bmatrix} 0 & \mathscr{E}_1(t) & 0 \\ \mathscr{W}_1(t) & 0 & \mathscr{W}_2(t) \\ \mathscr{E}_2(t) & 0 & 0 \end{bmatrix}.$$

The notation $\mathscr{E}_1, \mathscr{E}_2$ stands for exponential distributions such that

$$\mathscr{E}_i(t) = 1 - \exp\{-\lambda_i t\}, \quad t \geq 0,$$

with parameters λ_1, λ_2 being respectively equal to 0.1 and 0.04. We also denote by $\mathscr{W}_1, \mathscr{W}_2$ some Weibull distributions such that

$$\mathscr{W}_i(t) = 1 - \exp\{-(t/\alpha_i)^{\beta_i}\}, \quad t \geq 0,$$

with parameters $(\alpha_i, \beta_i)_{i=1,2}$ being respectively equal to (8, 2) and (4, 0.5).

The whole estimation process sums-up as follows: denoting by $a_0 = a \times \mathbb{E}_\pi[X_t]$ with $\mathbb{E}_\pi$ the expectation regarding the stationary law π of X, the Bogolyubov's averaging principle leads to a very simple deterministic process defined by

$$\widetilde{z}_t = z\exp\{a_0 t\}.$$

Taking the log, we perform a simple least-squared analysis to estimate the parameter a_0 (see, [3] for details). Then, paths of X can be extracted, prior to the estimation of the kernel Q. Once every parameters in system (21) are known, we rely on the very same strategy as in Sect. 3 to compute the reliability, this time *with the estimated kernel* $\widehat{Q}$. The whole learning procedure takes in input some 100 paths of Z simulated

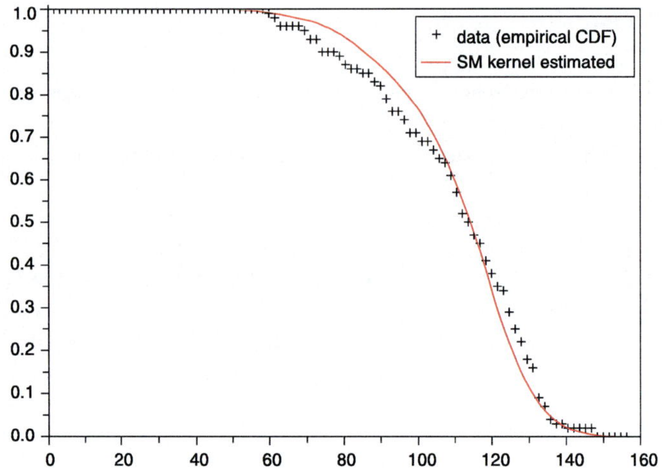

Fig. 8 Reliability evaluation and comparison to the testing set

according to (21), which consist the learning set. Some 100 other paths of Z are generated, consisting in the testing set, kept to evaluate the predictive performance of our inference strategy.

Figure 8 represents the reliability R of the system computed through the Markov renewal argument developed Sect. 3, using the estimated parameters (the kernel Q and the initial distribution law) as described in the current section. The empirical reliability which appears as an element of comparison in Fig. 8 has been computed on the test set, through

$$\widehat{R}(t) = \frac{1}{K} \sum_{k=1}^{K} \mathbb{1}_{\{Z_t^k < \Delta\}}.$$

A good fit is obtained, since the curves are quite closed: the slight discrepancy observed is likely to be due to the numerical discretization of the time interval and of the state space interval $[z, \Delta]$ of Z_t.

Remark 4.3. Note that, as future work, we plain to deeply investigate the numerical consistency of the empirical estimator of the kernel Q. It has been made for the corresponding estimator of the infinitesimal generator in the jump Markov case for X (see [3]). We also plan to take into account the dependency in τ in the likelihood maximization, since a closed-form of f_τ can be obtained through Markov renewal argument as developed in Sect. 3.

5 Concluding Remarks

Motivated by the fatigue crack-growth propagation problem, the point of view adopted in this chapter to model degradation processes does not include diffusion processes. So, we considered that the changes result from small or very small jumps. We have consequently developed a semi-Markov piecewise deterministic process as underlying model to achieve this goal.

As stated previously, this study was initially motivated and supported by the French Nuclear Power Plan Authority where we considered a Markov perturbing process. Here we considered a semi-Markov perturbing process which is much more general than the Markov one.

For a detailed modeling of reliability of SMPs the interested reader could find results in [17], for the discrete state space case, and in [16] for the general state space case. For estimation results of reliability and more general of dependability of semi-Markov systems see [18] and references therein.

Acknowledgments This chapter is dedicated to *Professor Shunji Osaki* for his 70th Birthday. It deals with semi-Markov processes and reliability which is the main area of expertise of Prof. Osaki. He has done many contributions to this field, a lot of them being included in his book "Stochastic System Reliability Modeling" (see Ref. [21] below), which is used as a basic reference by many researchers by the world including the present authors.

References

1. Bagdonavicius V, Bikelis A, Kazakevicius V, Nikulin M (2006) Non-parametric estimation in degradation-renewal-failure models. In: Probability, statistics and modelling in public health. Springer
2. Bogolyubov NN, Mitropol'skii YA (1961) Asymptotics methods in the theory of nonlinear oscillations.
3. Chiquet J, Limnios N (2006) Estimating stochastic dynamical systems driven by a continuous-time jump Markov process. Methodol Comput Appl Probab 8:431–447
4. Chiquet J, Limnios N (2008) A method to compute the reliability of a piecewise deterministic Markov process. Statist Probab Lett 78:1397–1403
5. Chiquet J, Limnios N, Eid M (2009) Piecewise deterministic Markov processes applied to fatigue crack growth modelling. J Stat Planning Inf 139:1657–1667
6. Çinlar E (1969) Markov renewal theory. Adv Appl Probab 1:123–187
7. Costa O, Dufour F (2008) Stability and ergodicity of piecewise deterministic Markov processes. SIAM J Control Optim 47(2):1053–1077
8. Davis MHA (1993) Markov models and optimization. In: Monographs on statistics and applied probability, vol 49. Chapman & Hall
9. Devooght J (1997) Dynamic reliability. Adv Nucl Sci Technol 25:215–278
10. Fridlyand J, Snijders A, Pinkel D, Albertson DG, Jain AN (2004) Hidden Markov models approach to the analysis of array CGH data. J Multivar Anal 90(1)
11. Iosifescu M, Limnios N, Oprişan G (2010) Introduction to stochastic models. Iste, J. Wiley, London
12. Korolyuk VS, Limnios N (2005) Stochastic systems in merging phase space. World Scientific, Singapore

13. Kotz S, Lumelskii Y, Pensky M (2003) The stress-strength model and its generalizations. World Scientific, New Jersey
14. Lebarbier E (2005) Detecting multiple change-points in the mean of Gaussian process by model selection. Sig Process 85
15. Lehmann A (2006) Degradation-threshold-shock models. In: Probability, statistics and modelling in public health. Springer
16. Limnios N (2011) Reliability measures of semi-Markov systems with general state space. Meth Comput Appl Probab. doi: 10.1007/s11009-011-9211-5
17. Limnios N, Oprişan G (2001) Semi-Markov processes and reliability. Birkhäuser, Boston
18. Limnios N, Ouhbi B (2003) Empirical estimators of reliability and related functions for semi-Markov systems. In: Lindqvist B, Doksum K (eds) Mathematical and statistical methods in reliability. World Scientific, Singapore
19. Lin YK, Yang JN (1985) A stochastic theory of fatigue crack propagation. AIAA J 23:117–124
20. Muller CH, Giunkel C, Denecke L (2011) Statistical analysis of damage evolution with a new image tool. Fatigue Fract Eng Mater Struct 34(7):510–520
21. Osaki S (1985) Stochastic system reliability modeling. World Scientific, Singapore
22. Paris PC, Erdogan F (1963) A critical analysis of crack propagation laws. J Basic Eng 85:528–534
23. Picard F (2005) Process segmentation/clustering. Application to the analysis of CGH microarray data. PhD. Thesis
24. Pyke R (1961) Markov renewal processes: definitions and preliminary properties. Ann Math Statist 32:1231–1242
25. Shurenkov VM (1984) On the theory of Markov renewal. Theory Probab Appl 29:247–265
26. Sobczyk K (1982) On the markovian models for fatigue accumulation. Journal de Mécanique Théorique et Appliquée (special issue), pp 147–160
27. Virkler DA, Hillberry BM, Goel PK (1979) The statistical nature of fatigue crack propagation. J Eng Mater Technol ASME 101:148–153

Customer-Perceived Software Reliability Predictions: Beyond Defect Prediction Models

Kazu Okumoto

Abstract In this chapter, we propose a procedure for implementing customer-perceived software reliability predictions, which address customer's concern about service-impacting outages and system stability. Data requirements are clearly defined in terms of test defects and field outages to ensure a good data collection process. We incorporate the effect of operational profile to demonstrate the changes in defect find rate from internal tests through precutover test and in-service operation. A software reliability growth model is a necessary key step, but not sufficient for addressing customer-perceived reliability measures. The proposed approach is a result of in-depth investigations of test defect data and field outage data over many years. It has been successfully demonstrated with actual field data and applied to a variety of software development projects.

1 Introduction

In recent years, many product suppliers have been implementing complex software-controlled systems with a large number of software functions or features for delivery on a short development schedule. A majority of field problems are associated with software. Although software does not physically break or wear out over time in a persistent way that can be easily examined with an optical or electron microscope, it does fail or crash. While hardware wears out over time, software does not. Customers are concerned about service-impacting outages and system stability. Customer-perceived software reliability and availability have become the common practice to be included in customer reviews and internal project reviews as key product quality metrics.

One of the critical customer operational issues has been on system performance, especially in terms of system outages impacting the service availability for their end

---

K. Okumoto (✉)
Alcatel-Lucent Technologies, Naperville, IL60563, USA
e-mail: kazu.okumoto@alcatel-lucent.com

T. Dohi and T. Nakagawa (eds.), *Stochastic Reliability and Maintenance Modeling*, Springer Series in Reliability Engineering 9, DOI: 10.1007/978-1-4471-4971-2_11,

users. It is becoming a common practice for service providers to ask their product suppliers for quality measurements and predictions such as software reliability and availability, representing customer views. For example, five-nines system availability (or equivalently 5.26 min/year/system) may be required.

For telecommunication products, TL 9000 [9] specifically requires outage measurements in the field to meet a customer's concern about service-impacting outages and system stability. The TL 9000 is a quality management system standard (QMS) which standardizes the quality system requirements for the design, development, delivery, installation, and maintenance of telecommunication products and services. It defines the customer-perceived reliability in terms of SO3 (service outage frequency) and SO4 (service outage duration) metrics.

As a product supplier, there is a need for predicting software field performance in terms of outage frequency and duration prior to software delivery. Analogous to hardware reliability predictions, we can look for a software product with similar functionality, complexity, and size. Unfortunately, most of the time field data do not match with the predictions. This happens mainly because every software development is different from others. It is essential to institutionalize internal software quality metrics through requirement/design document review, code inspection, and test defect density, so that appropriate corrective and preventive actions can be taken to improve the software quality. We will be using test defect data for predicting software field reliability.

The proposed approach is relatively new and based on customer views. It has been validated with actual data for release over release and successfully applied to various telecommunication products such as base station controller, radio network controller, and core network. It is not only a practical approach for tracking software reliability through defect data from internal test and field, but also a valuable tool for determining whether a software product is ready for delivery. It helps assure the delivery of highly reliable software products.

To build a common understanding of the subject, we can consider the following scenario as illustrated in Fig. 1, where a team of system engineers, developers, testers, project managers, and quality/reliability engineers is producing a set of software features to meet customers' need. As various new features are integrated into one software release, it goes through an intensified test program. Toward the end of the test program, approaching the committed delivery date, our management and customers typically start to ask "Are we done yet?" As a quality/reliability engineer, we need to answer the question based on quality versus delivery commitment [8].

In the following sections, we will clarify our objectives and assumptions, followed by an overview of our proposed customer-perceived reliability process.

1.1 Objectives

A key objective of this chapter is to establish a practical software reliability program for predicting customer-perceived software reliability and availability based on test

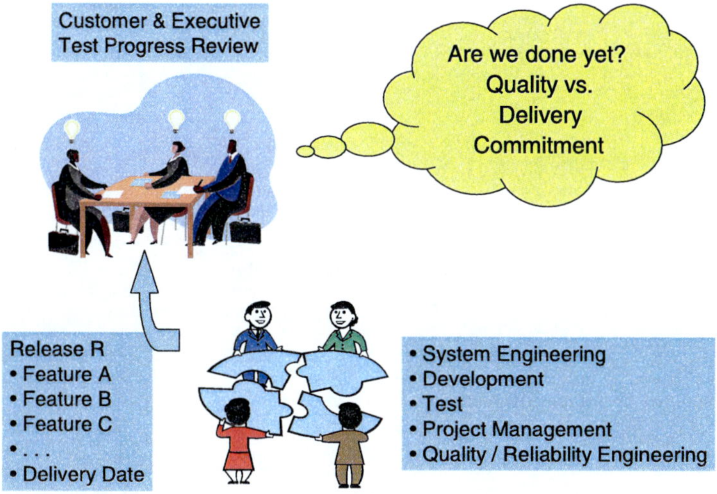

Fig. 1 A typical scenario of software reliability prediction

defect data. As a critical first step for the proposed approach, data requirements are clearly specified to distinguish test defects and field outages. A defect prediction model is used to predict the number of residual defects at the software delivery.

We will demonstrate that the defect find rate changes from internal test through precutover test and in-service operation due to the changes in operational profile. Additional steps beyond the traditional defect prediction model are needed to derive customer-perceived software reliability metrics. With the proposed approach, we can predict both outage frequency and duration prior to product delivery. The prediction results are validated with actual field data which are collected for each release for many years.

1.2 Assumptions

The following terms are used interchangeably here for simplicity: outages = failures and defects = faults = errors. A software failure is defined as a system outage caused by a software defect. Software reliability is defined as the rate of software failures (e.g., outages/year/system) and software availability is defined in terms of service downtime (e.g., minutes/year/system) due to software failures.

Software defect prediction models, which are typically called software reliability growth models, assume that there are a finite number of software defects in a release that can be exposed when subjected to a particular operational profile. Thus, as residual defects are discovered and removed, there are fewer defects left to be exposed in a particular operational profile. Fewer critical residual software defects should

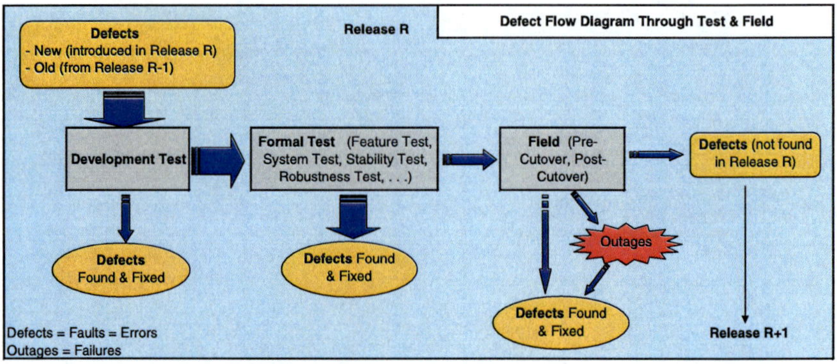

Fig. 2 Software defect flow diagram through test and field

be encountered less frequently in normal operation, and thus should yield a lower software failure rate. All defects found in a release are assumed to be fixed in the same release. This is referred to as software reliability growth through a find-fix process during test.

A diagram of defect flow through various test phases into the field is depicted in Fig. 2. At the start of test, there are a certain number of new defects introduced through software design and construction for a set of new features and functions in a release, plus old or base defects which were not found in a previous release and carried over into the current release. Although some defects are found and removed during development test, they are typically not reported precisely. In our analysis, we focus on defects found by testers. It is often called a formal test, which includes network element test, feature test, deliverable test, system test, cluster test, and network level test, stability test, and performance test, just name a few, since many different test names are used by different projects. Most of the defects are expected to be found and removed during formal test. Some residual defects at the end of formal test will be found in the field and a few of them will result in outages causing system downtime. There are always some defects which will not be found in the release. They will become old or base problems in the next release.

We focus on high severity defects representing severity 1 and 2 defects, as defined in TL 9000. Figure 3 provides further detail descriptions of high severity definitions. They are highly correlated to system stability and may trigger a failover or a reboot. Software failure rate is a function of residual high severity defects, known but not yet fixed high severity defects, and operational profile and configuration of deployed systems.

1.3 Software Reliability Tests

There are typically two types of software reliability testing (robustness and stability tests) to be performed to ensure highly reliable software prior to software delivery.

TL 9000 High Severity Definitions

- **Severity 1 (Critical):** Conditions that severely affect the primary functionality of the product and because of the business impact to the customer requires non-stop immediate corrective action, regardless of time of day or day of the week as viewed by a customer on discussion with the organization such as
  - product inoperability (total or partial outage),
  - a reduction in the capacity capability, that is, traffic/data handling capability, such that expected loads cannot be handled,
  - any loss of emergency capability (for example, emergency 911 calls), or
  - safety hazard or risk of security breach.
- **Severity 2 (Major):** Product is usable, but a condition exists that seriously degrades the product operation, maintenance or administration, etc., and requires attention during pre-defined standard hours to resolve the situation. The urgency is less than in critical situations because of a lesser immediate or impending effect on problem performance, customers and the customer's operation and revenue such as
  - reduction in product's capacity (but still able to handle the expected load),
  - any loss of administrative or maintenance visibility of the product and/or diagnostic capability,
  - repeated degradation of an essential component or function, or
  - degradation of the product's ability to provide any required notification of malfunction.

Fig. 3 TL 9000 high severity definitions

Results of robustness and stability tests should contribute to the validation of reliability requirements and the predictions of system availability and reliability.

Robustness testing (also called 'negative testing', 'adversarial testing', or 'fault insertion testing') confronts systems with plausible failure scenarios to assure that automatic failure detection, isolation, and recovery mechanisms work rapidly and reliably. Some examples of categories of robustness tests are:

- Software-related failures
  - Memory exhaustion/failure
  - Process/thread failure
  - File system exhaustion failure
  - Database/data structures
  - Application/platform software failures
  - Local and remote interprocess communication
  - Network communication failure
  - Timer failure
  - Overload condition
- Hardware-related failures
  - Disk system failure
  - Board level hardware fault insertion and recovery

  - Insertion and removal of field replaceable unit (FRU)
  - Cluster/processor/blade
- Procedure-related failures
  - Management and provisioning errors and failures
  - Software upgrade/install failures-Rollback/backouts

Stability (or 'endurance') testing uses a heavy, mixed traffic load against a system for an extended period, typically at least 72 hours. The stability run is performed to track whether new software features and code have impacted the overall network/solutions stability. Stability tests should be performed on final software release and after all feature testing is complete. This will assure that testing is against final product. A best practice is to simulate both heavy end user and Operation, Administration, Maintenance and Provisioning (OAM&P) activity. It is important to select the operational profile to replicate the actual operating environment in testing. Quantitative estimation of parameters is only meaningful if the operational profile during testing is representative of the field environment. In Sect. 3.3.1, we will discuss more details of defect data resulting from stability test.

1.4 Overview of Software Reliability Prediction Process

An overview of the proposed software reliability prediction process is illustrated in Fig. 4. Two sets of high severity defect data are needed from internal tests and field.

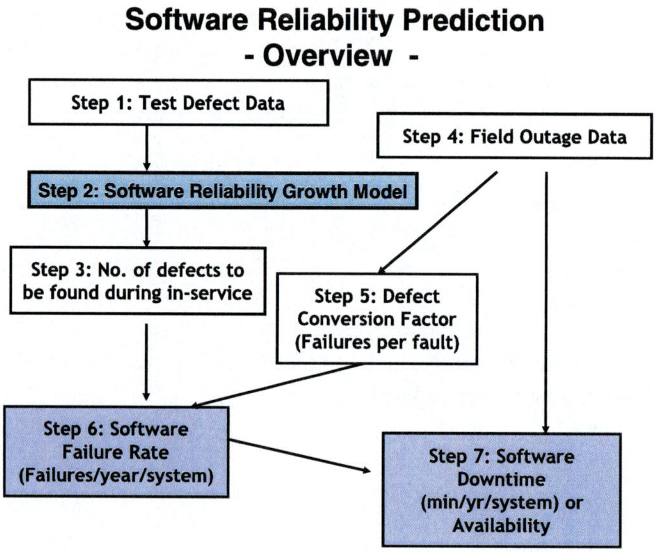

Fig. 4 Software reliability prediction-overview

Using test defect data and a software reliability growth model, we can identify the total number of high severity defects in a release. And, historical defect data from previous releases are used to derive the number of high severity defects to be found during an in-service operational phase. Combining field outage data and high severity defects found in an operational phase, we can also identify a defect conversion factor (i.e., outages per defect), which is used to convert high severity defects into outages. We can then derive software failure rate and downtime, and compare the predictions against TL 9000 frequency (SO3) and downtime (SO4).

In the following sections, we will address each of the steps in details.

2 Data Requirements and Analysis

A lack of high-quality data on test defects and field outages has been a main obstacle in implementing a software reliability program. In this section, we will address data requirements and provide some basic data analysis.

2.1 Defect Data

2.1.1 Defect Data from Stability Test

Software defect data should be normalized against test effort (e.g., stability run duration hours, tester-days) rather than calendar time, as pointed out first by Musa [5]. It will eliminate nonuniform effort over test interval, variations in staffing levels, weekends, holidays, etc.

Typical software defect data are shown in Fig. 5, where defect data are plotted against stability run duration in hours and calendar time in days, respectively. The data were taken from a relatively large-scale software development with over 2 million lines of code for a next generation radio network control system. The defect data and run hours were collected on a weekly basis. We can observe the defect find rate continuously leveling off with stability run hours as expected. This is referred to as software reliability growth through a find-fix process. As residual defects are discovered and removed, there are fewer defects left to be exposed. However, continuously leveling off behavior is not obvious for the calendar time-based defect find rate. It will be shown in Sect. 3 that the defect data based on stability run duration hours can be well represented by an exponential model.

Further examination of the test effort data is illustrated in Fig. 6, where the test execution rate is low at the startup period. This is mainly due to system stability problems or tools/lab environment problems. Once the early critical issues were removed, the execution rate became constant. Toward the end of the stability test, the effect is diminishing, i.e., the defect find rate is sufficiently low, indicating the

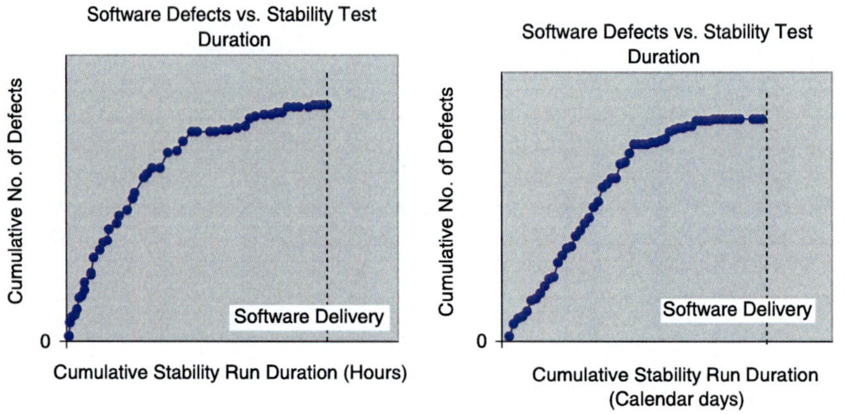

Fig. 5 Defect data based on calendar time versus test run duration hours

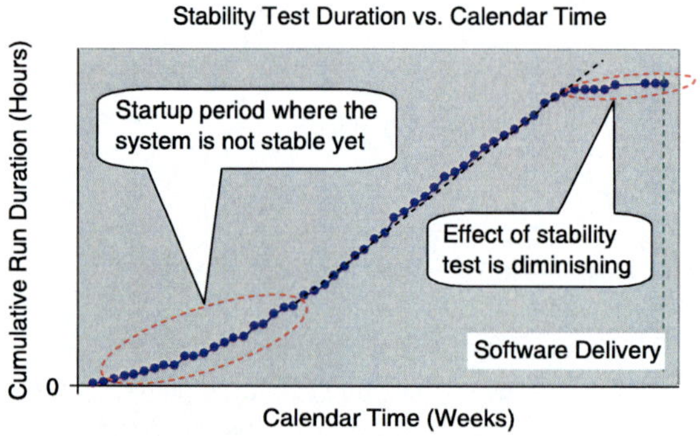

Fig. 6 Stability run duration hours versus calendar time

readiness of the software delivery. In fact, the project decided to move most of the test resources to the next release.

2.1.2 Defect Data from Formal Test

From a practical view point, test defect data are usually sorted by calendar time. Since not all new features are ready for test at the startup of test phase, defect rate is usually low in the beginning. This is the main reason why a cumulative defect find curve often exhibits an S-shaped curve. However, there are no specific trends in early test phase even within the same project, as illustrated in Fig. 7. Each release contains a different set of features, complexity, resource allocations, and test plans. We will address how to deal with the situation later in this section.

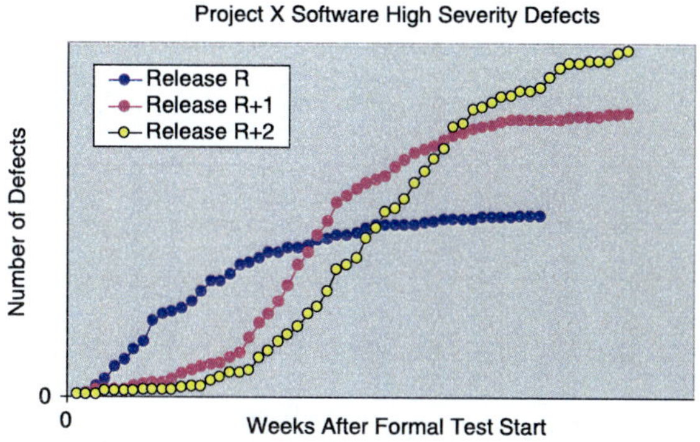

Fig. 7 Sample defect data from various releases within a same project

As described in Sect. 1.2, we need high severity defect data. Software defects should be unique, found in a release, and sorted by each test phase such as formal test, precutover test, and in-service operation. There are often multiple solutions, including carry-forward and carry-backward, in fixing a defect. In this analysis, we will count only unique defects, not solutions, found in the same release. Duplicate defects will not be counted here.

In addition, only defects found by testers, not by developers, should be counted. The defect data from precutover test, customer acceptance test, or in-service operation should also be tracked separately, so that we will be able to identify the percent of defects to be found during in-service based on release-over-release data.

Figure 8 illustrates the above requirements for defect data. It should be pointed out that not all features are ready for test in the beginning for typical projects. It results in a slow increase in defect find rate at the startup of test phase. This is the main reason why a cumulative defect find curve often exhibits an S-shaped curve. In this example, actual defects found after software delivery are mostly found through internal test. There are a very small portion of defects found during in-service. To meet customer's need, we often deliver a few more features after the first delivery.

Note that the defect find rate slows down as it moves from formal test, precutover test, and to in-service. This is due to the changes in operational profile where the intensity of test significantly changes. Test cases are developed to induce potential defects in the formal test in a highly simulated test environment or a heavy traffic condition. The precutover tests are designed to validate the functional requirements at customer sites. Lastly, in-service is in a normal operation environment.

Figure 9 further illustrates the effect of the operational profile changes in terms of weekly defect find rate from formal test to precutover, and to in-service phases. The concept of operational profile was first introduced by Musa [5] and applied by many others later, e.g., Okamura et al. [7], Jeske et al. [3], Zhang and Pham [14]. However, it is not an easy task to accurately predict operational software reliability based on test defect data, as reported.

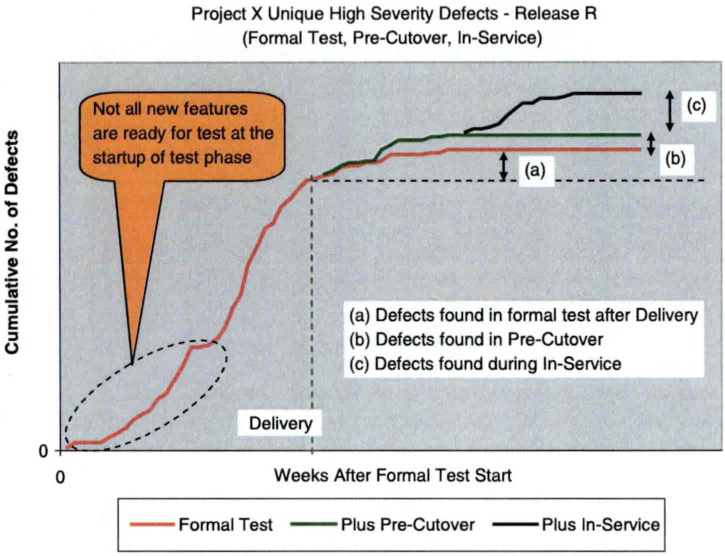

Fig. 8 Weekly high severity defect data in formal test, precutover, and in-service

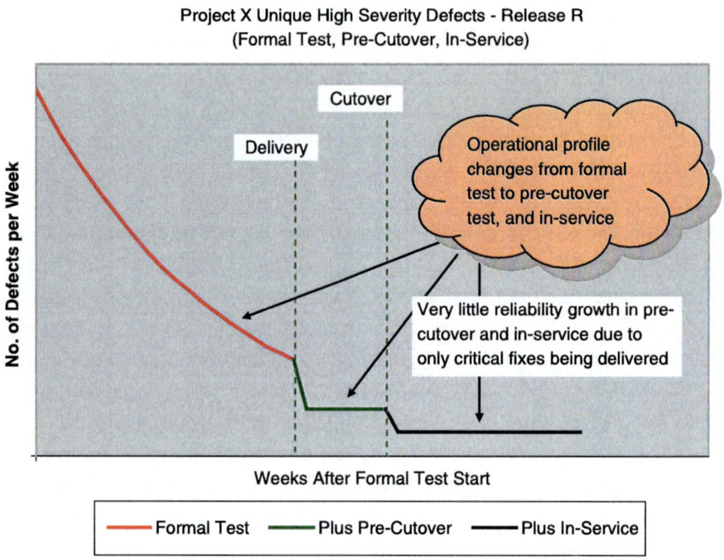

Fig. 9 Weekly defect find rate from formal test to precutover and in-service phases

Most of the projects under study demonstrated a trend similar to Fig. 9 with the significant changes in these phases. It also generally shows a constant defect rate in precutover and in-service phases, respectively. It means very little reliability growth in these phases mainly because only critical fixes are delivered during these phases.

In Sect. 2.1.2, we will further demonstrate the constant failure rate in the field based on outage data. This is the main reason why defect prediction models cannot appropriately describe the failure trend in the operational phase.

2.2 Field Outage Data

Software failure rate depends on operational profile and configuration of deployed systems. Operational profile is characterized by the system and solution configuration, usage and traffic mix, and other operational context that the system operates within. A particular system can often be used in several different operational profiles, each of which may stress the system in slightly different ways, thus exposing somewhat different residual defects. System test should reflect the operational profile(s) that the deployed system will experience to assure that the vast majority of design and residual defects are discovered and corrected before the field deployment. Differences between tested operational profiles and field operational profiles present gaps in testing that undiscovered software defects can escape to the field through. In addition, not all high severity defects will result in software failures in the field.

As part of TL 9000 metrics, the following outage data should be readily available in terms of the number of outages per month, prorated outage duration per incident, and the number of in-service systems per month. Figure 10 illustrates monthly service outage rate (commonly known as SO3 in terms of TL 9000 metrics), where the number of outages is normalized by the number of systems in service for each month.

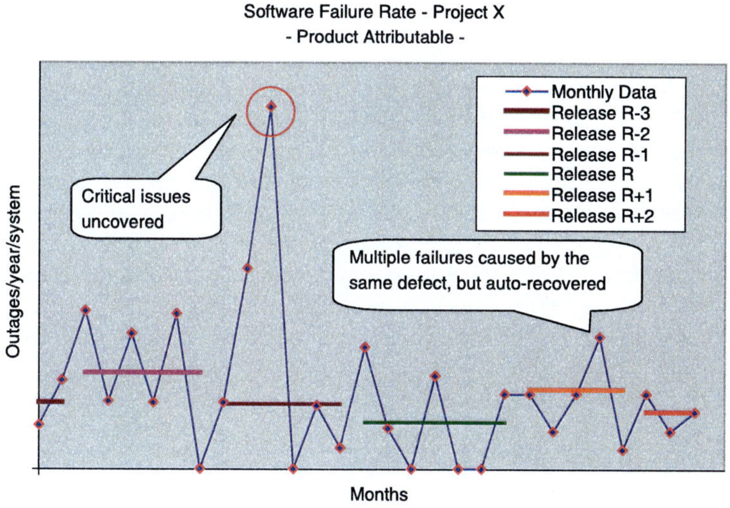

Fig. 10 Monthly failure rate and release-based failure rate

It typically displays a constant outage rate for each release with some variation from month to month.

In order to determine the outage rate for each release, we can take an average over the time period in which each release is actually deployed in the field. Multiple failures caused by the same defect are occasionally observed. We can perform a similar analysis for outage duration in minutes/year/system (commonly known as SO4 in terms of TL 9000 metrics), where outage duration is prorated based on the impact. According to the TL 9000 counting rules, an outage can be excluded if its impact is less than 10 %.

2.3 Other Reliability Related Data

2.3.1 Defect Conversion Factor

A defect conversion factor (i.e., outages per high severity defect) is used to convert high severity software defects into failure rate. The conversion factor can be derived from historical release data on high severity defects and field outages. Since it varies from project to project in the range of 0.4–0.9, use of historical data is highly recommended for each project. That is,

$$[\text{Defect Conversion Factor}] = [\text{No. of In-service Outages}]/[\text{No. of In-service Defects}]$$

2.3.2 Coverage Factor

A coverage factor is used to properly separate uncovered failures and covered failures. It is defined as the probability that the system diagnostic mechanism detects a failure, and therefore automatic recovery (typically through a reboot or a switch-over to a standby system) is triggered to bring the system back to normal operation. Some failures escape the system diagnoses, and are known as silent failures or uncovered failures. Silent failures usually take a longer time to be detected, which lead to longer outage durations. Figure 11 illustrates the impact of a coverage factor. In the test environment, coverage factor can be measured through a fault insertion test; however, it would be desirable if this parameter could be estimated from field data. It is typically in the range of 90–99 %, depending on the product maturity.

2.3.3 Outage Recovery Time

Outage recovery time per incident is an important measurement in calculating the system downtime or system availability. According to TL 9000 counting rules,

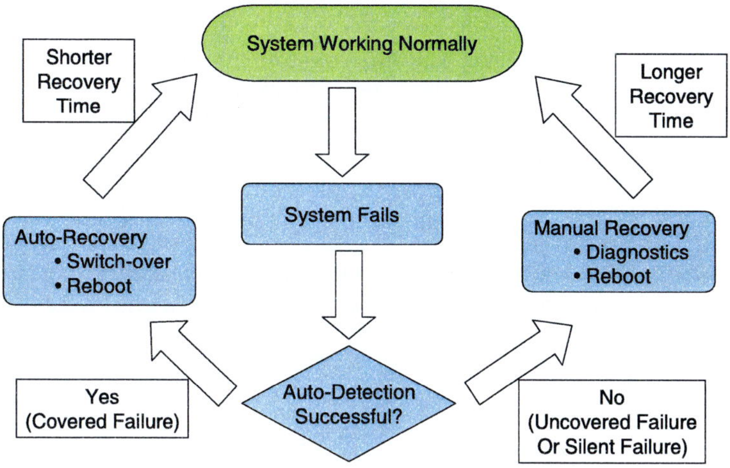

Fig. 11 Illustration of a coverage factor

outages with less than 15 s duration are excluded in the downtime calculation. "100 %" service availability may be declared despite critical software failures, but they were all automatically detected, isolated, and recovered by the system in seconds, so no outage was reported. Note that this counting rule may not be applicable to some customers.

3 Software Defect Prediction Models

In this section, we describe a method for predicting the number of high severity defects to be found during in-service.

3.1 Historical Model Development

Software defect prediction models are known as software reliability growth models. A large number of prediction models have been proposed and investigated over the last four decades. There are many references documenting and comparing various models in detail (e.g., Musa et al. [6], Lyu [4], Wallace and Coleman [12]), so we will not go into a comprehensive discussion here.

As defects are found and removed, encountering additional severe defects is less likely. Goel and Okumoto [1] first formulated this defect find process as a stochastic process in terms of defect find interval or time-varying defect find rate. Since we count defects as they are exposed, it seems logical to statistically formulate the number of high severity defects found during test as a Poisson process with a time-varying mean value function, which is known as a nonhomogeneous Poisson process (NHPP). That is, for a defect find process, N(t), the probability of finding n high severity defects by time t is expressed as a Poisson distribution with the mean value function, m(t), as:

$$P\{N(t) = n\} = m(t)^n \exp\{-m(t)\}/n! \quad (1)$$

It is typically assumed that there are a finite number of severe defects in any piece of software. Most of those frequently used models can be systematically sorted in terms of the shape of the mean value functional (i.e., an exponential curve or an S-shaped curve).

3.2 Exponential Models Versus S-Curve Models

Some representatives of an exponential curve are Jelinski and Morand [2], Schneidewind [11], Musa basic execution-time [5], Goel and Okumoto [1] while those of an S-shaped curve are Schick and Wolverton [10], Yamada et al. S-shaped [13], Weibull, Gamma, and logistic. S-curve models have flexibility in describing different shapes of the trend since they have more than two parameters. On the other hand, exponential models are simple with only two parameters. In next sections, we will present the reason why an exponential model is used in the proposed approach.

3.3 Software Defect Predictions

In this section we use an exponential model, which is simple and proven to provide predictions as accurate as, if not more than, S-shaped models. This will be also confirmed in the following analyses.

An exponential model is usually represented as an NHPP with the mean value function:

$$m(t) = \mathbf{a}\{1 - \exp(-\mathbf{b} * t)\}, \quad (2)$$

where m(t) = cumulative number of defects found at time t, **a** = total defects in the software, and **b** = rate at which each defect is exposed or found.

The corresponding defect intensity function or defect rate can be derived as the derivative of the mean value function:

$$\lambda(t) = \mathbf{a}\,\mathbf{b}\exp(-\mathbf{b} * t). \quad (3)$$

The maximum likelihood method is a commonly used statistical method for estimating the parameters, **a** and **b,** for a given set of defect data. A specific estimation procedure is described in Appendix A.

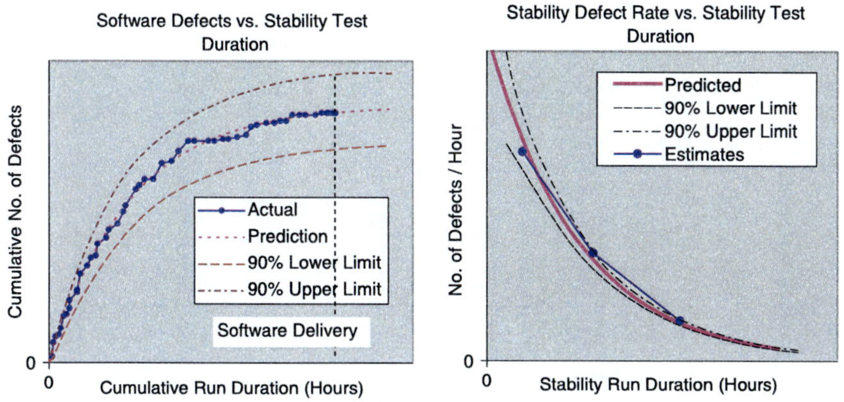

Fig. 12 Stability defect data with the defect prediction model

3.3.1 Execution Time Data

Using the stability run duration data shown in Fig. 5, we have obtained the maximum likelihood estimates for **a** and **b**, substituted them in the exponential function (2), and overlaid the predicted curve with the actual data, as illustrated in Fig. 12. It demonstrates a remarkably good representation of the defect data. Using the Poisson distribution, we also provided the 90 % lower and upper limits. It helps validate that actual defect data follows a Poisson process with the exponential mean value function.

Similarly, substituting the maximum likelihood estimates for **a** and **b** into (3), we can obtain the predicted defect rate. The 90 % limits can be easily converted from the cumulative curves. The defect intensity was estimated based on the actual defect data. Each data interval contained sufficiently large number of defects to avoid possible small sample size problems.

The estimated defect intensity data are shown in Fig. 3, along with the predicted defect rate with the 90 % limits. It demonstrates that the actual defect rate point estimates are within the 90 % limits.

In addition, following the procedure in Appendix A we have derived 90 % confidence limits for **a** and **b**. In Fig. 13, we illustrate the confidence limits along with the maximum likelihood estimates and the relationship between **a** and **b**, where we can observe they are negatively related in a nonlinear way. The normal approximation seems reasonable.

3.3.2 Calendar Time Data

In Sect. 3.3.1, we demonstrated that defect data based on stability run duration is a better measure for defect predictions. From a practical view point, however, test

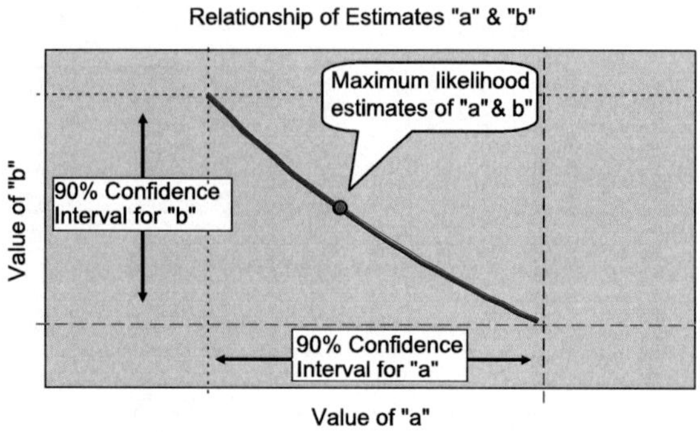

Fig. 13 90 % confidence intervals for *a* and *b*

defect data is usually collected by calendar time such as daily or weekly. In this section, we discuss how to deal with defect data based on calendar time.

In addition to early stability problems, not all new features are ready for test at the startup of test phase, and hence, defect rate is usually low in the beginning. This is the main reason why a cumulative defect find curve often exhibits an S-shaped curve as illustrated in Fig. 7.

Defect prediction should be based on the software release with a complete set of new features delivered for testing, i.e., based on defects found closer to the software delivery. The most recent data points are the most valuable. The earlier the data point, the less valuable it is as there are many factors contributing to defect discovery. This is the reason why we are removing defects found at a startup of test phase. By removing early defects from the analysis, the defect curve will look like an exponential curve rather than an S-curve. One of our goals for reliability prediction is to accurately predict the number of defects after the delivery date, not to accurately describe the defect trend during the entire test phase. Some experience will be required to determine where the start of the curve fitting will be.

It should be pointed out that a main reason why the find rate drops off is due to a lessening in test intensity (i.e., the find rate is related to test hours rather than calendar hours), and also because there are fewer bugs to find, so it takes longer to find them. An opposite effect could be said when we have increase in testing (people working overtime as we approach the delivery date, for example). However, it is worth stating that by combining the different test phases up to the delivery date, the overall intensity remains approximately constant up to the delivery date and that is why we project the curve at the delivery date, even though the find rate often drops significantly after the delivery date (due to the change in test intensity).

As described above, we are typically removing defects found at a startup of test phase. The mean value function (2) will be modified by adding another parameter, $\mathbf{t_0}$, which is often referred to as a shift parameter, allowing the curve to be shifted to the right by $\mathbf{t_0}$. In practice, $\mathbf{t_0}$ is chosen slightly after the inflection point of the cumulative curve. It often requires some experience to identify the starting point, $\mathbf{t_0}$.

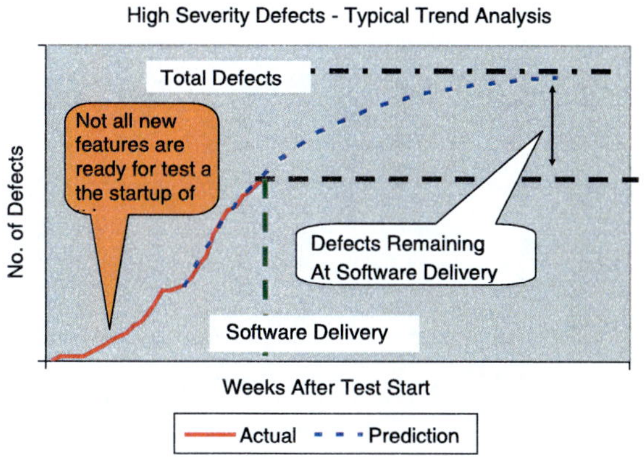

Fig. 14 Exponential model with actual defect data

The modified curve is expressed as:

$$m_0(t) = \mathbf{a}\{1 - \exp[-\mathbf{b} * (t - \mathbf{t_0})]\}. \tag{4}$$

The maximum likelihood method is commonly used for estimating the parameters, **a** and **b**, for a give set of weekly defect data. Specific equations for deriving the maximum likelihood estimates of **a** and **b** are provided in Appendix A.

Solving Eqs. (A.1) and (A.3) in Appendix A for the data set shown in Fig. 8, we have derived the estimates for **a** and **b** and substituted them in the modified exponential function. In Fig. 14, we overlaid the predicted curve with the actual data. It also indicates the total number of high severity defects in the release, and hence residual defects at the software delivery. As explained earlier, we used the defect data slightly after the inflection point of the cumulative curve for this analysis.

When we overlay actual defect data after the delivery date, we generate Fig. 15. We should always see a gap between total defects expected and actual defects found. This gap represents the number of defects to be carried over to the next release as base problems.

Next, we need to determine the percent of each defect category based on release over release data. Sample breakdown numbers are provided in Fig. 16 for illustration purposes. They significantly vary from project to project.

The percent of high severity in-service defects will be useful for determining the number of high severity defects to be found during in-service for this release. That is, once the number of residual defects is predicted at the delivery date from the reliability growth model, we can derive the following:

$$\begin{aligned}&[\text{No. of High Severity Defects to Be Found During In-Service}]\\&\quad = [\text{No. of Residual High Severity Defects at Delivery}]\\&\qquad \times [\%\text{High Severity Defects In-Service}]\end{aligned} \tag{5}$$

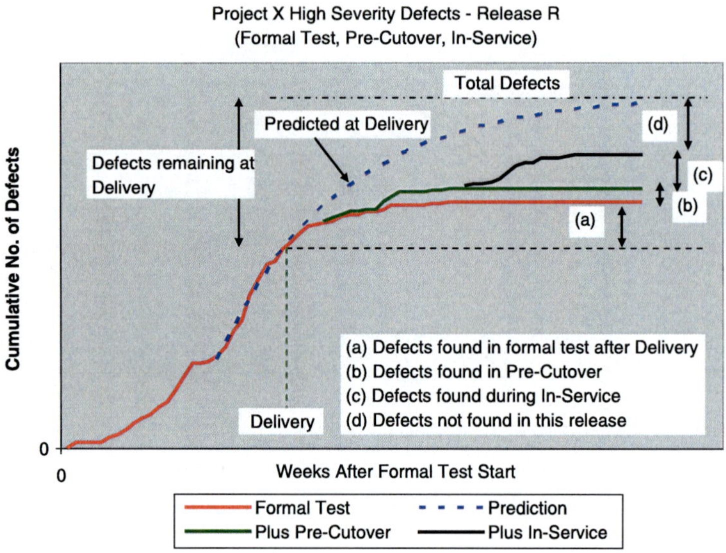

Fig. 15 Exponential model with actual defect data, including after the delivery

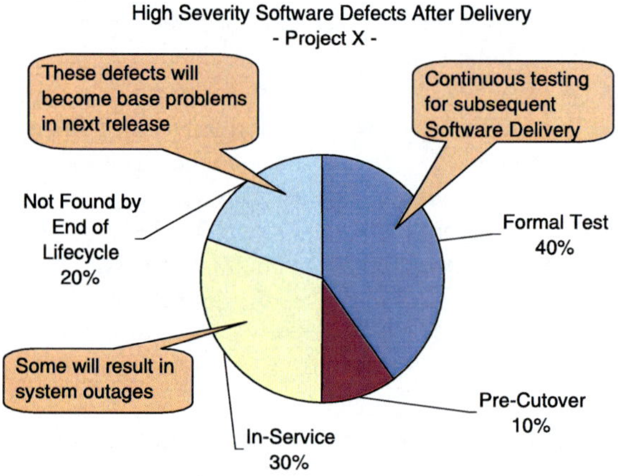

Fig. 16 Example of a breakdown of residual high severity software defects

3.3.3 Residual Defects at Software Delivery

In Sect. 3.3.2, we described how to predict high severity defects to be found during in-service using a software reliability growth model. In addition, we also need to assure that all defects found in a release are fixed in the same release, so that no known defects will be delivered. This is one of the assumptions described in Sect. 1.2.

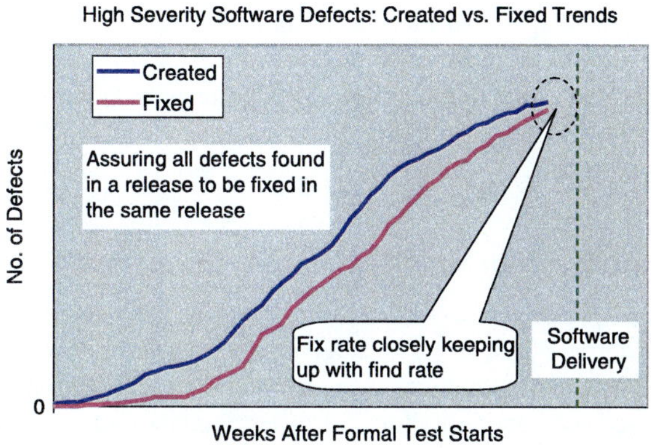

Fig. 17 Example of a find-fix tracking process

To confirm this assumption, it is a commonly used practice to track the number of defects found versus fixed during the formal test. Figure 17 illustrates the find-fix trends, where the fix rate is closely keeping up with the find rate at the delivery date. It is another indication that the release is ready for delivery. Most projects with tight development schedule show a similar trend as shown. In Sect. 6.5, we will address how to build confidence limits to monitor a gap between found and fixed defects.

4 Customer-Perceived Software Reliability Predictions

In Sect. 3.3.2, we described a method for predicting the number of high severity defects to be found during in-service. This section will address how to convert the high severity defects into field outages, i.e., customer-perceived software reliability and availability.

4.1 Software Failure Rate Predictions

One of the data requirements specified in Sect. 2.3.1 is a defect conversion factor. It is used to map the number of high severity defects into the number of field outages. That is,

$$[\text{No. of Outages}] = [\text{No. of High Severity Defects during in-service}] \times [\text{Defect Conversion Factor}]. \quad (6)$$

The number of outages can then be normalized by the number of systems in service and the number of months for the software release in service. The software failure rate is derived as the number of outages per year per system, as described in TL 9000.

4.2 Software Downtime and Availability Predictions

The failure rate derived in Sect. 4.1 is called an uncovered (or observed) failure rate, since it is based on reported outages. A coverage factor is used to properly separate uncovered failures and covered failures. As described in Sect. 2.3.2, it is defined as the probability that the system detects a failure and therefore automatic recovery, typically a reboot or a switch-over to a standby system, is triggered to bring the system back to normal operation. The coverage factor plays an important role if the recovery time (via either a reboot or a switch-over) is not trivial. It could be a significant contributor to the system downtime. In this section, we will illustrate how to incorporate the coverage factor in the software reliability and availability predictions.

Once the software failure rate is identified, we can calculate the system downtime using the average recovery time per incident. A coverage factor needs to be included in the calculation for a system without a redundant configuration, since the reboot time due to autorecovery is typically not negligible.

Customer-perceived software downtime can be calculated as follows:

$$\begin{aligned}[\text{Software Downtime}] = {} & [\text{Uncovered Software Failure Rate}] \\ & \times \big[\text{Manual Recovery Time}\big] \\ & + [\text{Covered Software Failure Rate}] \times [\text{Reboot Time}]\end{aligned} \tag{7}$$

Note that the covered and uncovered software failure rates are defined as:

$$\begin{aligned}&[\text{Covered Software Failure Rate}] \\ &\quad = [\text{Overall Software Failure Rate}] \times [\text{Coverage Factor}],\end{aligned} \tag{8}$$

and

$$\begin{aligned}&[\text{Uncovered Software Failure Rate}] \\ &\quad = [\text{Overall Software Failure Rate}] \times (1 - [\text{Coverage Factor}]).\end{aligned} \tag{9}$$

This will provide annual software downtime in minutes per year per system, which can then be converted to software availability as:

Table 1 Sample calculation of customer-perceived software availability

Software availability calculation		
Software metrics	Formula	Unit
Failure rate	(a)	Failures/year
Probability of successful auto-detection	(b)	Percentage
Auto-detection time	(c)	Seconds
Manual detection/recovery time	(d)	Minutes
Software reboot time or failover time	(e)	Minutes
Mean time to restore	(f) = (b) * ((c)/60 + (e)) + (1 − (b)) * (d)	Minutes/Failure
Software downtime	(g) = (a) * (f)	Minutes/Year
Software availability	(h) = 1 − (g)/(60 * 24 * 365)	Percentage

$$[\text{Software Availability}] = 1 - [\text{Software Downtime}]/(60 \times 24 \times 365). \qquad (10)$$

Table 1 illustrates the above availability calculation, where the system is recovered through either a failover to a standby system or a system reboot.

This predicted software availability is now compared against a customer requirement such as five-nines availability (or 99.999 %). It will help identify potential areas for improvement if it does not meet the requirement. In the next section, we will discuss a procedure for validating reliability and availability predictions against actual data.

5 Validation of Reliability Predictions with Field Outage Data

Having successfully predicted software reliability and availability, we will now demonstrate that the predictions based on the proposed approach are remarkably in line with actual field measurements.

To accomplish this task we will have to perform the reliability prediction procedure for several releases as shown in Fig. 18.

We have applied the validation process to our projects and summarized the results from one project as shown in Fig. 18. It overlays predicted values against actual field outage data for each release. We can observe predictions remarkably in line with TL 9000 metrics in terms of outage rate (SO3) and the prediction accuracy continues to improve as we incorporate more recent data. This results from our continuous refinement of the model parameters from release to release as the product becomes mature. The data set was chosen to show a more realistic situation. For example, one release encountered multiple outages resulting from the same defect, which distorted the data. In fact, if we remove the multiple outages from the data, actual data is now in line with the prediction. Predicting multiple outages like these is not an easy task.

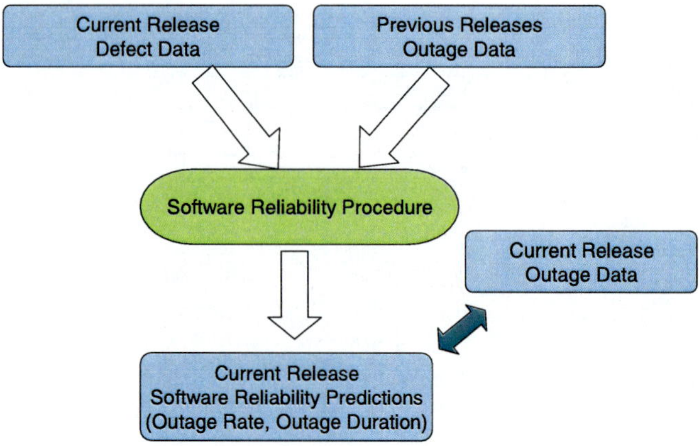

Fig. 18 Validation process for software reliability predictions versus outage data

A similar analysis was performed for software downtime data. The results are also shown in Fig. 19, where predictions are again remarkably in line with actual data. Note that actual data is consistently below the predictions even though the model parameters were continuously refined with recent data. This is due to the recovery time being consistently improved more than the previous release. It is one of the benefits resulting from periodic reviews of the quality metrics with project team. It turns out that the support team was making a special effort for reducing the recovery time to improve the metrics.

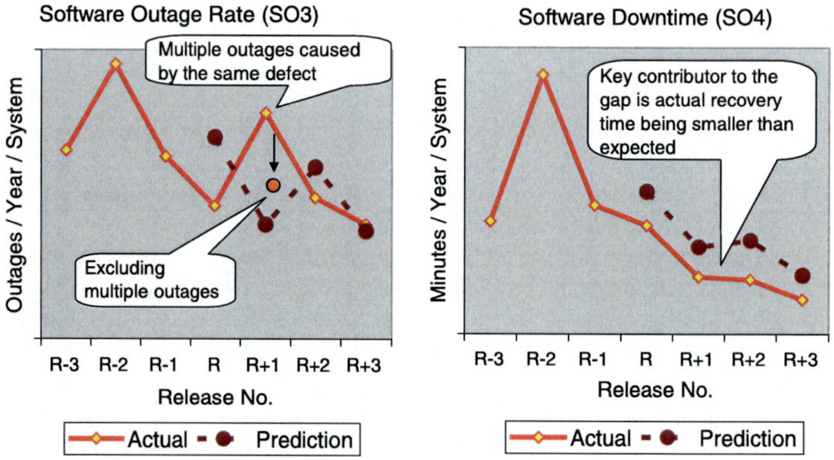

Fig. 19 Example of customer-perceived software reliability predictions versus actual data

6 Practical Uses of Software Reliability Models

In this section, we will introduce confidence limits for tracking and monitoring test defects to ensure the predicted reliability in the field. We also address how to estimate a defect curve at an early test phase where only a few data points are available. In addition, a few other practical uses of this approach will be discussed.

6.1 Tracking and Monitoring Test Defect Data

As discussed in Sect. 3.3, the defect prediction is a critical step for assuring the accuracy of customer-perceived reliability and availability predictions. During the course of various test phases in formal test, we need to provide a way for continuously tracking and monitoring actual defect data against the predicted curve.

To accomplish this monitoring process, we will build confidence limits around the mean value function, which was estimated based on previous few months data.

Assume t_e represents the last time at which the model parameters (**a** and **b**) were estimated, where $\mathbf{t_0}$ is the starting data point. That is, the mean value function is provided as:

$$m_0(t) = \mathbf{a}\{1 - \exp[-\mathbf{b} * (t - \mathbf{t_0})]\}. \tag{11}$$

The future trend from t_e can be described as a conditional NHPP for $t > t_e$. Assuming $m_e(t)$ represents the incremental mean value function from $m_0(t_e)$, i.e.,

$$m_e(t) = m_0(t) - m_0(t_e), \tag{12}$$

we can describe the conditional NHHP as follows:

$$\Pr\{N(t) = n|N(t)_e = n_e\} = m_e(t)^{(n-n_e)}\exp\{-m_e(t)\}/(n - n_e)! \tag{13}$$

For example, using this property, we can calculate 90 % limits. If the underlying defect find process follows the NHHP with the estimated mean value function, actual data should fall within the limits with 90 % confidence. If actual data starts falling outside the limits, we will have to refine the mean value function with the data set.

Figure 20 illustrates this defect tracking and monitoring process. Actual data shows close to the lower limit most of the time and then started to go up to the mean value function the last few data points. Once actual data starts to fall outside of the limits, we will need to refine the predictions with the new data set.

6.2 Predictions in Early Test Phase

Although more data points yield more accuracy in prediction, we often need to start predicting the defect find process with early limited data. After examining

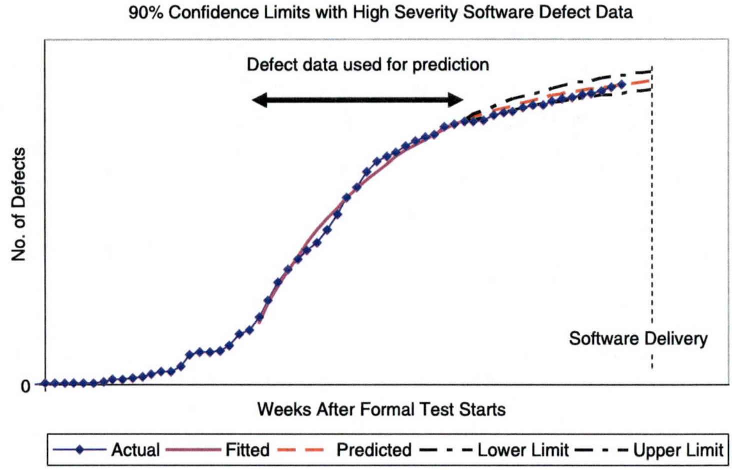

Fig. 20 Tracking and monitoring a defect find process

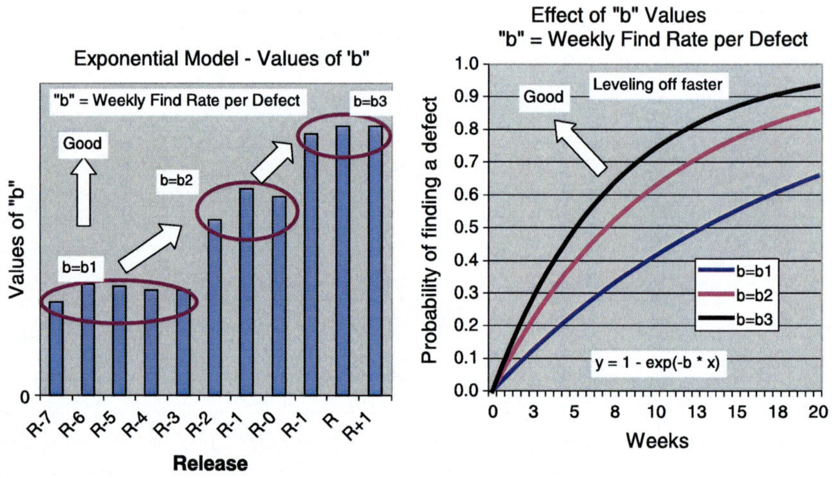

Fig. 21 Values of "b" versus test effectiveness

defect data from several releases, we recognize some consistency in the value of the parameter, **b**, which represents per-defect find rate. It is related to test effectiveness. It seems reasonable, considering that the test environment and test plans are relatively consistent from release to release for the same project.

In Fig. 21, we illustrate the above statement based on the project data used in Sect. 5, where "b" values are relatively consistent for a few releases until major changes took place in test plan in terms of improvements in lab environment and test scenarios.

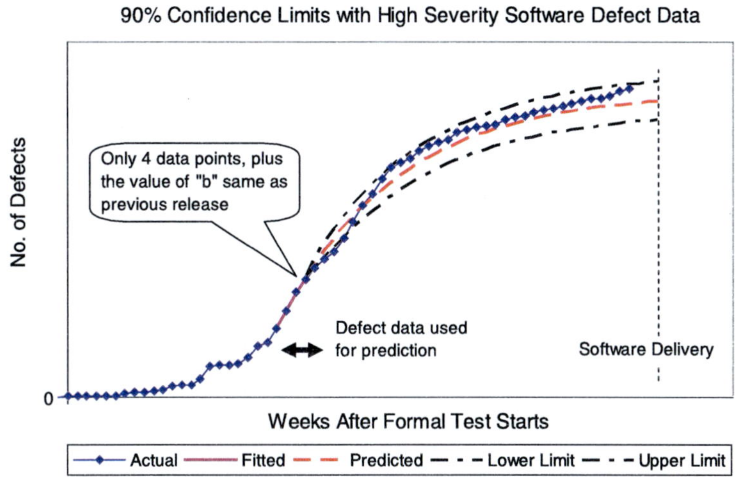

Fig. 22 Predictions in early test phase

In the following example, we assumed that there are no major changes in test plan from the previous release. Figure 22 illustrates the reasonable assumption of a constant value for **b**, where only four data points were used in estimating the total defects, **a**, and the shift parameter, $\mathbf{t}_0$, while the value for **b** is taken from the previous release. The 90 % confidence limits are also shown to demonstrate the validity of the early prediction against actual data. The release over release data plays an important role in predicting defect trends in early test phase.

6.3 Determination of Model Parameters Stability

In this section, we will address how to determine whether the estimates are stable enough to be used for predictions. Estimates of the model parameters are typically updated as a new data set becomes available.

To illustrate the procedure, we used stability test defect data based on run duration, shown in Fig. 5. We calculated parameter estimates (**a** and **b**) every week and monitored the changes in the estimates for every new data point, as shown in Fig. 23.

In order to illustrate a way to determine the stability of the parameter estimates, we used the parameter, **a**, and calculated the median of the estimates from the beginning to the latest data point. The median was used to avoid the influence by possible outliers. Figure 24 shows the median over the sample size, overlaid with actual estimates. It can be easily seen that the median of the estimates became stable when actual estimates stabilized.

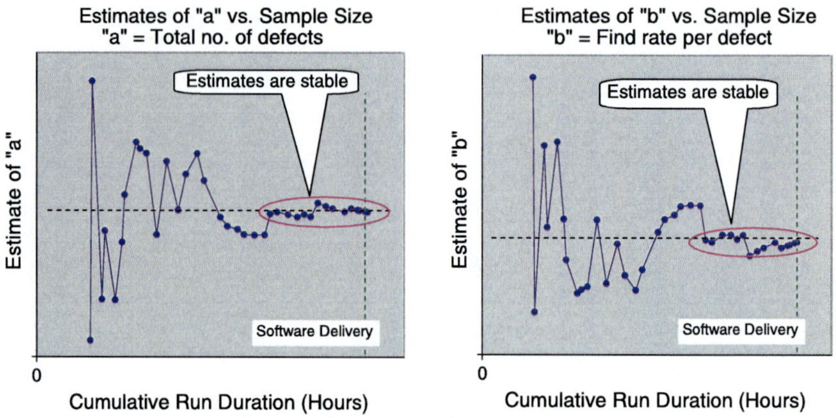

Fig. 23 Maximum likelihood estimates of **a** and **b** with sample size

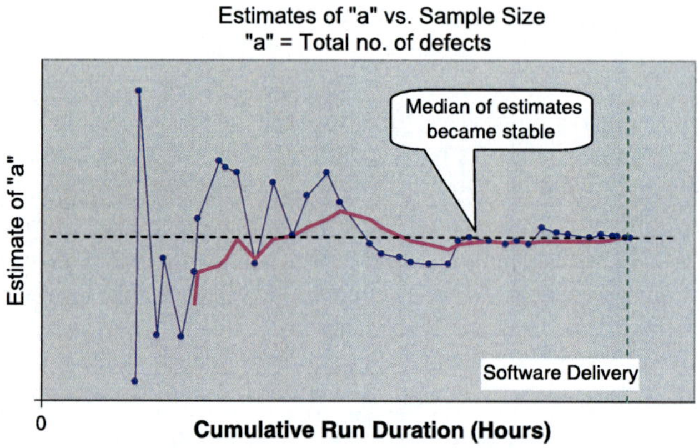

Fig. 24 Median of estimates of **a** and actual estimates with sample size

6.4 Determination of Additional Tests Needed

In addition to predicting the residual defects, we also need to know how much more testing is needed to meet the required reliability. As mentioned in Sect. 2.3.1, not all defects will result in failures or outages. Once the reliability requirement is set in terms of failures per year, it can be easily determined the corresponding weekly defect rate using the defect conversion factor. In this analysis, we assume that all defects found in stability testing will be fixed prior to release in order to meet the reliability requirement.

To illustrate the procedure, we used the same stability data set as used in Sect. 6.3 for consistency. Substituting the estimates of **a** and **b** into the defect intensity

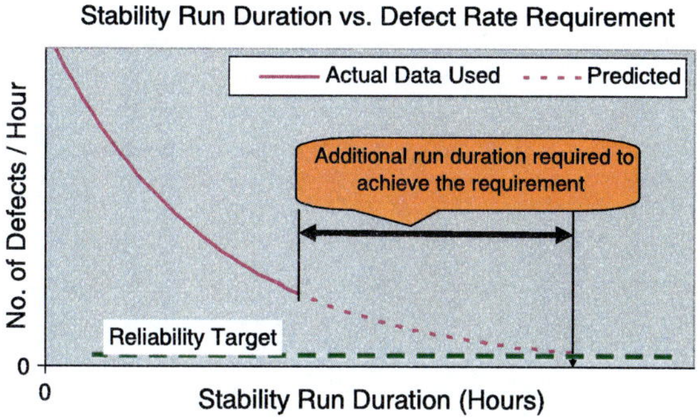

Fig. 25 Determination of additional tests required

function (3), we can derive the required test duration hours to reach the required defect rate, as shown in Fig. 25.

From (3) we can derive the run duration to achieve the required defect rate, λ_r, as follows:

$$t_r = -\ln\{\lambda_r/(\mathbf{a} * \mathbf{b})\}/\mathbf{b} \tag{14}$$

The additional run duration is thus obtained as $t_r - t_e$, where the prediction was made at t_e.

6.5 Tracking and Monitoring "Not Fixed" Defects

In Sect. 3.3.3, we discuss the importance of tracking the number of defects found versus fixed during formal test. It helps to assure that all defects found in a release are fixed in the same release, so that no known defects will be delivered. We will now address how to track and monitor defects which are found but not yet fixed.

To illustrate our approach, we will use the data shown in Fig. 17, where both defects found and closed are available. We are proposing to use an exponential model for defects closed after having investigated several other projects. We have applied the model to defects closed, similar to what we did for defects found. Both actual and predicted curves are overlaid in Fig. 26, where we can see the two curves are getting closer as the software delivery time approaches. In reality, the gap represents the time required to fix new defects although it is relatively small.

Next, we will explain how to develop a control chart (similar to a hardware quality control chart) for monitoring actual open defects on a weekly basis. A control chart will help the gap get smaller and actual data will be within the limits. For this purpose, we assume that defects between created and closed will follow a Poisson distribution with the mean, provided by [the predicted created curve]–[the predicted closed curve]. Then, we can easily construct confidence limits as shown in Fig. 27.

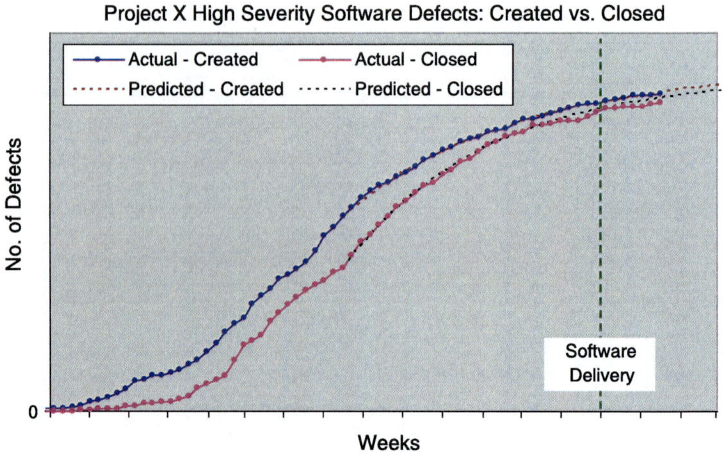

Fig. 26 Defects created versus closed with predicted curves

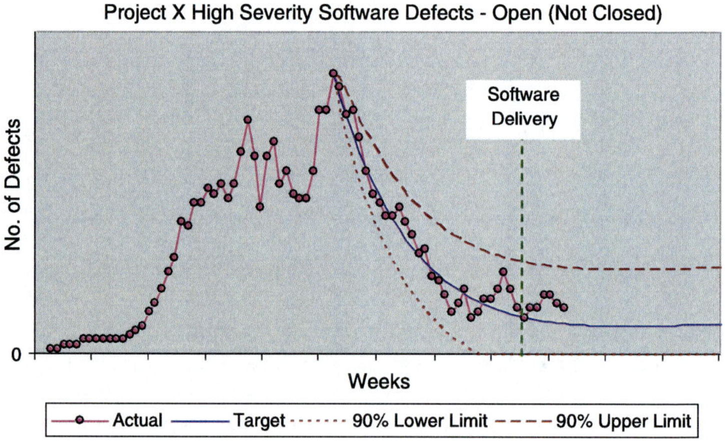

Fig. 27 Sample control chart for tracking "not yet closed" defects

7 Conclusions

We have demonstrated the proposed software reliability program with clear definition of data requirements. The specific procedures were illustrated with many years of actual defect and outage data. The proposed approach was designed to meet customer's expectations on software reliability in terms of field outages. It is not a simple extension of the traditional software reliability growth model, but it involves additional steps to predict customer-perceived reliability prior to software delivery. The proposed approach has been extensively applied to various software development projects over the several years. One of key lessens learned is that every software development is different from others, so that most of the model parameters have to be adjusted accordingly.

A large number of software reliability growth models have been proposed in the last four decades. There are still some hesitation and reluctance in applying to actual projects. This is mainly due to a lack of data requirement specification and not focusing on field outages as expected by most customers. The proposed approach is not a mathematical exercise but to address customer's reliability expectations through many years of in-depth analysis on test defect and field outage data from various software development projects.

Acknowledgments I am deeply honored and grateful to Professor Shunji Osaki (then at Hiroshima University, Japan) for his guidance and support throughout my professional career. He provided me the opportunity to pursue my graduate degree under Professor Amrit Goel at Syracuse University, NY. Professor Goel not only supported my Ph. D. program but also introduced me to new research in software reliability, which was supported by the Rome Air Force Development Center. It was my pleasure working with John Musa at Bell Labs in Whippany, NJ, including our coauthored book. He was instrumental in developing a software reliability engineering program. My special thanks to Doris Ryan, Ken Ng, and Heather Brown at Alcatel-Lucent Technologies for their management support and constructive feedback over the last decade. Their guidance and directions have been extremely helpful for applying the software quality/reliability approach to various projects.

Appendix A: Derivation of Maximum Likelihood Estimates for an Exponential Model

In this section, we will consider a case where defect data are available on a grouped basis such as weekly. The equations for deriving estimates of **a** and **b** for an exponential model will be provided using the maximum likelihood estimation method. Additional details are available from Musa et al. [6].

Let $y_i\,(i = 1, \ldots, p)$ be the number of defects found in $(0, x_i)$. Then the likelihood function of **a** and **b**, given the defect data set $y_i\,(i = 1, \ldots, p)$, is derived from (1) as:

$$L(\mathbf{a}, \mathbf{b}; y_1, \ldots, y_p) = \prod_{i=1}^{p} \mathrm{m}(x_i - x_{i-1})^{y_i - y_{i-1}} \exp\{-\mathrm{m}(x_i - x_{i-1})\}/(y_i - y_{i-1})! \tag{A.1}$$

where $x_0 = y_0 = 0$, and, the mean value function $m(t)$ is given by (2). After some algebra by taking partial derivatives of the log-likelihood function of (A.1) with respect to **a** and **b** and setting to zeros, we have the following two equations:

$$\mathbf{a} = y_p/[1 - \exp(-\mathbf{b}x_p)] \tag{A.2}$$

and

$$\sum_{i=1}^{p}(\mathrm{A}_i/\mathrm{B}_i) - \mathrm{C} = 0 \tag{A.3}$$

where A_i, B_i, and C are, respectively, given by:

$$\mathrm{A}_i = (y_i - y_{i-1})[x_i\exp(-\mathbf{b}x_i) - x_{i-1}\exp(-\mathbf{b}x_{i-1})] \tag{A.4}$$

$$\mathrm{B}_i = \exp(-\mathbf{b}x_{i-1}) - \exp(-\mathbf{b}x_i) \tag{A.5}$$

$$\mathrm{C} = x_p y_p/[\exp(\mathbf{b}x_p) - 1]. \tag{A.6}$$

Maximum likelihood estimates of **a** and **b** can be obtained by solving Eqs. (A.2) and (A.3). Note that Eq. (A.2) implies **a** and **b** satisfy the last data point (x_p, y_p). That is, the mean value function with the maximum likelihood estimates of **a** and **b** always goes through the first data point (x_0, y_0) and last data points point (x_p, y_p). It should be pointed out that Eq. (A.3) is nonlinear but can be easily implemented in a spreadsheet with the use of a built-in function such as "solver".

In order to obtain confidence intervals for **a** and **b**, we take a second derivative with respect to **b** and substitute the estimate of **b** into the negative of the second derivative. Since the inverse of the above quantity is considered as the variance of estimate **b**, the 90 % confidence interval for **b** can be constructed using a normal approximation. The 90 % confidence interval for **a** can be obtained using (A.2) for each limit of **b**.

References

1. Goel AL, Okumoto K (1979) Time-dependent error-detection rate model for software reliability and other performance measures. IEEE Trans Reliab 206–211
2. Jelinski Z, Moranda PB (1972) Software reliability research. In: Feiberger W (ed) Statistical computer performance evaluation. Academic, New York, pp 465–484
3. Jeske DR, Zhang X, Pham L (2005) Adjusting software failure rates that are estimated from test data. IEEE Trans Reliab 107–114
4. Lyu MR (1995) Handbook of Software Reliability Engineering. Computer Society Press, McGraw-Hill, Los Alamitos, New York
5. Musa JD (1993) Operational profiles in software-reliability engineering. IEEE Softw 14–32
6. Musa JD, Iannino A, Okumoto K (1987) Software Reliability: Measurement, Prediction, Application. McGraw-Hill, New York
7. Okamura H, Dohi T, Osaki S (2001) A reliability assessment method for software products in operational phase—proposal of an accelerated life testing model. Electron Commun Japan 25–33
8. Okumoto K (2010) Software reliability predictions—Are we done yet? QuEST Americas best practices conference, Atlanta, GA
9. QuEST Forum's TL 9000 (2007) Measurements Handbook Release 4.0
10. Schick GJ, Wolverton RW (1973) Assessment of software reliability. In: Proceedings of operations research, Physica-Verlag, Wurzburg-Wien, pp 395–422

11. Schneidewind NF (1975) Analysis of error processes in computer software. In: Proceedings of the international conference on reliable software, IEEE Computer Society, pp 337–346
12. Wallace D, Coleman C (2001) Application and improvement of software reliability models. Hardware and software reliability. Software Assurance Technology Center (SATC), pp 323–08
13. Yamada S, Ohba M, Osaki S (1983) S-shaped reliability growth modeling for software error detection. IEEE Trans Reliab 475–478
14. Zhang X, Pham H (2006) Software field failure rate prediction before software deployment. J Syst Softw 291–300

Recent Developments in Software Reliability Modeling and its Applications

Shigeru Yamada

Abstract Management technologies for improving software reliability are very important for software total quality management (TQM). The quality characteristics of software reliability are that computer systems can continue to operate regularly without the occurrence of failures on software systems. In this chapter, we describe several recent developments in software reliability modeling and its applications as quantitative techniques for software quality/reliability measurement and assessment. That is, a quality engineering analysis of human factors affecting software reliability during the design review phase, which is the upper stream of software development, and software reliability growth models based on stochastic differential equations (SDEs) and discrete calculus during the testing-phase, which is the lower one, are discussed. Finally, we discuss quality-oriented software management analysis by applying the multivariate analysis method and the existing software reliability growth models to actual process monitoring data.

1 Introduction

At present, it is important to assess the reliability of software systems because of increasing demands on quality and productivity in social systems. Moreover, they may cause serious accidents affecting people's lives. Against such a background, software reliability technologies for the purpose of producing quality software systems efficiently, systematically, and economically have been developed and researched energetically. Especially, comprehensive use of technologies and methodologies in software engineering is needed for improving software quality/reliability.

S. Yamada (✉)
Department of Social Management Engineering, Graduate School of Engineering,
Tottori University, Tottori-shi 680-8552, Japan
e-mail: yamada@sse.tottori-u.ac.jp

T. Dohi and T. Nakagawa (eds.), *Stochastic Reliability and Maintenance Modeling*,
Springer Series in Reliability Engineering 9, DOI: 10.1007/978-1-4471-4971-2_12,

A computer-software is developed by human work, therefore many software faults may be introduced into the software product during the development process. These software faults often cause breakdowns in computer systems. Recently, it has become more difficult for developers to produce highly reliable software systems efficiently because of the diversified and complicated software requirements. Therefore, it is necessary to control the software development process in terms of quality and reliability. Note that *software failure* is defined as an unacceptable departure of program operation caused by a *software fault* remaining in the software system.

First, in this chapter, we focus on a software design-review process which is more effective than other processes in the upper stream of software development for elimination and prevention of software faults. Then conducting a design-review experiment, we discuss a quality engineering approach for analyzing the relationships among the quality of the design-review activities, i.e., software reliability, and human factors to clarify the fault-introduction process in the design-review process.

Basically, software reliability can be evaluated by the number of detected faults or the software failure-occurrence time in the testing-phase which is the last phase of the development process, and it can be also estimated in the operational phase. Especially, *software reliability models* which describe software fault-detection or failure-occurrence phenomena in the system testing-phase are called *software reliability growth models* (*SRGMs*). The SRGMs are useful to assess the reliability for quality control and testing-process control of software development. Most of the SRGMs proposed till date treat the event of fault-detection in the testing and operational phases as a counting process. However, if the size of the software system is large, the number of faults detected during the testing-phase become large, and the change in the number of faults which are detected and removed through debugging activities becomes sufficiently small compared with the initial fault content at the beginning of the testing phase.

Then, in this chapter, we model the fault-detection process as a stochastic process with a continuous state space for reliability assessment in an open source solution developed under several open source softwares (OSSs) to consider the active state of the open source projects and the collision among the open source components. We propose a new SRGM describing the fault-detection process by applying a mathematical technique of stochastic differential equations of Itô-type.

Further, based on discrete analogs of nonhomogeneous Poisson process (NHPP) models as SRGMs, which have exact solutions in terms of solving the hypothesized differential equations, we propose two discrete models described by difference equations derived by transforming the continuous testing-time into a discrete one. Thus we show that such a difference calculus enables us to assess software reliability more accurately than conventional discrete models.

Finally, we discuss quality-oriented software management through statistical analysis of process monitoring data. Based on the desired software management models, we obtain the significant process factors affecting quality, cost, and delivery (QCD) measures. At the same time, we propose a method of software reliability assessment as process monitoring evaluation with actual data for the process monitoring progress ratio and the pointed-out problems (i.e., detected faults).

2 Human Factors Analysis

In this chapter, we discuss an experiment study to clarify human factors [1–3] and their interactions affecting software reliability by assuming a model of human factors which consist of inhibitors and inducers. In this experiment, we focus on the software design-review process which is more effective than the other processes in the elimination and prevention of software faults. For an analysis of experimental results, a quality engineering approach base on a *signal-to-noise ratio* (defined as SNR) [4] is introduced to clarify the relationships among human factors and software reliability measured by the number of seeded faults detected by review activities, and the effectiveness of significant human factors judged by the design of experiment [5] is evaluated. As a result, applying the orthogonal array $L_{18}(2^1 \times 3^7)$ to the human factor experiment, we obtain the optimal levels for the selected inhibitors and inducers.

2.1 Design-Review and Human Factors

The inputs and outputs for the design-review process are shown in Fig. 1. The design-review process is located in the intermediate process between design and coding phases, and have software requirement-specifications as inputs and software design-specifications as outputs. In this process, software reliability is improved by detecting software faults effectively [6].

The attributes of software designers and design process environment are mutually related to the design-review process (see Fig. 1). Then, influential human factors for the design-specifications as outputs are classified into two kinds of attributes in the following [7–9] (see Fig. 2):

(1) Attributes of the design reviewers (*Inhibitors*): Attributes of the design reviewers are those of software engineers who are responsible for design-review work. For example, they are the degree of understanding of software requirement-specifications and software design-methods, the aptitude of programmers, the

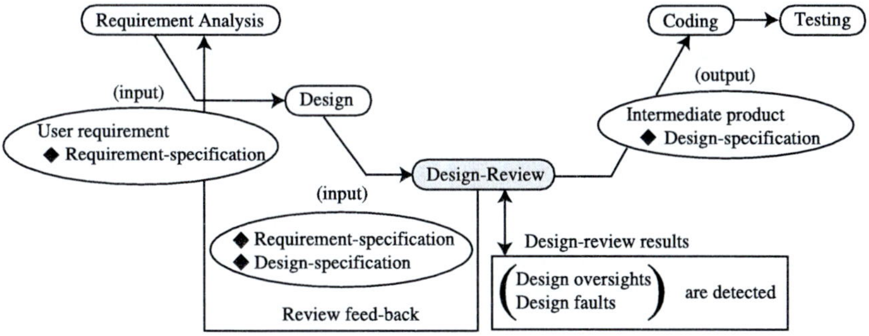

Fig. 1 Inputs and outputs in the software design process

experience and capability of software design, the volition of achievement of software design, etc. Most of them are psychological human factors which are considered to contribute directly to the quality of software design-specification.

(2) Attributes of environment for the design-review (*Inducers*): In terms of design-review work, many kinds of influential factors are considered such as the education of software design-methods, the kind of software design methodologies, the physical environmental factors in software design work, e.g., temperature, humidity, noise, etc. All of these influential factors may affect indirectly the quality of software design-specification.

2.2 Design-Review Experiment

In order to find the relationships among the reliability of software design-specification and its influential human factors, we have performed the design of experiment by selecting five human factors as shown in Table 1.

In this experiment, we conduct an experiment to clarify the relationships among human factors affecting software reliability and the reliability of design-review work by assuming a human factor model consisting of inhibitors and inducers as shown in Fig. 2. The actual experiment has been performed by 18 subjects based on the same design-specification of a triangle program which receives three integers representing the sides of a triangle and classifies the kind of triangle such sides form [10]. We measured the 18 subjects' capability of both the degrees of understanding of design-method and requirement-specification by the preliminary tests before the design of experiment. Further, we seeded some faults in the design-specification intentionally. Then, we have executed such a design-review experiment in which the 18 subjects detect the seeded faults.

Table 1 Human factors in the design-review experiment

	Human factor	Level		
		1	2	3
A[b]	BGM of classical music in the review work environment	A1:yes	A2:no	-
B[b]	Time duration of software design work (minute)	B1:20 min	B2:30 min	B3:40 min
C[a]	Degree of understanding of the designmethod (R-Net technique)	C1:high	C2:common	C3:low
D[a]	Degree of understanding of requirement-specifications	D1:high	D2:common	C3:low
E[b]	Check kist (indicating the matters that require attention in review work)	E1:detailed	E2:common	E3:nothing

[a]Inhibitors
[b]Inducers

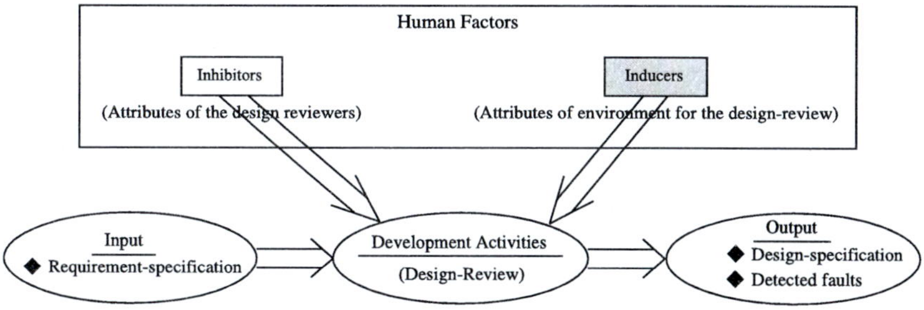

Fig. 2 A human factor model including inhibitors and inducers

We have performed the experiment by using the five human factors with three levels as shown in Table 1, which are assigned to the orthogonal-array $L_{18}(2^1 \times 3^7)$ of the design of experiment as shown in Table 3. We distinguish the design parts as follows to be pointed out in the design-review as detected faults into the descriptive-design and symbolic-design parts.

- *Descriptive-design faults* The descriptive-design parts consist of words or technical terminologies which are described in the design-specification to realize the required functions. In this experiment, the descriptive-design faults are algorithmic ones, and we can improve the quality of design-specification by detecting and correcting them.
- *Symbolical-design faults* The symbolical-design parts consist of marks or symbols which are described in the design-specification. In this experiment, the symbolical-design faults are notation mistakes, and the quality of the design-specification cannot be improved by detecting and correcting them.

For the orthogonal-array $L_{18}(2^1 \times 3^7)$ as shown in Table 3, setting the classification of detected faults as outside factor R and the human factors A, B, C, D, and E as inside factors, we perform the design-review experiment. Here, the outside factor R has two levels such as descriptive-design parts (R_1) and symbolical-design parts (R_2).

2.3 Analysis of Experimental Results

We define the efficiency of design-review, i.e., the reliability, as the degree that the design reviewers can accurately detect correct and incorrect design parts for the design-specification containing seeded faults. There exist the following relationships among the total number of design parts, n, the number of correct design parts, n_0, and the number of incorrect design parts containing seeded faults, n_1:

$$n = n_0 + n_1. \tag{1}$$

Table 2 Input and output tables for two kinds of error

	Output		
	0 (true)	1 (false)	Total
Input			
(i)Observed values			
0 (true)	n_{00}	n_{11}	n_0
1 (false)	n_1	n_{01}	n_{10}
Total	r_0	r_1	n
(ii)Error rates			
0 (true)	$1-p$	p	1
1 (false)	q	$1-q$	1
Total	$1-p+q$	$1-q+p$	2

Therefore, the design parts are classified as shown in Table 2 by using the following notations:

n_{00} = the number of correct design parts detected accurately as correct design parts,

n_{01} = the number of correct design parts detected by mistake as incorrect design parts,

n_{10} = the number of incorrect design parts detected by mistake as correct design parts,

n_{11} = the number of incorrect design parts detected accurately as incorrect design parts,

where two kinds of error rates are defined by

$$p = \frac{n_{01}}{n_0}, \tag{2}$$

$$q = \frac{n_{10}}{n_1}. \tag{3}$$

Considering the two kinds of error rates, p and q, we can derive the *standard error rate*, p_0, [4] as

$$p_0 = \frac{1}{1+\sqrt{\left(\frac{1}{p}-1\right)\left(\frac{1}{q}-1\right)}}. \tag{4}$$

Then, the SNR based on Eq. (4) is defined by (see Ref. [4])

$$\eta_0 = -10\log_{10}\left\{\frac{1}{(1-2p_0)^2}-1\right\}. \tag{5}$$

The standard error rate, p_0, can be obtained by transforming Eq. (5) using the SNR of each control factor as

$$p_0 = \frac{1}{2}\left\{1 - \frac{1}{\sqrt{10^{\left(-\frac{\eta_0}{10}\right)} + 1}}\right\}. \tag{6}$$

The method of experimental design based on orthogonal-arrays is a special one that requires only a small number of experimental trials to help us discover main factor effects. On traditional researches [7, 11], the design of experiment has been conducted using orthogonal-array $L_{12}(2^{11})$. However, since the orthogonal-array $L_{12}(2^{11})$ has only two levels for grasp of factorial effect to the human factors experiment, the middle effect between two levels cannot be measured. Thus, in order to measure it, we adopt the orthogonal-array $L_{18}(2^1 \times 3^7)$ that can lay out one factor with 2 levels (1, 2) and 7 factors with 3 levels (1, 2, 3) as shown in Table 3, and dispense with $2^1 \times 3^7$ trials by executing experimental independent 18 experimental trials with each other. For example, as for the experimental trial No. 10, we executed the design-review work under the conditions A_2, B_1, C_1, D_3, and E_3, and obtained the computed SNRs as 0.583 (dB) for the descriptive-design faults from the observed values $n_{00} = 52$, $n_{01} = 0$, $n_{10} = 10$, and $n_{11} = 4$, and as 4.497 (dB) for the symbolical-design faults from the observed values $n_{00} = 58$, $n_{01} = 1$, $n_{10} = 1$, and $n_{11} = 3$.

Table 3 The orthogonal array $L_{18}(2^1 \times 3^7)$ with assigned human factors and experimental data

No.	Human factors					Observed values								SNR (dB)	
						R_1				R_2					
	A	B	C	D	E	n_{00}	n_{01}	n_{10}	n_{11}	n_{00}	n_{01}	n_{10}	n_{11}	R_1	R_2
1	1	1	1	1	1	52	0	2	12	58	1	0	4	7.578	6.580
2	1	1	2	2	2	49	3	8	6	59	0	2	2	−3.502	3.478
3	1	1	3	3	3	50	2	12	2	59	0	4	0	8.769	2.342
4	1	2	1	1	2	52	0	2	12	59	0	0	4	7.578	8.237
5	1	2	2	2	3	50	2	4	10	57	2	0	4	1.784	4.841
6	1	2	3	3	1	45	7	8	6	59	0	3	1	−7.883	0.419
7	1	3	1	2	1	52	0	2	12	59	0	2	2	7.578	3.478
8	1	3	2	3	2	47	5	6	8	59	0	2	2	−3.413	3.478
9	1	3	3	1	3	52	0	10	4	58	1	1	3	0.583	4.497
10	2	1	1	3	3	52	0	10	4	58	1	1	3	0.583	4.497
11	2	1	2	1	1	47	5	1	13	59	0	3	1	3.591	0.419
12	2	1	3	2	2	46	6	8	6	59	0	4	0	6.909	2.342
13	2	2	1	2	3	46	6	10	4	59	0	0	4	−10.939	8.237
14	2	2	2	3	1	49	3	11	3	59	0	4	0	8.354	2.342
15	2	2	3	1	2	46	6	10	4	59	0	0	4	−10.939	8.237
16	2	3	1	3	2	50	2	2	12	59	0	0	4	4.120	8.237
17	2	3	2	1	3	50	2	4	10	57	2	0	4	1.784	4.841
18	2	3	3	2	1	44	8	6	8	59	0	3	1	−5.697	0.419

2.4 Investigation of Analysis Results

We analyze the simultaneous effects of outside factor R and inside human factors A, B, C, D, and E. As a result of the analysis of variance by taking account of correlation among inside and outside factors discussed in Sect. 2.2, we obtain Table 4. There are two kinds of errors in the analysis of variance: e_1 is the error among experiments of the inside factors, and e_2 the mutual correlation error between e_1 and the outside factor. In this analysis, since there was no significant effect by performing F-test for e_1 with e_2, F-test for all factors was performed by e_2. As a result, significant human factors such as the degree of understanding of the design-method (Factor C), the degree of understanding of requirement-specification (Factor D), and the classification of detected faults (Factor R) were recognized. Figure 3 shows the factor effect for each level in the significant factors which affect design-review work.

As a result of analysis, in the inside factors, only Factors C and D are significant and the inside and outside factors are not mutually interacted. That is, it turns out that reviewers with the high degree of understanding of the design-method and the high degree of understanding of requirement-specification can exactly review the design-specification efficiently regardless of the classification of detected faults. Moreover, the result that outside factor R is highly significant, and the descriptive-design faults are detected less than the symbolic-design faults, can be obtained. That is, although it is a natural result, it is difficult to detect and correct the algorithmic faults which lead to improvement in quality rather than the notation mistakes. However, it is important

Table 4 The result of analysis of variance by taking account of correlation among inside and outside factors

Factor	f	S	V	F_0	$\rho(\%)$
A	1	37.530	37.530	2.497	3.157
B	2	47.500	23.750	1.580	3.995
C	2	313.631	156.816	10.435[b]	26.380
D	2	137.727	68.864	4.582[a]	11.584
E	2	4.684	2.342	0.156	0.394
A×B	2	44.311	22.155	1.474	3.727
e_1	6	38.094	6.460	0.422	3.204
R	1	245.941	245.941	16.366[b]	20.686
A×R	1	28.145	28.145	1.873	2.367
B×R	2	78.447	39.224	2.610	6.598
C×R	2	36.710	18.355	1.221	3.088
D×R	2	9.525	4.763	0.317	0.801
E×R	2	46.441	23.221	1.545	3.906
e_2	8	120.222	15.028	3.870	10.112
T	35	1188.909			100.0

[a] 5 % level of significant
[b] 1 % level of significant

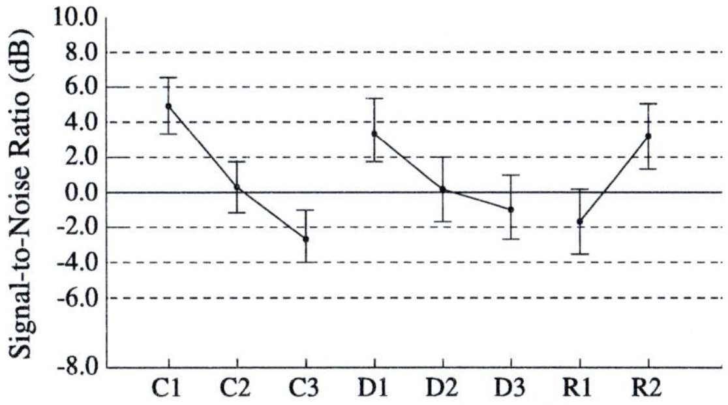

Fig. 3 The estimation of significant factors with correlation among inside and outside factors

to detect and correct the algorithmic faults as an essential problem of the quality improvement for design-review work. Therefore, in order to increase the rate of detection and correction of the algorithmic faults which lead to the improvement of quality, it is required before design-review work to make reviewers fully understand the design techniques used for describing design-specifications and the contents of requirement-specifications.

3 Stochastic Differential Equation Modeling

The software development environment has been changing into new development paradigms such as concurrent distributed development environment and the so-called open source project by using network computing technologies. Especially, such OSS systems which serve as key components of critical infrastructures in the society are still ever-expanding now [12].

The successful experience of adopting the distributed development model in such open source projects includes GNU/Linux operating system, Apache Web server, and so on [12]. However, the poor handling of the quality and customer support prohibits the progress of OSS. We focus on problems in the software quality, which prohibit the progress of OSS.

Especially, SRGMs [6, 13] have been applied to assess the reliability for quality management and testing-progress control of software development. On the other hand, the effective method of dynamic testing management for new distributed development paradigms as typified by the open source project has been presented by only a few works [14–17]. In case of considering the effect of the debugging process on the entire system for the development of a method of reliability assessment for OSS, it is necessary to grasp the situation of registration for bug tracking system, degree of maturation of OSS, and so on.

In this chapter, we focus on an open source solution developed under several OSSs. We discuss a useful software reliability assessment method in open source solution as a typical case of next-generation distributed development paradigm. Especially, we propose a software reliability growth model based on stochastic differential equations (SDEs) in order to consider the active state of the open source project and the component collision of OSS. Then, we assume that the software failure intensity depends on the time, and the software fault-report phenomena on the bug tracking system keeps an irregular state. Also, we analyze the actual software fault-count data to show numerical examples of software reliability assessment for the open source solution. Moreover, we compare our model with the conventional model based on SDEs in terms of goodness-of-fit for actual data. Then, we show that the proposed model can assist improvement of quality for an open source solution developed under several OSSs.

3.1 Stochastic Differential Equation Model

Let $S(t)$ be the number of detected faults in the open source solution by testing-time $t(t \geq 0)$. Suppose that $S(t)$ takes on continuous real values. Since latent faults in the open source solution are detected and eliminated during the operational phase, $S(t)$ gradually increases as the operational procedures go on. Thus, under common assumptions for software reliability growth modeling, we consider the following linear differential equation:

$$\frac{dS(t)}{dt} = \lambda(t)S(t), \tag{7}$$

where $\lambda(t)$ is the intensity of inherent software failures at operational time t and is a non-negative function.

Generally, it is difficult for users to use all functions in open source solution, because the connection state among open source components is unstable in the testing-phase of open source solution. Considering the characteristic of open source solution, the software fault-report phenomena keeps an irregular state in the early stage of testing-phase. Moreover, the addition and deletion of software components are repeated under the development of an OSS system, i.e., we consider that the software failure intensity depends on the time.

Therefore, we suppose that $\lambda(t)$ and $\mu(t)$ have irregular fluctuation. That is, we extend Eq. (7) to the following SDE [18, 19]:

$$\frac{dS(t)}{dt} = \{\lambda(t) + \sigma\mu(t)\gamma(t)\}S(t), \tag{8}$$

where σ is a positive constant representing a magnitude of the irregular fluctuation, $\gamma(t)$ a standardized Gaussian white noise, and $\mu(t)$ the collision level function of open source component.

We extend Eq. (8) to the following SDE of an Ito type:

$$dS(t) = \left\{\lambda(t) + \frac{1}{2}\sigma^2\mu(t)^2\right\} S(t)dt + \sigma\mu(t)S(t)\mathrm{d}\omega(t), \tag{9}$$

where $\omega(t)$ is a one-dimensional Wiener process which is formally defined as an integration of the white noise $\gamma(t)$ with respect to time t. The Wiener process is a Gaussian process and it has the following properties:

$$Pr[\omega(0) = 0] = 1, \tag{10}$$

$$\mathrm{E}[\omega(t)] = 1, \tag{11}$$

$$\mathrm{E}[\omega(t)\omega(t')] = \mathrm{Min}[t, t']. \tag{12}$$

By using Ito's formula [18, 19], we can obtain the solution of Eq. (8) under the initial condition $S(0) = \nu$ as follows [20]:

$$S(t) = \nu \cdot \exp\left(\left(\int_0^t \lambda(s)ds + \sigma\mu(t)\omega(t)\right)\right), \tag{13}$$

where ν is the number of detected faults for the previous software version. Using solution process $S(t)$ in Eq. (13), we can derive several software reliability measures.

Moreover, we define the intensity of inherent software failures, $\lambda(t)$, and the collision level function, $\mu(t)$, as follows:

$$\int_0^t \lambda(s)ds = (1 - \exp[-\alpha t]), \tag{14}$$

$$\mu(t) = \exp[-\beta t], \tag{15}$$

where α is an acceleration parameter of the intensity of inherent software failures, and β the growth parameter of the open source project.

3.2 *Method of Maximum-Likelihood*

In this section, the estimation method of unknown parameters α, β, and σ in Eq. (13) is presented. Let us denote the joint probability distribution function of the process $S(t)$ as

$$P(t_1, y_1; t_2, y_2; \ldots; t_K, y_K) \equiv \Pr[S(t_1) \le y_1, \ldots, S(t_K) \le y_K \mid S(t_0) = \nu], \tag{16}$$

where $S(t)$ is the cumulative number of faults detected up to the operational time $t (t \ge 0)$, and denote its density as

$$p(t_1, y_1; t_2, y_2; \ldots; t_K, y_K) \equiv \frac{\partial^K P(t_1, y_1; t_2, y_2; \cdots; t_K, y_K)}{\partial y_1 \partial y_2 \ldots \partial y_K}. \tag{17}$$

Since $S(t)$ takes on continuous values, we construct the likelihood function l for the observed data $(t_k, y_k)(k = 1, 2, \ldots, K)$ as follows:

$$l = p(t_1, y_1; t_2, y_2; \ldots; t_K, y_K). \tag{18}$$

For convenience in mathematical manipulations, we use the following logarithmic likelihood function:

$$L = \log l. \tag{19}$$

The maximum-likelihood estimates α^*, β^*, and σ^* are the values making L in Eq. (19) to maximize. These can be obtained as the solutions of the following simultaneous likelihood equations [20]:

$$\frac{\partial L}{\partial \alpha} = \frac{\partial L}{\partial \beta} = \frac{\partial L}{\partial \sigma} = 0. \tag{20}$$

3.3 Expected Number of Detected Faults

We consider the expected number of faults detected up to operational time t. The density function of $\omega(t)$ is given by:

$$f(\omega(t)) = \frac{1}{\sqrt{2\pi t}} \exp\left\{-\frac{\omega(t)^2}{2t}\right\}. \tag{21}$$

Information about the cumulative number of detected faults in the OSS system is important to estimate the situation of the progress on the software operational procedures. Since it is a random variable in our model, its expected value and variance can be useful measures. We can calculate the expected number of faults detected up to time t from Eq. (13) as follows [20]:

$$\mathrm{E}[S(t)] = \nu \cdot \exp\left(\int_0^t \lambda(s)ds + \frac{\sigma^2 \mu(t)^2}{2} t\right). \tag{22}$$

3.4 Numerical Illustrations

We focus on a large-scale open source solution based on the Apache HTTP Server [21], Apache Tomcat [22], MySQL [23], and JavaServer Pages (JSP). The fault-count data used in this chapter are collected in the bug tracking system on the website of each open source project. The estimated expected cumulative number of detected faults in Eq. (22) is shown in Fig. 4. Also, the sample path of the estimated numbers of detected faults in Eq. (13) is shown in Fig. 5, approximately.

We show the reliability assessment results for the other SDE model in terms of the performance evaluation of our model. The sample path of the estimated cumulative numbers of detected faults in the conventional SDE model for OSS [24] are shown in Fig. 6. Also, Fig. 7 is the sample path of the estimated numbers of remaining faults in the conventional SDE model [25]. From Figs. 6 and 7, we have found that the

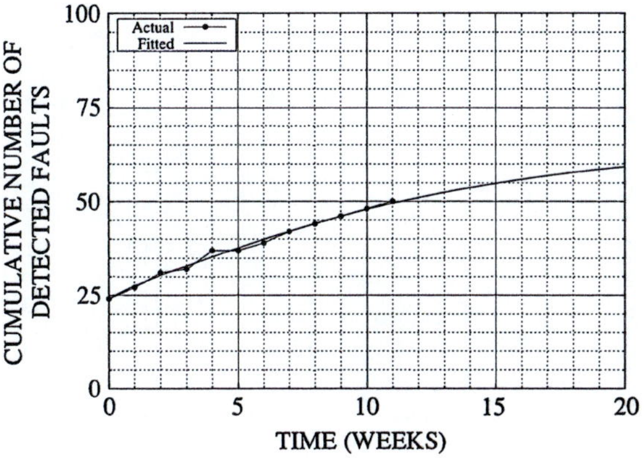

Fig. 4 The estimated cumulative number of detected faults, E[$S(t)$]

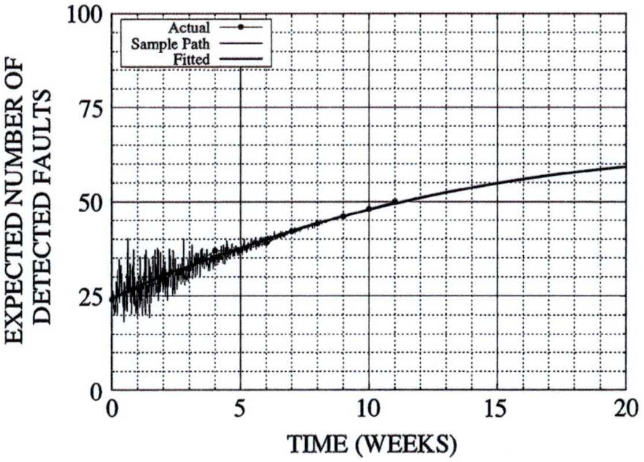

Fig. 5 The estimated path of the estimated number of detected faults

magnitude of the irregular fluctuation in the early phase of the proposed model is larger than those of the conventional SDE models, i.e., the irregular fluctuation in the proposed model depends on the time. Then, for the large-scale open source solution [26, 27], we may utilize the proposed model for assisting improvement of quality, in which it can describe actual fault-detection phenomena.

4 Discrete NHPP Modeling

In recent researches, Satoh [28] proposed a discrete Gompertz curve model, and Satoh and Yamada [29] suggested parameter estimation procedures for software reliability assessment of a discrete logistic curve model, and compared these models by using

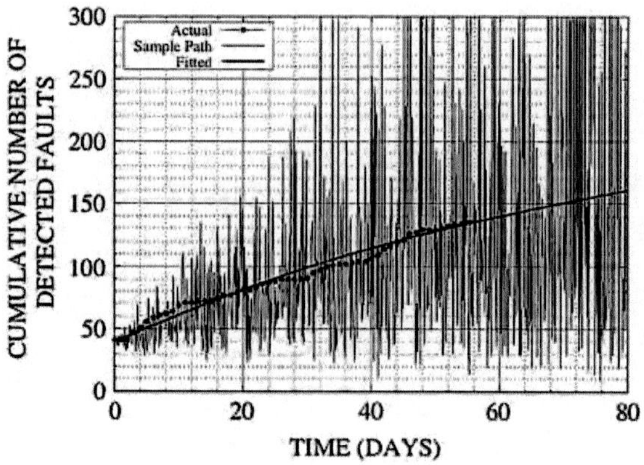

Fig. 6 The sample path of the estimated cumulative number of detected faults for SDE model for OSS

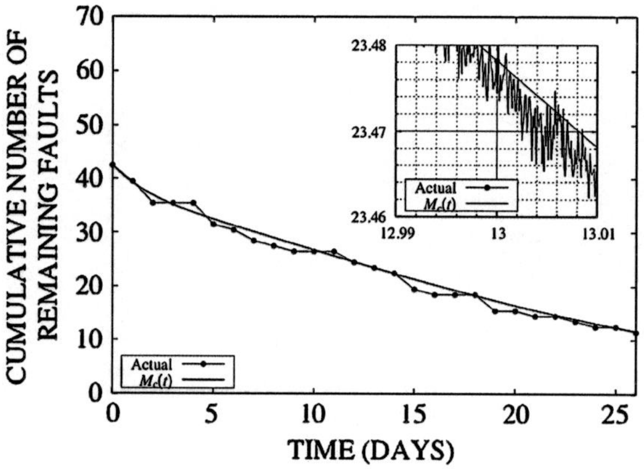

Fig. 7 The sample path of the estimated number of remaining faults for the conventional SDE model

a new proposed criterion. They reported that the discrete models as statistical data analysis models enable us to obtain accurate parameter estimates even with a small amount of observed data for particular applications.

In this chapter, we discuss the discrete NHPP models [6] derived by employing a difference method which conserves the gauge invariance from above results and high applicability of NHPP models point of view. The discrete NHPP models, that is, the discrete exponential SRGM and the discrete inflection S-shaped SRGM, have exact solutions. The difference equations and their exact solutions tend to the differential equations and their exact solutions. Therefore, the proposed models conserve the characteristics of the continuous NHPP models. The proposed models can be easily applied to regression equations to get accurate parameter estimates, and have more advantages in terms of numerical calculations than the maximum-likelihood estimation [30].

We assume a discrete counting process $\{N_n, n \geq 0\}(n = 0, 1, 2, \ldots)$ representing the cumulative number of faults detected by nth period from the test beginning. Then, the NHPP model with mean value function D_n representing the expected cumulative number of faults is formulated by

$$\Pr\{N_n = x\} = \frac{[D_n]^x}{x!}\exp[-D_n] \qquad (n, x = 0, 1, 2, \ldots). \tag{23}$$

We employ a difference method which conserves the gauge invariance because the proposed discrete NHPP models have to conserve the characteristic of the continuous NHPP models, i.e., the continuous NHPP models have exact solutions. With regard to parameter estimations, the difference equations can be easily applied to regression equations to get accurate parameter estimates, and these models have some advantages in terms of numerical calculations. Therefore, we can estimate unknown parameters by adopting the method of ordinary least-square procedures from the regression equations.

4.1 Discrete Exponential SRGM

We propose a discrete analog of the original exponential SRGM whose mean value function is of the simplest form in the SRGMs. The difference equation for this model has an exact solution. Let H_n denote the expected cumulative number of software faults detected by nth period from the test beginning. Then, we derive a discrete analog of the exponential SRGM from the hypotheses of the continuous NHPP model as follows:

$$H_{n+1} - H_n = \delta b(a - H_n). \tag{24}$$

Solving the above equation, an exact solution H_n in Eq. (24) is given by

$$H_n = a[1 - (1 - \delta b)^n] \qquad (a > 0,\ 0 < b < 1), \tag{25}$$

where δ represents the constant time-interval, a the expected total number of potential software failures occured in an infinitely long duration or the expected initial fault content, and b the fault-detection rate per fault. As $\delta \to 0$, Eq. (25) converges to an exact solution of the original exponential SRGM which is described by the differential equation.

We can derive a regression equation from Eq. (24) to estimate the model parameters. The regression equation is obtained as

$$Y_n = A + BH_n, \tag{26}$$

where

$$\begin{cases} Y_n = H_{n+1} - H_n \\ A = \delta ab \\ B = -\delta b. \end{cases} \tag{27}$$

Using Eq. (26), we can estimate $\hat{A}$ and $\hat{B}$ by using the observed data, which are the estimates of A and B. Therefore, we can obtain the parameter estimates $\hat{a}$ and $\hat{b}$ from Eq. (27) as follows:

$$\begin{cases} \hat{a} = -\hat{A}/\hat{B} \\ \hat{b} = -\hat{B}/\delta. \end{cases} \tag{28}$$

Y_n in Eq. (26) is independent of δ because δ is not used in calculating Y_n in Eq. (26). Hence, we can obtain the same parameter estimates of $\hat{a}$ and $\hat{b}$, respectively, when we choose any value of δ.

4.2 Discrete Inflection S-Shaped SRGM

We also propose a discrete analog of the original inflection S-shaped SRGM which is the continuous one. Let I_n denote the expected cumulative number of software faults detected by nth period from the test beginning. Then, we can derive a discrete analog of the inflection S-shaped SRGM from the hypotheses of the continuous NHPP model as follows:

$$I_{n+1} - I_n = \delta abl + \frac{\delta b(1-2l)}{2}[I_n + I_{n+1}] - \frac{\delta b(1-l)}{a} I_n I_{n+1}. \tag{29}$$

Solving the above difference equation, an exact solution I_n in Eq. (29) is given by

$$I_n = \frac{a\left[1 - \left(\frac{1-\frac{1}{2}\delta b}{1+\frac{1}{2}\delta b}\right)^n\right]}{1 + c\left(\frac{1-\frac{1}{2}\delta b}{1+\frac{1}{2}\delta b}\right)^n} \quad (a > 0,\ 0 < b < 1,\ c > 0,\ 0 \le l \le 1), \tag{30}$$

where δ represents the constant time-interval, a the expected total number of potential software failures occured in an infinitely long duration or the expected initial fault content, b the fault-detection rate per fault, and c the inflection parameter. The inflection parameter is specified as follows: $c = (1-l)/l$ where l is the inflection rate which indicates the ratio of the number of detectable faults to the total number of faults in the software system. As $\delta \to 0$, Eq. (30) converges to an exact solution of the original inflection S-shaped SRGM which is described by the differential equation.

Defining the difference operator as

$$\Delta I_n \equiv \frac{I_{n+1} - I_n}{\delta}. \tag{31}$$

We show that the inflection point occurs when

$$\bar{n} = \begin{cases} < n^* > & (\text{if } \Delta I_{<n^*>} \geq \Delta I_{<n^*>+1}) \\ < n^* > +1 & (\text{otherwise}), \end{cases} \tag{32}$$

where

$$n^* = -\frac{\log c}{\log\left(\frac{1-\frac{1}{2}\delta b}{1+\frac{1}{2}\delta b}\right)} - 1, \tag{33}$$

$$< n^* > = \{n | \max(n \leq n^*),\ n \in Z\}. \tag{34}$$

Moreover, we define t^* as

$$t^* = n^* \delta. \tag{35}$$

When n^* is an integer, we can show that t^* converges the inflection point of the inflection S-shaped SRGM which is described by the differential equation as $\delta \to 0$ as follows:

$$t^* = -\delta \frac{\log c}{\log\left(\frac{1-\frac{1}{2}\delta b}{1+\frac{1}{2}\delta b}\right)} - \delta \ \to \ \frac{\log c}{b} \text{ as } \delta \to 0. \tag{36}$$

By the way, the inflection S-shaped SRGM is regarded as a Riccati equation. Hirota [31] proposed a discrete Riccati equation which has an exact solution. A Bass model [32] which forecasts the innovation diffusion of products is also a Riccati equation. Satoh [33] proposed a discrete Bass model which can overcome the shortcomings of the ordinary least-square procedures in the continuous Bass model.

We can derive a regression equation to estimate the model parameters from Eq. (29). The regression equation is obtained as

$$Y_n = A + BK_n + CL_n, \tag{37}$$

where

$$\begin{cases} Y_n = I_{n+1} - I_n \\ K_n = I_n + I_{n+1} \\ L_n = I_n I_{n+1} \\ A = \delta abl \\ B = \delta b(1 - 2l)/2 \\ C = -\delta b(1 - l)/a. \end{cases} \tag{38}$$

Using Eq. (37), we can estimate $\hat{A}$, $\hat{B}$, and $\hat{C}$ by using the observed data, which are the estimates of A, B, and C, respectively. Therefore, we can obtain the parameter estimates $\hat{a}$, $\hat{b}$, and $\hat{l}$ from Eq. (38) as follows:

$$\begin{cases} \hat{a} = \hat{A}/(\sqrt{\hat{B}^2 - \hat{A}\hat{C}} - \hat{B}) \\ \hat{b} = 2\sqrt{\hat{B}^2 - \hat{A}\hat{C}}/\delta \\ \hat{l} = (1 - \hat{B}/\sqrt{\hat{B}^2 - \hat{A}\hat{C}})/2. \end{cases} \tag{39}$$

Y_n, K_n, and L_n in Eq. (37) are independent of δ because δ is not used in calculating Y_n, K_n, and L_n in Eq. (37). Hence, we can obtain the same parameter estimates $\hat{a}$, $\hat{b}$, and $\hat{l}$, respectively, when we choose any value of δ.

4.3 Model Comparisons

We show the result of goodness-of-fit comparisons in this section. We compare the four discrete models by using four data sets (DS1–DS4) observed in actual software testing. The four discrete models are as follows: two discrete NHPP models that were discussed in Sects. 4.1 and 4.2, a discrete logistic curve model [29, 30], and a discrete Gompertz curve model [28]. The data sets of DS1 and DS2 indicate exponential growth curves, and those of DS3 and DS4 indicate S-shaped growth curves, respectively. We employ the predicted relative error [6], the mean square errors (MSE) [6], and Akaike's Information Criterion (AIC) [6] as criteria of the model comparison in this chapter.

The predicted relative error is a useful criterion for indicating the relative errors between the predicted number of faults discovered by termination time of testing by using the part of observed data from the test beginning and the observed number of faults discovered by the termination time. Let $R_e[t_e]$ denote the predicted relative error at arbitrary testing-time t_e. Then, the predicted relative error is given as b

$$R_e[t_e] = \frac{\hat{y}(t_e; t_q) - q}{q}, \tag{40}$$

where $\hat{y}(t_e; t_q)$ is the estimated value of the mean value function at the termination time t_q using the observed data by the arbitrary testing-time $t_e (0 \leq t_e \leq t_q)$, and q is the observed cumulative number of faults detected by the termination time. We show Figs. 8, 9 and 10 which are the results of the model comparisons based on the predicted relative error for DS1 and DS3. MSE is obtained by using the sum of squared errors between the observed and estimated cumulative numbers of detected faults, y_k and $\hat{y}(t_k)$ during $(0, t_k]$, respectively. Getting N data pairs $(t_k, y_k)(k = 1, 2, \ldots, N)$, MSE is given by

$$\text{MSE} = \frac{1}{N}\sum_{k=1}^{N}[y_k - \hat{y}(t_k)]^2, \tag{41}$$

where $\hat{y}(t_k)$ denote the estimated value of the expected cumulative number of faults by using exact solutions of each model by the arbitrary testing-time $t_k (k = 1, 2, \ldots, N)$.

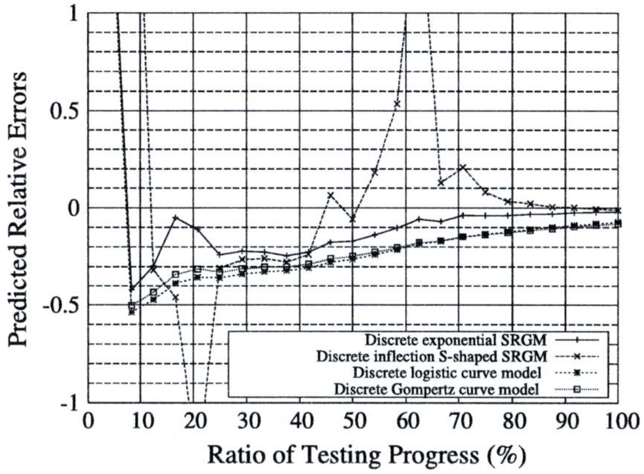

Fig. 8 The predicted relative for DS1

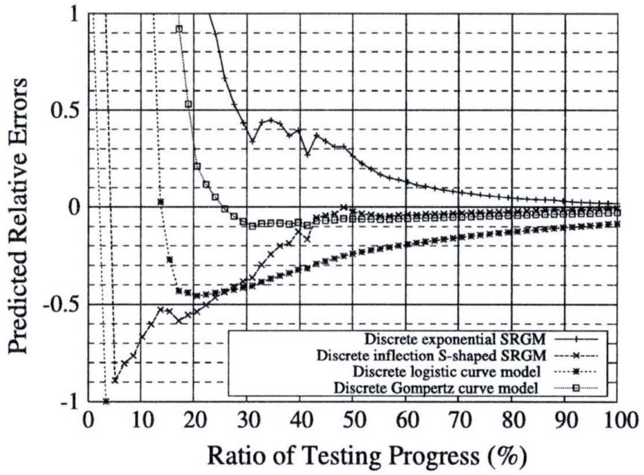

Fig. 9 The predicted relative error for DS3

Table 5 shows the result of model comparison based on MSE for each model. From Table 5, we conclude that the discrete inflection S-shaped SRGM fits better to all data sets except for DS2. However, the result of model comparison based on MSE depends on the number of model parameters of each model, e.g., the discrete exponential SRGM has two parameters and the discrete inflection S-shaped one has three parameters. Therefore, as a criterion of goodness-of-fit comparison for the two discrete models, i.e., the discrete exponential SRGM and the discrete inflection S-shaped one, we adopt the value of AIC. Table 6 shows the result of model comparison based on AIC. From Table 6, we can validate the above evaluation for MSE.

From these three results of goodness-of-fit comparison, we conclude that the discrete exponential SRGM is a more useful model for software reliability assessment for the observed data which indicate an exponential growth curve, and the

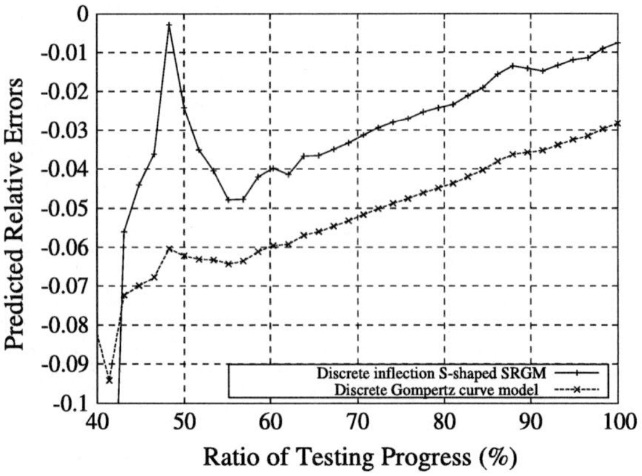

Fig. 10 The model comparison based on the predicted relative error for DS3 focusing on the discrete Gompertz curve model and the discrete inflection S-shaped SRGM

Table 5 The result of model comparison based on MSE

Data set	Discrete exponential SRGM	Discrete inflection S-shaped SRGM	Discrete logistic curve model	Discrete Gompertz curve model
DS1	39.643	**12.141**	101.92	72.854
DS2	**1762.5**	2484.0	27961	13899
DS3	25631	**9598.1**	149441	19579
DS4	11722	**438.59**	49741	27312

Table 6 The result of model comparison between the discrete exponential SRGM and discrete inflection S-shaped SRGM based on AIC

Data set	Discrete expo-nential SRGM	Discrete inflection S-shaped SRGM	Absolute value of difference
DS1	110.031	109.195	0.836
DS2	**115.735**	118.752	3.017
DS3	617.434	**606.132**	11.30
DS4	315.069	**274.818**	40.25

discrete inflection S-shaped SRGM is a more useful one for assessment after 60 % of the testing progress ratio for the observed data which indicate an S-shaped growth curve.

4.4 Software Reliability Assessment

We show useful quantitative measures for software reliability assessment by using the discrete NHPP models proposed in this chapter. We adopt DS1, i.e., the observed

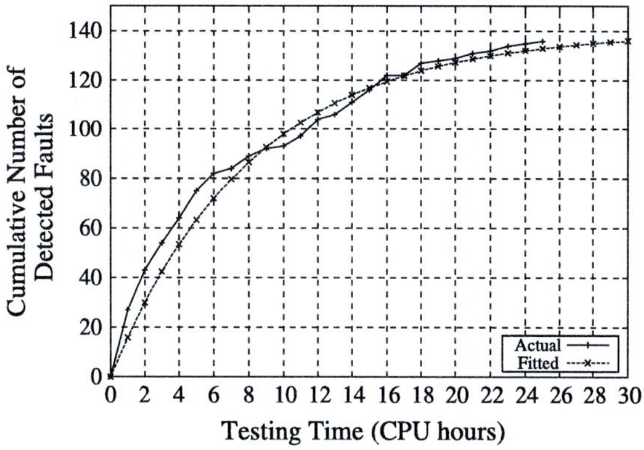

Fig. 11 The estimated discrete mean value function, $\hat{H}_n$, for DS1

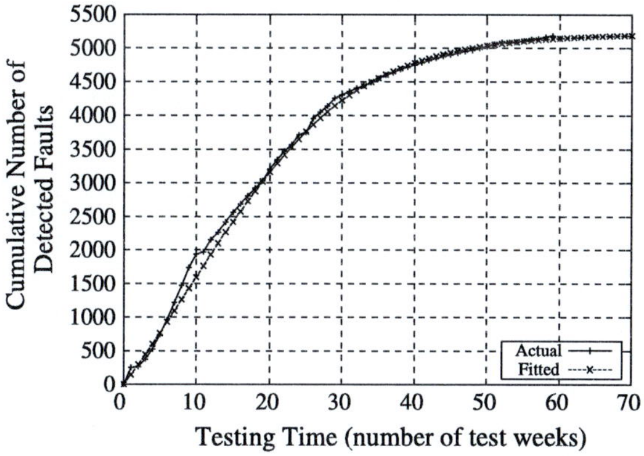

Fig. 12 The estimated discrete mean value function, $\hat{I}_n$, for DS3

Table 7 The estimated parameters of $\hat{H}_n$ for DS1 and $\hat{I}_n$ for DS3

	$\hat{a}$	$\hat{b}(\delta = 1)$	$\hat{c}$	n^*	$< n^* >$	$\bar{n}$
Hn	139.956	0.113				
In	5217.88	0.0906	2.350	8.383	8	9

25 pairs $(t_k, y_k)(k = 1, 2, \ldots, 25\ ;\ t_{25} = 25,\ y_{25} = 136)$ for the discrete exponential SRGM, and DS3, i.e., the observed 59 pairs $(t_k, y_k)(k = 1, 2, \ldots, 59\ ;\ t_{59} = 59,\ y_{59} = 5,186)$ for the discrete inflection S-shaped SRGM, where y_k is the cumulative number of faults detected by the execution of testing time t_k. The observation time unit of DS1 is CPU hours, and that of DS3 the number of weeks. We show the estimated mean value functions of H_n in Eq. (25) and I_n in Eq. (30) in Figs. 11 and 12, respectively, where several quantities are shown in Table 7.

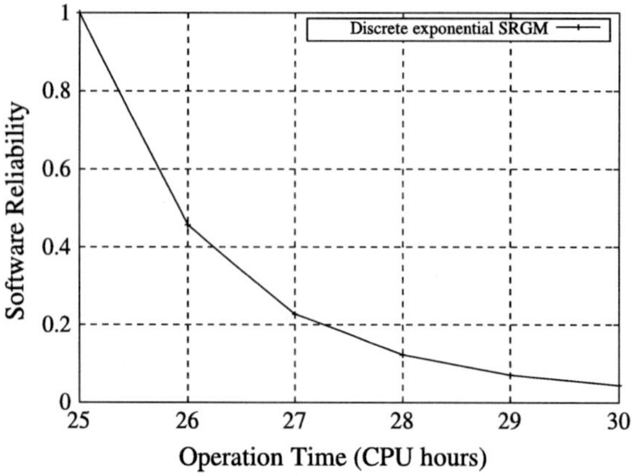

Fig. 13 The estimated software reliability function, $\hat{R}(25, h)$, for DS1

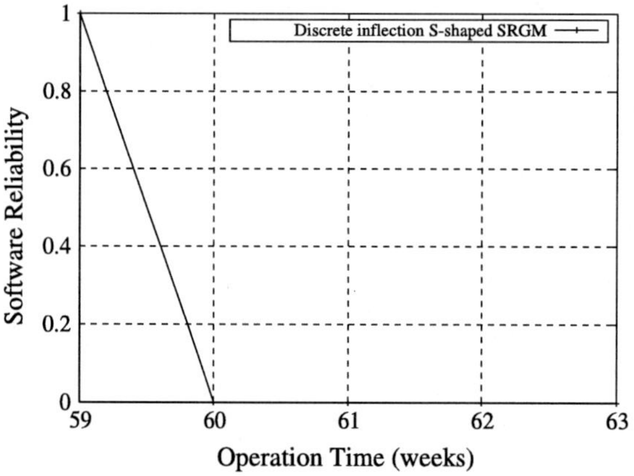

Fig. 14 The estimated software reliability function, $\hat{R}(59, h)$, for DS3

We can derive the software reliability function which is a useful measure for software reliability assessment. The software reliability function is obtained by Eq. (23) as follows:

$$
\begin{aligned}
R(n, h) &\equiv \Pr\{N_{n+h} - N_n = 0 | N_n = x\} \\
&= \exp[-\{D_{n+h} - D_n\}]. \qquad (42)
\end{aligned}
$$

Letting $\delta = 1$, the software reliability function for H_n after the termination time $n = 25$ (CPU hours), and for I_n after the termination time $n = 59$ (weeks), are shown in Figs. 13 and 14, respectively. After releasing the software systems at these time points, assuming that the software users operate these software systems under the same environment as the software testing one, we can estimate the software reliability $\hat{R}(25, 1.0)$ for H_n to be about 0.46. Also, we can estimate one $\hat{R}(59, 1.0)$ for I_n to be about 0.0.

5 Quality-Oriented Software Management Analysis

In this chapter, first, we conduct multivariate linear analyses by using process monitoring [34] data, derive effective process factors affecting the final products' quality, and discuss the significant process factors with respect to software management measures of QCD [35, 36]. Then, we analyze actual process monitoring data based on the derivation procedures of a process improvement model, i.e., software management model [37, 38] (as shown in Fig. 15). Then, we discuss project management on the significant process factors affecting the QCD measures, and show their effect on them. Second, we analyze the process monitoring data in a viewpoint of software reliability measurement and assessment in the process monitoring activities.

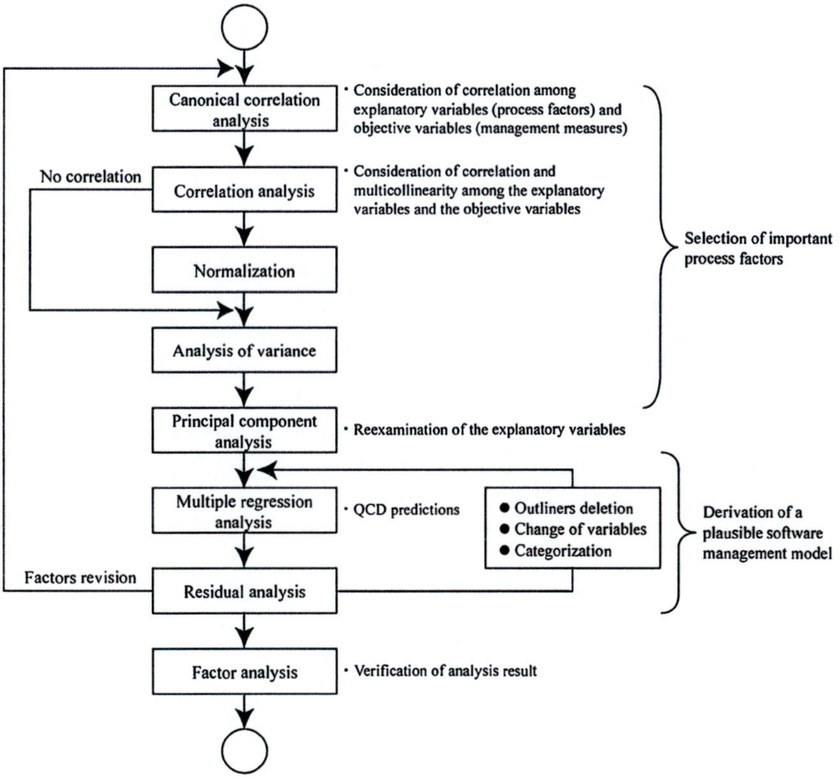

Fig. 15 Derivation procedures of software management model

5.1 Process Monitoring Data

We predict software management measures of QCD by using the process monitoring data as shown in Table 8. Five variables measured in terms of the number of faults (QCD problems) detected through the process monitoring, i.e., the contract review, the development planning review, the design completion review, the test planning review, and the test completion review phases are used as explanatory variables. The observed values of these five factors are normalized by each project development size (KLOC, 10^3LOC) in this chapter. Three variables, i.e., the number of faults detected during customer acceptance testing, the cost excess rate, and the number of delivery-delay days, are used as objective variables.

5.2 Factor Analysis Affecting QCD

Based on the canonical correlation analysis and the correlation analysis in Fig. 15, X_3 is selected as an important factor for estimating a software quality prediction model. Then, a single regression analysis is applied to the process monitoring data as shown in Table 8. Then, using X_3, we have the estimated single regression equation predicting the number of software faults, $\hat{Y}_q$, given by Eq. (43) as well as the normalized single regression expression, $\hat{Y}_q^N$, given by Eq. (44):

$$\hat{Y}_q = 11.761 \cdot X_3 + 0.998, \tag{43}$$

$$\hat{Y}_q^N = 0.894 \cdot X_3, \tag{44}$$

where the squared multiple correlation coefficient adjusted for degrees of freedom (adjusted R^2) is given by 0.758, and the derived linear quality prediction model is significant at 1 % level.

In a similar discussion to factor analysis affecting the number of faults above, as the result of canonical correlation analysis, correlation analysis, and principal component analysis, we can select X_1 and X_5 as the important factors for estimating the cost excess rate and delivery-delay days. Then, using X_1 and X_5, we have the estimated multiple regression equation predicting cost excess rate, $\hat{Y}_c$, given by Eq. (45) as well as the normalized multiple regression expression, $\hat{Y}_c^N$, given by Eq. (46):

$$\hat{Y}_c = 0.253 \cdot X_1 + 1.020 \cdot X_5 + 0.890, \tag{45}$$

$$\hat{Y}_c^N = 0.370 \cdot X_1 + 0.835 \cdot X_5, \tag{46}$$

where the adjusted R^2 is given by 0.917, and the derived cost excess prediction model is significant at 1 % level.

By the same way as the cost excess rate, using X_1 and X_5, we have the estimated multiple regression equation predicting the number of delivery-delay days, $\hat{Y}_d$, given

Table 8 Process monitoring data

Project No.	Contract review (X1) Number of faults per development size	Development planning (X2) Number of faults per development size	Design completion review (X3) Number of faults per development size	Test planning review (X4) Number of faults per development size	Test completion (X5) Number of faults per development size
1	0.591	1.181	0.295	0.394	0.394
2	0.323	0.645	0	0.108	0.108
3	0.690	0.345	0	0.345	0
4	0.170	0.170	0	0.085	0
5	0.150	0.451	0.301	0.075	0.075
6	1.186	0.149	0	0.037	0.037
7	0.709	0	0	0	0
	Quality (Yq) Number of faults detected during acceptance testing	Cost (Yc) Cost excess rate	Delivery (Yd) Number of delivery-delay days		
1	4	1.456	28		
2	1	1.018	3		
3	0	1.018	4		
4	2	0.953	0		
5	5	1.003	0		
6	0	1	−8		
7	2	1.119	12		

by Eq. (47) as well as the normalized multiple regression expression, $\hat{Y}_d^N$, given by Eq. (48):

$$\hat{Y}_d = 24.669 \cdot X_1 + 55.786 \cdot X_5 - 9.254, \tag{47}$$

$$\hat{Y}_d^N = 0.540 \cdot X_1 + 0.683 \cdot X_5, \tag{48}$$

where the adjusted R^2 is given by 0.834, and the derived delivery-delay prediction model is significant at 5 % level.

5.3 Analysis Results of Software Management Models

We have derived software management models by applying the methods of multivariate linear analysis to actual process monitoring data. Quantitative evaluation based on the derived prediction models about final product quality, cost excess, and delivery-delay, has been conducted with high accuracy. Then, it is very effective to promote software process improvement under Plan, Do, Check, Act (PDCA) management cycle by using the derivation procedures of software management models as shown in Fig. 15.

Further, the design completion review has an important impact on software quality. Then, it is possible to predict software product quality in the early-stage of software development project by using the result of the design completion review in process monitoring activities.

Next, the contract review and the test completion review processes have important impacts on the cost excess rate and the delivery-delay days. That is, it is difficult to predict cost excess and delivery-delay measures at the early stage of software development project, and it is found that the cost excess and delivery-delay measures can be predicted according to the same process monitoring factors.

5.4 Implementation of Project Management

5.4.1 Continuous Process Improvement

From the result of software management model analyses and factor analyses, it is found that the contract review has an important relationship with cost and delivery measures. Then, in order to improve the cost excess and delivery-delay, we perform suitable project management for the problems detected in the contract review.

The project management practices to be performed for the important problems detected in the contract review are:

- Early decision of the specification domain.
- Improvement of requirement specification technologies.

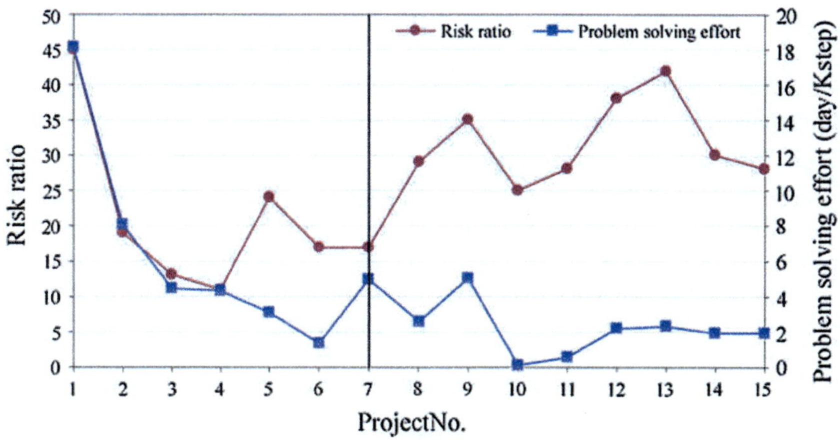

Fig. 16 Relationship between risk ratio and problem solving effort

- Early decision of development schedule.
- Improvement of project progress management.
- Improvement of testing technology.

As a result of carrying out project management and continuous process improvement, the relationship between the risk ratio measured at the initial stage of a project and the amount of problem solving effort (man-day) in the contract review become as shown in Fig. 16 where Projects 8–21 were monitored under process improvement based on the analysis results for Projects 1–7, and the risk ratio is given by

$$R = \sum_i \{\text{risk item}(i) \times \text{weight}(i)\}. \tag{49}$$

In Eq. (49), the risk estimation checklist has weight(i) in each risk item(i), and the risk ratio ranges between 0 and 100 points. Project risks are identified by interviewing based on the risk estimation checklist. From the identified risks, the risk ratio of a project is calculated by Eq. (49).

From Fig. 16, it is found that by performing suitable project management for the important problems in the contract review from Projects 8–15, the problem can be solved at the early stage of the software project even if the risk ratio is high.

5.4.2 Implementation of Design Quality Evaluation

In a similar fashion to cost and delivery measures, it is found that the design completion review has an important relationship with software quality. Then, in order to improve software quality, we decide to perform suitable project management called design evaluation in the design completion review.

The design evaluation assesses the following items based on the risk estimation checklist by the project manager, the designer, and the members of quality control department. Through the following design evaluation, we have to judge if the development can proceed to the next stage:

- After the requirements analysis, how many requirements are included in the requirement specifications? Are the requirements (function requirements and non-function requirements) suitably defined?
- After the elementary design, have the requirements (function requirements and non-function requirements) been taken over from the user requirements to the design documents without omission of the description items in the requirement specification?
- As for elementary design documents, is the elementary design included?

After implementation of the design evaluation, we have found that by performing design evaluation in the design completion review from Projects 17–21, the software quality has improved, and the cost excess rate and the delivery-delay days are also stable.

5.5 Software Reliability Assessment

Next, we discuss software reliability measurement and assessment based on the process monitoring data. A software reliability growth curve in process monitoring activities shows the relationship between the process monitoring progress ratio and the cumulative number of faults (QCD problems) detected during process monitoring. Then, we apply SRGMs based on an NHPP [6]. Table 9 shows the process monitoring data which are analyzed to evaluate software reliability, where Table 8 is derived from Table 9 for Projects 1–7, and Projects 8–21 were monitored under process improvement based on the analysis results for Projects 1–7. However, the collected process monitoring data have some missing values in metrics. Therefore, we apply collaborative filtering to the observed data to complement the missing values for assessing software reliability. The underlined values in Table 9 are the metrics values complemented by collaborative filtering.

We discuss software reliability growth modeling based on an NHPP because an analytic treatment of it is relatively easy. Then, we choose the process monitoring progress ratio as the alternative unit of testing-time by assuming that the observed data for testing-time are continuous.

In order to describe a fault-detection phenomenon at processing monitoring progress ratio $t\,(t \geq 0)$, let $\{N(t),\ t \geq 0\}$ denote a counting process representing the cumulative number of faults detected up to progress ratio t. Then, the fault-detection phenomenon can be described as follows:

$$\Pr\{N(t) = n\} = \frac{\{H(t)\}^n}{n!} \exp[-H(t)] \ (n = 0, 1, 2, \ldots), \tag{50}$$

Table 9 Process monitoring data for applying SRGM's

Project No.	Contact review (X_1)		Development planning review (X_2)		Design completion review (X_3)		Test planning review (X_4)		Test completion review (X_5)	
	Number of detected faults	Total days for desolving faults	Number of detected faults	Total days for desolving faults	Number of detected faults	Total days for desolving faults	Number of detected faults	Total days for desolving faults	Number of detected faults	Total days for desolving faults
1	6	184	12	223	3	109	4	49	4	132
2	3	75	6	97	0	0	1	7	1	5
3	4	26	2	14	0	0	2	47	0	0
4	2	51	2	14	0	0	1	8	0	0
5	2	41	6	158	4	39	1	6	1	5
6	5	36	4	122	0	0	1	27	1	5
7	1	7	0	0	0	0	0	0	0	0
8	3	12	9	188	0	0	3	20	0	0
9	3	42	7	161	1	21	4	43	2	25
10	4	4	3	15	1	24	3	3	4	4
11	2	15	3	15	1	20	4	8	1	18
12	5	27	5	40	1	20	6	30	1	18
13	6	32	5	51	1	20	6	33	1	18
14	3	15	4	25	1	20	4	22	1	18
15	2	13	2	20	1	18	3	12	0	0
16	6	107	4	104	1	19	2	39	0	0
17	3	12	5	100	1	20	2	2	1	6
18	2	30	3	42	1	22	0	0	1	6
19	1	56	1	2	1	18	0	0	0	0
20	3	54	3	6	3	20	0	0	0	0
21	1	1	4	8	0	0	0	0	0	0

where $H(t)$ represents the expected value of $N(t)$ called a mean value function of the NHPP. $\Pr\{A\}$ in Eq. (50) means the probability of event A. In this chapter, we apply three NHPP models [6], i.e., the exponential SRGM, the delayed S-shaped SRGM, and the logarithmic Poisson execution time model.

Software reliability assessment measures play an important role in quantitative software reliability assessment based on an SRGM. The expected number of remaining faults, $n(t)$, represents the number of faults latent in the software system by arbitrary testing-time t, and is formulated as

$$n(t) \equiv \mathrm{E}[N(\infty) - N(t)] = \mathrm{E}[N(\infty)] - H(t), \tag{51}$$

where E[A] represents the expected value for random variable A. And an instantaneous mean time between software faults (MTBF) is formulated as

$$\mathrm{MTBF}_I(t) = \frac{1}{\mathrm{d}H(t)/\mathrm{d}t}, \tag{52}$$

which is one of the substitute measures of the MTBF for the NHPP model.

Further, a software reliability function represents the probability that a software failure does not occur in the time-interval $(t,\ t+x]$ $(t \geq 0,\ x \geq 0)$ given that the testing or the user operation has been going up to time t. Then, if the counting process $\{N(t),\ t \geq 0\}$ follows the NHPP with mean value function $H(t)$, the software reliability function is derived as

$$R(x \mid t) = \exp[-\{H(t+x) - H(t)\}]. \tag{53}$$

We have found that the logarithmic Poisson execution time model shows the best goodness-of-fit in Projects 11–14 in which the test completion review's missing values are complemented by collaborative filtering. We have also found that the

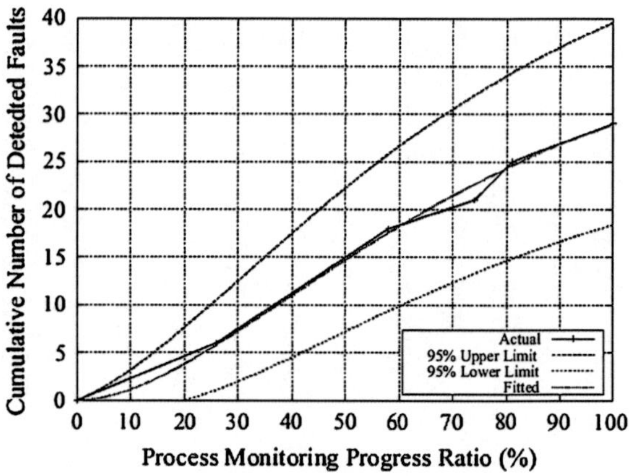

Fig. 17 The estimated mean value function for Project 1

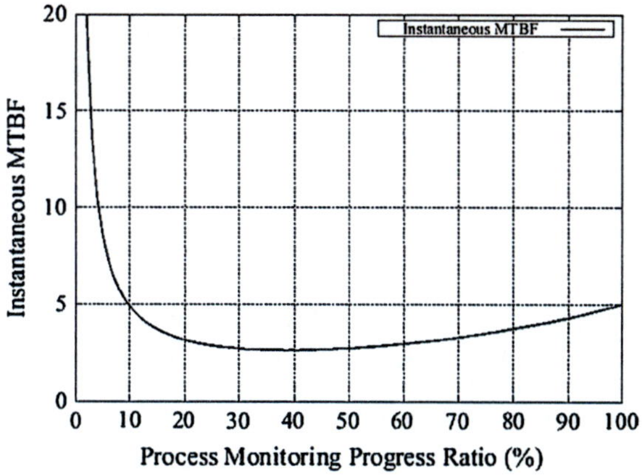

Fig. 18 The estimated instantaneous MTBF for Project 1

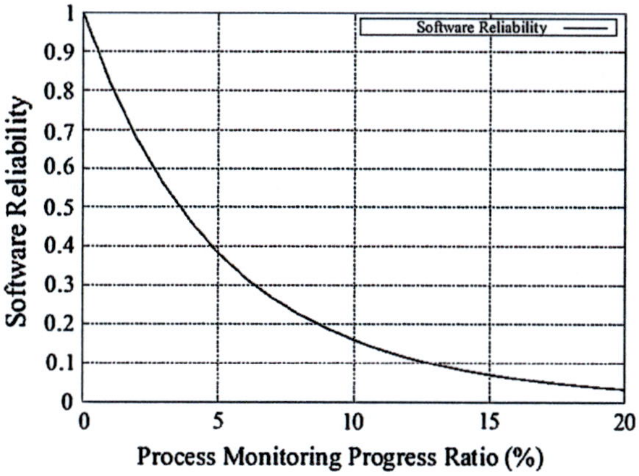

Fig. 19 The estimated software reliability function for Project 1

delayed S-shaped SRGM shows suitability in all projects. Therefore, if we select the process monitoring progress ratio as the unit of testing-time for SRGMs based on an NHPP, then the delayed S-shaped SRGM becomes a very useful one for quantitative software reliability assessment based on the process data derived from software process monitoring activities.

Further, we show numerical illustration of software reliability assessment by using the delayed S-shaped SRGM for Project 1. Figure 17 shows the estimated mean value function and its 95 % confidence limits, where the parameter estimates are obtained as $\hat{a} = 39.67$ and $\hat{b} = 0.0259$. We can find that there are 10 remaining faults at the end of test completion review phase. Figure 18 shows the estimated instantaneous MTBF in Eq. (52). From Fig. 18, we can estimate the instantaneous MTBF at the finishing test completion review phase to be about 5 days. Figure 19 shows the estimated software reliability at process monitoring progress ratio $t = 100$ (%). From Fig. 19, if the process monitoring progress ratio is 120 %, we can find that a software failure will occur with high probability.

6 Concluding Remarks

In this chapter, we have discussed several recent developments in software reliability modeling and its applications, i.e., quality engineering approach based on the human factor model in design-review process, SDE modeling for OSS projects, NHPP modeling with discrete calculus, and software project evaluation based on quality-oriented software management models. The first human factor analysis is very important to promote software quality/reliability management during the upper stream of development process by controlling the effective inhibitors and inducers. The latter two SRGMs enable us to obtain plausible results of software reliability assessment more than ever. The last quality-oriented software management analysis enables us to manage software projects quantitatively for successful project management in terms of QCD.

Acknowledgments The author is very grateful to Professor Shunji Osaki who stimulated his interest in software reliability theory through many interesting discussions.

References

1. Basili VR, Reiter RW Jr (1979) An investigation of human factors in software development. IEEE Comput Mag 12(12):21–38
2. Curtis B (ed) (1985) Tutorial : Human factors in software development. IEEE Computer Society Press, Los Alamitos, CA
3. Nakajo T, Kume H (1991) A case history analysis of software error cause-effect relationships. IEEE Trans Softw Eng 17(8):830–838
4. Taguchi G (ed) (1998) Signal-to-Noise raito for quality evaluation (in Japanese). Japanese Standards Association, Tokyo
5. Taguchi G (1976) A method of design of experiment (the First volume (2nd edn)) (in Japanese). Maruzen, Tokyo
6. Yamada S (2011) Elements of software reliability : modeling approach (in Japanese). Kyoritsu-Shuppan, Tokyo
7. Esaki K, Yamada S, Takahashi M (2001) A quality engineering analysis of human factors affecting software reliability in software design review process (in Japanese). Trans IEICE Japan J84–A(2):218–228
8. Yamada S (2008) Early-stage software product quality prediction based on process measurement data. In: Misra KB (ed) Springer handbook of performability engineering. Springer, London, pp 1227–1237 chapter 74
9. Yamada S (2006) A human factor analysis for software reliability in design-review process. Intern J Performability Eng 2(3):223–232
10. Miyamoto I (1982) Software engineering—Current status and perspectives- (in Japanese). TBS Publishing, Tokyo

11. Esaki K, Takahashi M (1997) A software design review on the relationship between human factors and software errors classified by seriousness (in Japanese). J Qual Eng Forum 5(4):30–37
12. E-Soft Inc., Internet Research Reports. (Online). Available:http://www.securityspace.com/s_survey/data/index.html
13. Yamada S (2002) Software reliability models. In: Osaki S (ed) Stochastic models in reliability and maintenance. Springer, Berlin, pp 253–280 chapter 10
14. MacCormack A, Rusnak J, Baldwin CY (2006) Exploring the structure of complex software designs: An empirical study of open source and proprietary code. Inf J Manage Sci 52(7):1015–1030
15. Kuk G (2006) Strategic interaction and knowledge sharing in the KDE developer mailing list. Inf J Manage Sci 52(7):1031–1042
16. Zhoum Y, Davis J (2005) Open source software reliability model: an empirical approach. In: Proceedings workshop on open source software engineering (WOSSE), vol 30(4), pp 67–72
17. Li P, Shaw M, Herbsleb J, Ray B, Santhanam P (2004) Empirical evaluation of defect projection models for widely-deployed production software systems. In: Proceedings of 12th international symposium foundations of software engineering (FSE-12), pp 263–272
18. Arnold L (1974) Stochastic differential equations-theory and applications. Wiley, New York
19. Wong E (1971) Stochastic processes in information and systems. McGraw-Hill, New York
20. Yamada S, Kimura M, Tanaka H, Osaki S (1994) Software reliability measurement and assessment with stochastic differential equations. IEICE Trans Fundam E77–A(1):109–116
21. The Apache HTTP Server Project, The Apache Software Foundation. (Online). Available: http://httpd.apache.org/
22. Apache Tomcat, The Apache Software Foundation. (Online). Available: http://tomcat.apache.org/
23. PostgreSQL, PostgreSQL Global Development Group. (Online). Available: http://www.postgresql.org/
24. Tamura Y, Yamada S (2007) Software reliability growth model based on stochastic differential equations for open source software. In: Proceedings of 4th IEEE international conference on mechatronics, CD-ROM (ThM1-C-1)
25. Tamura Y, Yamada S (2006) A flexible stochastic differential equation model in distributed development environment. Eur J Operl Res 168(1):143–152
26. Tamura Y, Yamada S (2009) Optimisation analysis for reliability assessment based on stochastic differential equation modeling for open source software. Int J Syst Sci 40(4):429–438
27. Tamura Y, Yamada S (2011) Reliability assessment based on hazard rate model for an embedded OSS porting phase. Softw Test, Verification Reliab, vol 21, to be published
28. Satoh D (2000) A discrete Gompertz equation and a software reliability growth model. IEICE Trans Inf Syst E83–D(7):1508–1513
29. Satoh D, Yamada S (2001) Discrete equations and software reliability growth models. In: Proceedings of 12th international symposium on software reliability engineering (ISSRE'01), pp 176–184
30. Inoue S, Yamada S (2007) Generalized discrete software reliability modeling withe effect of program size. IEEE Trans Sys, Man, Cybern (Part A) 37(2):170–179
31. Hirota R (1979) Nonlinear partial difference equations. V. Nonlinear equations reducible to linear equations. J Phys Soc Japan 46(1):312–319
32. Bass FM (1969) A new product growth model for consumer durables. Manage Sci 15:215–227
33. Satoh D (2001) A discrete Bass model and its parameter estimation. J Oper Res Soc Japan 44(1):1–18
34. Kasuga K, Fukushima T, Yamada S (2006) A practical approach software process monitoring activities (in Japanese). In: Proceedings of 25th JUSE software quality symposium, pp 319–326
35. Yamada S, Fukushima T (2007) Quality-oriented software management (in Japanese). Morikita-Shuppan, Tokyo
36. Yamada S, Takahashi M (1993) Introduction to software management model (in Japanese). Kyoritsu-Shuppan, Tokyo

37. Yamada S, Kawahara A (2009) Statistical analysis of process monitoring data for software process improvement. Int J Reliab, Qual Saf Eng 16(5):435–451
38. Yamada S, Yamashita T, Fukuta A (2010) Product quality prediction based on software process data with development-period estimation. Int J Syst Assur Eng Manage 1(1):69–73

Application of EM Algorithm to NHPP-Based Software Reliability Assessment with Ungrouped Failure Time Data

Hiroyuki Okamura and Tadashi Dohi

Abstract This chapter presents computation procedures for maximum likelihood estimates (MLEs) of software reliability models (SRMs) based on nonhomogeneous Poisson processes (NHPPs). The idea behind our methods is to regard usual failure time data as incomplete data. This leads to quite simple computation procedures for NHPP-based SRMs based on the EM (expectation–maximization) algorithm, and these algorithms overcome a problem arising in practical use of SRMs. In this chapter, we discuss the algorithms for 10 types of NHPP-based SRMs. Numerical examples show that the proposed EM algorithms help us to reduce computational efforts in the parameter estimation of NHPP-based SRMs.

1 Introduction

Since Jelinski and Moranda [12] and Goel and Okumoto [8] exhibited software reliability models (SRMs) based on stochastic processes, a number of SRMs have been proposed to quantitatively assess the reliability of software products [19, 21, 22, 31, 36]. In particular, nonhomogeneous Poisson processes (NHPPs) have much popularity for the software reliability modeling based on observed software failure data due to their mathematical tractability.

In general, NHPPs are defined by *mean value functions*, which are the expected numbers of failures with respect to testing time or efforts. Therefore, NHPP-based SRMs are classified by types of mean value functions. Goel and Okumoto [8], Goel [6], Musa and Okumoto [23], Ohba [25, 26], Yamada, Ohba and Osaki [38], Zhao and

H. Okamura (✉) · T. Dohi
Department of Information Engineering, Graduate School of Engineering,
Hiroshima University, 1–4–1 Kagamiyama, 739-8527 Higashi-Hiroshima, Japan
e-mail: okamu@rel.hiroshima-u.ac.jp

T. Dohi
e-mail: dohi@rel.hiroshima-u.ac.jp

T. Dohi and T. Nakagawa (eds.), *Stochastic Reliability and Maintenance Modeling*, Springer Series in Reliability Engineering 9, DOI: 10.1007/978-1-4471-4971-2_13,

Xie [39] and Pham [32] proposed NHPP-based SRMs whose mean value functions represent typical nonlinear curves. Their formulation of NHPP-based SRMs is based on only the dynamics of mean value functions. That is, they derived differential equations for the mean value functions from several modeling assumptions, and developed NHPP-based SRMs by solving the differential equations. In this modeling framework, the number of failures can be divided into two essential components; a trend curve and a noise process. This framework is almost same as those of regression models.

On the other hand, almost all NHPP-based SRMs can be described by the stochastic frameworks based on Markov processes. In recent years, many researchers focus on the stochastic classification and unification of NHPP modelings rather than differential equation-based modeling. These provide rich properties from statistical point of view. Shanthikumar [34] provided a bridge between a time-homogeneous Markov processes and NHPPs based on a binomial distribution. Langberg and Singpurwalla [18] also presented that a class of NHPP-based SRMs could be unified from the Bayesian point of view. Chen and Singpurwalla [3] showed that almost all SRMs belonged to subclasses of self-exiting point processes. Gokhale et al. [9] proposed the similar unification approach to Langberg and Singpurwalla [18], the framework was introduced from the concept of test coverage. Joe [14] and Miller [20] also proposed a modeling framework based on exponential order statistics, and the modeling framework can be classified as an extension of Langberg and Singpurwalla's work. In fact, all the NHPP-based SRMs can be described by either general order statistics or record value statistics of the software failure time data under the assumption that the failure times are mutually independent random variables [17].

This chapter focuses on model parameter estimation of NHPP-based SRMs. The estimation of model parameters is needed to quantitatively assess the software reliability from observed failure data. The commonly used method is the maximum likelihood (ML) estimation. The ML estimation is to find the maximum of likelihood function of observed software failure data. Since ML estimates (MLEs) have rational properties like asymptotic efficiency, MLEs are expected to be suitable even in the parameter estimation for NHPP-based SRMs. Knafl and Morgan [16] presented a method to solve systematically the likelihood equations of SRMs with two model parameters. Joe [14] also discussed confidence intervals of MLEs. Zhao and Xie [39] derived the MLEs for an extended Goel and Okumoto model. Jeske and Pham [13] discovered empirically that the MLEs in Goel and Okumoto model are not statistically consistent.

Although ML estimation allows us to compute statistically proper estimates for model parameters, we occasionally encounter several difficulties for the parameter estimation. In general, MLEs are obtained by maximizing log-likelihood functions (LLFs) or by solving nonlinear equation called likelihood equation which is derived from the first derivative of LLF. However, even for Goel and Okumoto model, we cannot obtain closed forms of MLEs. In other words, we employ numerical approaches to find MLEs based on LLFs.

The common approaches to find MLEs are Newton's method, quasi-Newton's method and Nelder-Mead method. In fact, they were applied to obtain MLEs of

NHPP-based SRMs in many chapters. As is well-known, though Newton's method is a powerful tool to calculate MLEs, it has the local convergence property and may fail to get the solution due to unsuitable initial guesses. The Nelder-Mead method is one of the direct search methods, which is more stable for initial guesses than Newton's method, but a few design parameters, such as expansion rate, must be manually adjusted before executing the algorithm.

These are fatal problems when we develop and implement the software reliability assessment tool. In general, users of software reliability assessment tools are not expert at numerical optimization. Then tools must not ask the users to select appropriate initial guesses and appropriate design parameters, since most of the tool users or practitioners cannot judge if the resulting estimates are reliable or not. In fact, much effort will be wasted to obtain the reliable solutions in parameter estimation. Unfortunately, such a computational problem has not been studied sufficiently in the software reliability engineering community.

Recently, we developed an alternative parameter estimation algorithm based on the expectation–maximization (EM) principle [5, 35] and applied it to the software reliability assessment based on the NHPP-based SRMs [27–30]. As another examples of EM algorithms in SRMs, Kimura and Yamada [15], and Ando et al. [2] attempt to use the EM algorithms to estimation of imperfect debugging models [7] and architecture-based SRMs [4]. Their models are based on the continuous-time Markov chain, and are closely related to Markov-modulated Poisson processes and/or Markovian arrival processes. Thus, their EM algorithms are developed on the completely different ideas from this chapter. Our key idea here is to regard the underlying software failure data as incomplete data. It was shown that the EM algorithm can be applied to typical NHPP-based SRMs and can give much advantages on global convergence and reduction of computation efforts. It is worth noting that the EM algorithm has to be carefully designed for individual SRM. The main purpose of this chapter is to figure out the EM algorithms for some typical NHPP-based SRMs as well as their variations. Here, we deal with 10 NHPP-based SRMs under ungrouped failure data, i.e., failure time data, and design the concrete EM algorithms. This chapter summarizes the earlier results by the same authors [27–30] and extends them with aim of practical use. We believe that the results in this chapter are directly applicable to the actual software reliability assessment practice and useful to implement in the software reliability assessment tools [33].

This chapter is organized as follows. In Sect. 2, we describe the basic concept of software reliability modeling and introduce 10 typical NHPP-based SRMs. In Sect. 3, we derive the fundamental formulas of our EM algorithm for NHPP-based SRMs. Section 4 focuses on the specific EM algorithms for ten NHPP-based SRMs. Some practical remarks in use of the EM algorithms are given in Sect. 5, where termination condition in numerical calculation, initial guesses, model selection, and extension to maximum a posterior estimation are discussed. A numerical experiment is given in Sect. 6. We compare the proposed EM algorithm and Newton's method from the viewpoint of convergence properties. Finally, we conclude the chapter with remarks in Sect. 7.

2 NHPP-Based SRMs

2.1 Model Description

Let $\{X(t), t \geq 0\}$ denote the number of software failures experienced before time t. According to Langberg and Singpurwalla [18], this chapter considers the following model assumptions:

Assumption A: Software failures occur at mutually independent random times. The probability distributions of all failure times are identical. The probability density and cumulative distribution functions (p.d.f. and c.d.f.) are given by $f(t)$ and $F(t)$, respectively.
Assumption B: The number of inherent software faults causing failures is given by a Poisson random variable.

Under the assumption that the number of inherent faults is fixed as N, the probability mass function (p.m.f.) of the cumulative number of failures experienced by time t is given by

$$P(X(t) = n) = \binom{N}{n} F(t)^n \overline{F}(t)^{N-n}, \tag{1}$$

where $\overline{F}(\cdot) = 1 - F(\cdot)$. When N is a Poisson random variable with mean ω, the cumulative number of software failures before time t has the following p.m.f.

$$P(X(t) = n) = \frac{(\omega F(t))^n}{n!} e^{-\omega F(t)}. \tag{2}$$

Equation (2) is equivalent to the probability mass function of NHPP with mean value function $\omega F(t)$.

2.2 Specific Models

In the modeling framework of Eq. (2), NHPP-based SRMs are defined as respective failure time distributions. This chapter deals with typical five types of failure time distributions; gamma distribution, normal distribution, logistic distribution, and maximum/minimum extreme value distributions. In addition, two different failure time distributions are derived from each of normal, logistic, and maximum/minimum extreme value distributions. Concretely, these distributions are defined on the range $(-\infty, \infty)$, and thus we apply truncation and logarithm techniques to change their domain to the range $[0, \infty)$. Table 1 presents NHPP-based SRMs and corresponding failure time distributions treated in this chapter.
GAMMA (EXP): GAMMA is the NHPP model whose failure time distribution is the following gamma distribution:

Table 1 NHPP-based SRMs and their failure time distributions

Model abbr.	Failure time dist.	References
GAMMA (EXP)	Gamma	[1, 8, 38, 39]
TNORM	Truncated normal	—
LNORM	Log normal	[1]
TLOGIS	Truncated logistic	[25, 28]
LLOGIS	Log logistic	[10, 28]
TEVMAX	Truncated maximum extreme value	[27, 37]
LEVMAX	Log maximum extreme value	[27]
TEVMIN	Truncated minimum extreme value	[27]
LEVMIN	Log minimum extreme value	[6, 27]

$$F(t) = \int_0^t \frac{\beta^\alpha u^{\alpha-1} \exp(-\beta u)}{\Gamma(\alpha)} du, \quad \alpha > 0, \quad \beta > 0, \tag{3}$$

where α and β are shape and scale (rate) parameters for gamma distribution, and $\Gamma(\cdot)$ is the standard gamma function. GAMMA was discussed as the delayed S-shaped model [38] and its generalized models [39]. In particular, since the gamma distribution includes the exponential distribution, GAMMA also includes Goel and Okumoto model (EXP).

TNORM: The failure time distribution of TNORM is given by the truncated normal distribution:

$$F(t) = \Phi\left(\frac{t-\mu}{\sigma}\right) / \{1 - \Phi(-\mu/\sigma)\}, \quad -\infty < \mu < \infty, \quad \sigma > 0, \tag{4}$$

where $\Phi(\cdot)$ is the c.d.f. of the standard normal distribution function

$$\Phi(t) = \int_{-\infty}^t \phi(u)\, du, \tag{5}$$

$$\phi(t) = \frac{1}{\sqrt{2\pi}} \exp\left(-\frac{1}{2}t^2\right). \tag{6}$$

LNORM: LNORM was built from the log-normal distribution:

$$F(t) = \Phi\left(\frac{\log t - \mu}{\sigma}\right), \quad -\infty < \mu < \infty, \sigma > 0. \tag{7}$$

Achcar et al. [1] introduced LNORM as the log-normal order statistics model.

TLOGIS: TLOGIS is the model whose failure time distribution is given by the truncated logistic distribution:

$$F(t) = \Psi\left(\frac{t-\mu}{\psi}\right) / \{1 - \Psi(-\mu/\psi)\}, \quad -\infty < \mu < \infty, \quad \psi > 0, \tag{8}$$

where $\Psi(\cdot)$ is the c.d.f. of standard logistic distribution

$$\Psi(t) = \frac{1}{1 + \exp(-t)}. \tag{9}$$

TLOGIS is same as the inflection S-shaped SRM by Ohba [25]. Although Ohba [25] derived TLOGIS from differential equations, Okamura and Dohi [28] revealed that TLOGIS can also be built from the logistic order statistics model.
LLOGIS: LLOGIS is the model whose failure time distribution is the log-logistic distribution:

$$F(t) = \Psi\left(\frac{\log t - \mu}{\psi}\right), \quad -\infty < \mu < \infty, \quad \psi > 0. \tag{10}$$

LLOGIS was proposed by Gokhale and Trivedi [10].
TEVMAX: TEVMAX is built from the truncated maximum extreme value type I distribution (Gumbel distribution):

$$F(t) = \Theta\left(\frac{t-\mu}{\theta}\right) / \{1 - \Theta(-\mu/\theta)\} \quad -\infty < \mu < \infty, \quad \theta > 0, \tag{11}$$

where $\Theta(\cdot)$ is the c.d.f. of standard Gumbel distribution is defined as

$$\Theta(t) = \exp\{-\exp(-t)\}. \tag{12}$$

TEVMAX is same as the modified Gompertz model by Yamada [37] and its mean value function draws Gompertz curve.
LEVMAX: LEVMAX is derived by exponentially transformed Gumbel random variables, so-called samples from a log-Gumbel distribution:

$$F(t) = \Theta\left(\frac{\log t - \mu}{\theta}\right), \quad -\infty < \mu < \infty, \quad \theta > 0. \tag{13}$$

It is well-known that the above c.d.f. can be reduced to Fréchet distribution with positive support.
TEVMIN: TEVMIN is the model whose failure time distribution is the truncated minimum extreme value type I distribution. By using the survival function of standard Gumbel distribution, we have

$$F(t) = \overline{\Theta}\left(\frac{t+\mu}{\theta}\right) / \left\{1 - \overline{\Theta}(\mu/\theta)\right\} \quad -\infty < \mu < \infty, \quad \theta > 0, \tag{14}$$

where $\overline{\Theta}(t) = 1 - \Theta(-t)$.

LEVMIN: LEVMIN is the model whose failure time distribution is the Weibull distribution. It is known that logarithmic transformation of Weibull random variables provides Gumbel random variables. For the notational convenience, this chapter uses the survival function of standard Gumbel distribution to express the failure time distribution of LEVMIN:

$$F(t) = \overline{\Theta}\left(\frac{\log t + \mu}{\theta}\right) \quad -\infty < \mu < \infty, \theta > 0. \tag{15}$$

Since the above failure time distribution is equivalent to Weibull distribution, LEVMIN gives the generalized exponential model by Goel [6].

3 EM Algorithm for NHPP-Based SRMs

3.1 Maximum Likelihood Estimation

In the software reliability assessment, we should estimate model parameters of NHPP-based SRMs from observed data. The most commonly used technique to parameter estimation is ML estimation. Let $D_T = (t_1, \ldots, t_K)$ be a set of failure times experienced by time T. Without loss of generality, we assume $0 < t_1 < \cdots < t_K$. For the observed data D_T, the LLF for NHPP-based SRMs is given by

$$\mathcal{L}(\omega, \boldsymbol{\lambda}; D_T) = \log p(D_T; \omega, \boldsymbol{\lambda}), \tag{16}$$

$$p(D_T; \omega, \boldsymbol{\lambda}) = \omega^K \prod_{k=1}^{K} f(t_k; \boldsymbol{\lambda}) \exp\left(-\omega F(T; \boldsymbol{\lambda})\right), \tag{17}$$

where $\boldsymbol{\lambda}$ is a parameter vector for the failure time distribution, $f(\cdot; \boldsymbol{\lambda})$ is a density function of $F(\cdot; \boldsymbol{\lambda})$ and $p(\cdot)$ is an appropriate probability mass or density function. The MLE is to find the parameters maximizing the LLF, so-called maximum likelihood estimates (MLEs). In general, MLEs of NHPP-based SRMs cannot be expressed as closed forms, even for the simplest model, i.e., Goel and Okumoto model. That is, we need to utilize any iterative methods for numerical optimization such as Newton's method, quasi-Newton's method, Fisher's scoring method, and Nelder-Mead method. However, it is well-known that it is difficult to choose appropriate initial parameters in these iterative methods. In addition, some methods require appropriate design parameters such as reflection and expansion rates in Nelder-Mead method. This property adversely affects making software reliability assessment tools. In the reliability assessment tool with Newton's or quasi-Newton's method, the users should change initial parameters and other design parameters depending on observed failure data.

3.2 EM Algorithm

The EM algorithm is an iterative method for computing ML estimates with incomplete data [5, 35]. Let D and Z be observable and unobservable data vectors, respectively, and we generally estimate a model parameter vector $\boldsymbol{\lambda}$ from only the observable data vector D. In the context of ML estimation, the problem corresponds to finding a parameter vector that maximizes a marginal LLF:

$$\hat{\boldsymbol{\lambda}}_{ML} = \underset{\boldsymbol{\lambda}}{\operatorname{argmax}}\, \mathcal{L}(\boldsymbol{\lambda}; D), \tag{18}$$

$$\mathcal{L}(\boldsymbol{\lambda}; D) = \log p(D; \boldsymbol{\lambda}) = \log \int p(D, Z; \boldsymbol{\lambda}) dZ, \tag{19}$$

where $p(\cdot)$ is any probability density or mass function.

Taking account of the posterior distribution of unobservable data vector with the parameter vector $\boldsymbol{\lambda}'$ and Jensen's inequality, we have

$$\begin{aligned} \mathcal{L}(\boldsymbol{\lambda}; D) &= \log \int p(D, Z; \boldsymbol{\lambda}) dZ \\ &= \log \int \frac{p(D, Z; \boldsymbol{\lambda})}{p(Z|D; \boldsymbol{\lambda}')} p(Z|D; \boldsymbol{\lambda}') dZ \\ &\geq \int p(Z|D; \boldsymbol{\lambda}') \log \frac{p(D, Z; \boldsymbol{\lambda})}{p(Z|D; \boldsymbol{\lambda}')} dZ \\ &\equiv \mathcal{Z}(\boldsymbol{\lambda}; \boldsymbol{\lambda}'). \end{aligned} \tag{20}$$

The posterior distribution for unobservable data can be obtained from Bayes theorem:

$$p(Z|D; \boldsymbol{\lambda}') = \frac{p(D, Z; \boldsymbol{\lambda}')}{\int p(D, Z; \boldsymbol{\lambda}') dZ}. \tag{21}$$

Equation (20) yields

$$\mathcal{L}(\boldsymbol{\lambda}; D) - \mathcal{Z}(\boldsymbol{\lambda}; \boldsymbol{\lambda}') = D_{KL}(p(Z|D; \boldsymbol{\lambda}') || p(Z|D; \boldsymbol{\lambda})), \tag{22}$$

where $D_{KL}(P||Q)$ is the Kullback-Leibler distance from the distribution P to the distribution Q. Hence the difference $\mathcal{L}(\boldsymbol{\lambda}; D) - \mathcal{L}(\boldsymbol{\lambda}'; D)$ is given by

$$\mathcal{L}(\boldsymbol{\lambda}; D) - \mathcal{L}(\boldsymbol{\lambda}'; D) = \mathcal{Z}(\boldsymbol{\lambda}; \boldsymbol{\lambda}') - \mathcal{Z}(\boldsymbol{\lambda}'; \boldsymbol{\lambda}') + D_{KL}(p(Z|D; \boldsymbol{\lambda}') || p(Z|D; \boldsymbol{\lambda})). \tag{23}$$

Since $D_{KL}(\cdot||\cdot) \geq 0$, Eq. (23) implies that the maximization of lower bound results in the maximization of marginal LLF.

Let $Q(\boldsymbol{\lambda}|\boldsymbol{\lambda}')$ denote the conditional expected LLF with respect to the complete data vector (D, Z) using the posterior distribution for unobservable data vector with provisional parameter vector $\boldsymbol{\lambda}'$:

$$\begin{aligned} Q(\boldsymbol{\lambda}|\boldsymbol{\lambda}') &= \mathrm{E}[\log p(D, Z; \boldsymbol{\lambda})|D; \boldsymbol{\lambda}'] \\ &= \int p(Z|D; \boldsymbol{\lambda}') \log p(D, Z; \boldsymbol{\lambda}) dZ. \end{aligned} \tag{24}$$

Then Eq. (20) is rewritten in the form:

$$\mathcal{Z}(\boldsymbol{\lambda}; \boldsymbol{\lambda}') = Q(\boldsymbol{\lambda}|\boldsymbol{\lambda}') - \int p(Z|D; \boldsymbol{\lambda}') \log p(Z|D; \boldsymbol{\lambda}') dZ. \tag{25}$$

Since the second term of the above equation is constant, the maximization of $Q(\boldsymbol{\lambda}|\boldsymbol{\lambda}')$ with respect to $\boldsymbol{\lambda}$ is directly reduced to the maximization of $\mathcal{Z}(\boldsymbol{\lambda}; \boldsymbol{\lambda}')$ with respect to $\boldsymbol{\lambda}$.

Based on the above discussion, the EM algorithm consists of E-step and M-step. E-step computes the conditional expected LLF with respect to the complete data vector (D, Z) using the posterior distribution for unobservable data vector with provisional parameter vector $\boldsymbol{\lambda}'$, i.e., $Q(\boldsymbol{\lambda}|\boldsymbol{\lambda}')$. In M-step, we find a new parameter vector $\boldsymbol{\lambda}''$ that maximizes the expected LLF:

$$\boldsymbol{\lambda}'' := \operatorname*{argmax}_{\boldsymbol{\lambda}} Q(\boldsymbol{\lambda}|\boldsymbol{\lambda}'), \tag{26}$$

and $\boldsymbol{\lambda}''$ becomes a provisional parameter vector at the next E- and M-steps. These steps surely increase the marginal LLF. The E- and M-steps are repeatedly executed until the parameters converge to ML estimates.

3.3 Fundamental EM-Step Formulas for NHPP-Based SRMs

Consider the EM algorithm for NHPP-based SRMs with the following p.m.f.

$$P(N(t) = x) = \frac{(\omega F(t; \boldsymbol{\lambda}))^x}{x!} \exp\left(-\omega F(t; \boldsymbol{\lambda})\right). \tag{27}$$

It is obvious that failure times after time T are not observable. Then we define the complete data as $0 < T_1 < T_2 < \ldots < T_N$, where N is the total number of inherent faults and T_k is the k-th ordered failure time. In this case, $Z_T = (T_{K+1}, \ldots, T_N, N)$ is unobserved data. Since N is a Poisson distributed random variable and T_k obeys $F(\cdot; \boldsymbol{\lambda})$, the complete LLF is given by

$$\log p(D_T, Z_T; \omega, \boldsymbol{\lambda}) = N \log \omega - \omega + \sum_{k=1}^{N} \log f(T_k; \boldsymbol{\lambda}). \tag{28}$$

From the standard argument of MLEs, the MLEs of ω and $\boldsymbol{\lambda}$ can be derived as

$$\hat{\omega} = N \tag{29}$$

and

$$\hat{\boldsymbol{\lambda}} = \underset{\boldsymbol{\lambda}}{\operatorname{argmax}} \sum_{k=1}^{N} \log f(T_k; \boldsymbol{\lambda}), \tag{30}$$

respectively. This implies that the estimation problem of NHPP-based SRMs under the complete data can be decomposed into separate data fitting problems for Poisson distribution and failure time distribution with independent and identically distributed (IID) samples.

From Eq. (26), we have the following M-step formulas

$$\omega: = \mathrm{E}[N | D_T; \omega', \boldsymbol{\lambda}'] \tag{31}$$

and

$$\boldsymbol{\lambda}: = \underset{\boldsymbol{\lambda}}{\operatorname{argmax}}\, \mathrm{E}\left[\sum_{i=1}^{N} \log f(T_i; \boldsymbol{\lambda}) \middle| D_T; \omega', \boldsymbol{\lambda}' \right], \tag{32}$$

where ω' and $\boldsymbol{\lambda}'$ are provisional parameters. On the other hand, by applying Bayes theorem, we have

$$p(N|D_T; \omega, \boldsymbol{\lambda}) \propto \omega^K e^{-\omega} \prod_{k=1}^{K} f(t_k; \boldsymbol{\lambda}), \quad N = K \tag{33}$$

and

$$p(N|D_T; \omega, \boldsymbol{\lambda}) \propto \omega^N e^{-\omega} \prod_{k=1}^{K} f(t_k; \boldsymbol{\lambda})$$
$$\times \int_{T_{N-1}}^{\infty} \cdots \int_{T}^{\infty} \prod_{k=K}^{N} f(T_k; \boldsymbol{\lambda}) dT_{K+1} \cdots dT_N, \quad N \geq K + 1. \tag{34}$$

Then the posterior distribution of N becomes the Poisson p.m.f. with mean $\omega \overline{F}(T; \boldsymbol{\lambda})$:

$$p(N|D_T; \omega, \boldsymbol{\lambda}) = \frac{\left(\omega \overline{F}(T; \boldsymbol{\lambda})\right)^{N-K}}{(N-K)!} \exp\left(-\omega \overline{F}(T; \boldsymbol{\lambda})\right), \quad N \geq K. \tag{35}$$

Therefore, the EM-step formula for ω can be obtained as

$$\omega: = K + \omega' \overline{F}(T; \boldsymbol{\lambda}'). \tag{36}$$

4 Concrete EM-Step Formulas for NHPP-Based SRMs

4.1 EM-Step Formulas for EXP

The failure time distribution of EXP is given by the following exponential distribution:

$$F(t; \beta) = 1 - e^{-\beta t}. \tag{37}$$

Also, since the MLE of exponential distribution under IID ordered samples $T_1, \ldots, T_N$ is given by

$$\hat{\beta} = \frac{N}{\sum_{k=1}^{N} T_k}. \tag{38}$$

Thus, from Eq. (32), we have the following M-step formula

$$\beta := \frac{\mathrm{E}\left[N | D_T; \omega', \beta'\right]}{\mathrm{E}\left[\sum_{k=1}^{N} T_k | D_T; \omega', \beta'\right]}. \tag{39}$$

In general, for a measurable function $h(\cdot)$, the expected value with the posterior distribution can be computed as follows (see Appendix).

$$\mathrm{E}\left[\sum_{k=1}^{N} h(T_k) \middle| D_T; \omega, \boldsymbol{\lambda}\right] = \sum_{k=1}^{K} h(t_k) + \omega \int_T^{\infty} h(u) dF(u; \boldsymbol{\lambda}). \tag{40}$$

Applying the above formula, the EM-step formula for parameter β is given by

$$\beta := \frac{K + \omega' e^{-\beta' T}}{\sum_{k=1}^{K} t_k + \omega'(T + 1/\beta') e^{-\beta' T}}. \tag{41}$$

4.2 EM-Step Formulas for GAMMA

The failure time distribution of GAMMA is the following gamma distribution:

$$F(t; \alpha, \beta) = \int_0^t \frac{\beta^{\alpha} u^{\alpha-1} \exp(-\beta u)}{\Gamma(\alpha)} du. \tag{42}$$

According to the ordinary ML estimation for gamma distribution, the MLEs for complete data are given by parameters satisfying the following likelihood equations:

$$\log \hat{\alpha} - \psi(\hat{\alpha}) = \log\left(\frac{1}{N}\sum_{i=1}^{N} T_i\right) - \frac{1}{N}\sum_{i=1}^{N} \log X_i, \tag{43}$$

$$\hat{\beta} = \frac{\hat{\alpha} N}{\sum_{k=1}^{N} T_k}, \tag{44}$$

where $\psi(\cdot)$ is the digamma function, i.e., $\psi(\alpha) = d\log\gamma(\alpha)/d\alpha$. Thus, the EM-step formulas are given by

$$\alpha := \inf\left\{\alpha > 0; \log\alpha - \psi(\alpha) = \log\left(\frac{T^{(1)}}{N^{(1)}}\right) - \frac{T^{(2)}}{N^{(1)}}\right\}, \tag{45}$$

$$\beta := \frac{\alpha N^{(1)}}{T^{(2)}}, \tag{46}$$

where

$$N^{(1)} = K + \omega' \overline{F}(T; \alpha', \beta'), \tag{47}$$

$$\begin{aligned} T^{(1)} &= \sum_{k=1}^{K} t_k + \omega' \int_T^{\infty} u dF(u; \alpha', \beta') \\ &= \sum_{k=1}^{K} t_k + \omega' \frac{\alpha'}{\beta'} \overline{F}(T; \alpha' + 1, \beta'), \end{aligned} \tag{48}$$

$$T^{(2)} = \sum_{k=1}^{K} \log t_k + \omega' \int_T^{\infty} \log u \, dF(u; \alpha', \beta'). \tag{49}$$

Note that numerical root-finding and integration algorithms are required for Eqs. (45) and (49), respectively. As a special case, GAMMA with fixed shape parameter includes several important NHPP-based SRMs such as the delayed S-shaped SRM. The EM-step for GAMMA with fixed shape parameter becomes simpler than that for general GAMMA. Given the shape parameter α, the EM-step formulas can be more simplified to

$$\beta := \frac{\alpha N^{(1)}}{T^{(1)}}, \tag{50}$$

$$N^{(1)} = K + \omega' \overline{F}(T; \alpha, \beta'), \tag{51}$$

$$T^{(2)} = \sum_{k=1}^{K} t_k + \omega' \frac{\alpha}{\beta'} \overline{F}(T; \alpha + 1, \beta'). \tag{52}$$

4.3 EM-Step Formulas for TNORM

For the EM-step formulas, we suppose that samples generated before time 0 are also unobserved. Let $T_{-\tilde{N}} < \cdots < T_{-1} < 0$ be unobserved failure times before time 0, where $\tilde{N}$ is the number of failures already experienced by $t = 0$. Then unobserved data can be rewritten by $Z_T = (T_{-\tilde{N}}, \ldots, T_{-1}, T_{K+1}, \ldots, T_N, \tilde{N}, N)$. Under the assumption, we obtain the M-step formulas from EM principle.

$$\tilde{\omega} := \mathrm{E}[N + \tilde{N} | D_T; \tilde{\omega}', \boldsymbol{\lambda}'], \tag{53}$$

$$\boldsymbol{\lambda} := \underset{\boldsymbol{\lambda}}{\operatorname{argmax}} \, \mathrm{E}\left[\sum_{k=-\tilde{N}}^{N} \log f(T_k; \boldsymbol{\lambda}) \middle| D_T; \tilde{\omega}', \boldsymbol{\lambda}' \right], \tag{54}$$

where $\tilde{\omega}$ is a Poisson parameter for the total number of failures on the range $(-\infty, \infty)$. Then we should change $\tilde{\omega}$ into the mean number of failures experienced on the positive support $[0, \infty)$ after the EM algorithm is finished, i.e., $\omega := \tilde{\omega}\overline{F}(0)$.

Similar to the posterior distribution of N, we have

$$\mathrm{E}[N + \tilde{N} | D_T; \tilde{\omega}', \boldsymbol{\lambda}'] = K + \tilde{\omega}'\overline{F}(T; \boldsymbol{\lambda}') + \tilde{\omega}' F(0; \boldsymbol{\lambda}'). \tag{55}$$

Also, for any measurable function $h(\cdot)$, the following equation holds (see Appendix):

$$\mathrm{E}\left[\sum_{k=-\tilde{N}}^{N} h(T_k) \middle| D_T; \tilde{\omega}', \boldsymbol{\lambda}' \right] = \sum_{k=1}^{K} h(t_k) + \tilde{\omega}' \int_{-\infty}^{0} h(u) dF(u; \boldsymbol{\lambda}') + \tilde{\omega}' \int_{T}^{\infty} h(u) dF(u; \boldsymbol{\lambda}'). \tag{56}$$

It is worth noting that the above equation includes a left-truncated term as well as a right-truncated term.

According to the standard argument of parameter estimation of normal distribution, MLEs of TNORM under the complete data are given by

$$\hat{\tilde{\omega}} = N + \tilde{N}, \tag{57}$$

$$\hat{\mu} = \frac{\sum_{k=-\tilde{N}}^{N} T_k}{\tilde{N} + N}, \tag{58}$$

$$\hat{\sigma} = \sqrt{\frac{\sum_{k=-\tilde{N}}^{N} T_k^2}{\tilde{N} + N} - \hat{\mu}^2}. \tag{59}$$

Define the following functions based on the standard normal distribution:

$$\overline{\Phi}^{(1)}(z) = \frac{1}{\sigma}\int_{\sigma z+\mu}^{\infty} u\phi(u)du = \sigma\phi(z) + \mu\overline{\Phi}(z), \tag{60}$$

$$\overline{\Phi}^{(2)}(z) = \frac{1}{\sigma}\int_{\sigma z+\mu}^{\infty} u^2\phi(u)du = (\sigma^2 z + 2\mu\sigma)\phi(z) + (\sigma^2 + \mu^2)\overline{\Phi}(z). \tag{61}$$

By using $\overline{\Phi}^{(1)}(z)$ and $\overline{\Phi}^{(2)}(z)$, the EM-step formulas for TNORM can be obtained as follows.

$$\tilde{\omega} := N^{(1)}, \tag{62}$$

$$\mu := \frac{T^{(1)}}{N^{(1)}}, \tag{63}$$

$$\sigma := \sqrt{\frac{T^{(2)}}{N^{(1)}} - \left(\frac{T^{(1)}}{N^{(1)}}\right)^2}, \tag{64}$$

where

$$N^{(1)} = K + \tilde{\omega}'\left(1 - \overline{\Phi}\left(-\frac{\mu'}{\sigma'}\right) + \overline{\Phi}\left(\frac{T-\mu'}{\sigma'}\right)\right), \tag{65}$$

$$T^{(1)} = \sum_{k=1}^{K} t_k + \tilde{\omega}'\left(\mu' - \overline{\Phi}^{(1)}\left(-\frac{\mu'}{\sigma'}\right) + \overline{\Phi}^{(1)}\left(\frac{T-\mu'}{\sigma'}\right)\right), \tag{66}$$

$$T^{(2)} = \sum_{k=1}^{K} t_k^2 + \tilde{\omega}'\left(\sigma'^2 + \mu'^2 - \overline{\Phi}^{(2)}\left(-\frac{\mu'}{\sigma'}\right) + \overline{\Phi}^{(2)}\left(\frac{T-\mu'}{\sigma'}\right)\right). \tag{67}$$

After the EM algorithm converges to MLEs, we take $\omega := \tilde{\omega}\overline{\Phi}(-\mu/\sigma)$.

4.4 EM-Step Formulas for LNORM

The idea behind the EM-step formulas for LNORM is to take logarithm of samples, i.e., we consider samples $\log T_1 < \cdots < \log T_K$ instead of original samples. Since $\log T_1 < \cdots < \log T_K$ follow a normal distribution, we obtain the EM-step formulas for LNORM in a similar manner to TNORM:

$$\omega := N^{(1)}, \tag{68}$$

$$\mu := \frac{T^{(1)}}{N^{(1)}}, \tag{69}$$

$$\sigma := \sqrt{\frac{T^{(2)}}{N^{(1)}} - \left(\frac{T^{(1)}}{N^{(1)}}\right)^2}, \tag{70}$$

where

$$N^{(1)} = K + \omega' \overline{\Phi}\left(\frac{\log T - \mu'}{\sigma'}\right), \tag{71}$$

$$T^{(1)} = \sum_{k=1}^{K} \log t_k + \omega' \overline{\Phi}^{(1)}\left(\frac{\log T - \mu'}{\sigma'}\right), \tag{72}$$

$$T^{(2)} = \sum_{k=1}^{K} (\log t_k)^2 + \omega' \overline{\Phi}^{(2)}\left(\frac{\log T - \mu'}{\sigma'}\right). \tag{73}$$

4.5 EM-Step Formulas for TLOGIS

Similar to TNORM, we suppose that failure times are observed only on the range $[0, T]$. Each of failure time in TLOGIS follows a logistic distribution:

$$F(t; \mu, \psi) = \Psi\left(\frac{t - \mu}{\psi}\right), \tag{74}$$

$$\Psi(t) = \frac{1}{1 + \exp(-t)}. \tag{75}$$

Let $T_{-\tilde{N}} < \cdots < T_{-1} < 0$ and $T_{K+1} < \cdots < T_N$ denote unobservable failure times experienced before time 0 and after time T, respectively, under the observed data $T_1 < \cdots < T_K$. For the complete samples $T_{-\tilde{N}} < \cdots < T_{-1} < T_1 < \cdots < T_N$, MLEs of logistic distribution are given by the parameters satisfying the following likelihood equation:

$$\sum_{k=-\tilde{N}}^{N} \overline{\Psi}\left(\frac{T_k - \hat{\mu}}{\hat{\psi}}\right) = \frac{\tilde{N} + N}{2}, \tag{76}$$

$$\sum_{k=-\tilde{N}}^{N} \frac{T_k - \hat{\mu}}{\hat{\psi}}\left(1 - 2\overline{\Psi}\left(\frac{T_k - \hat{\mu}}{\hat{\psi}}\right)\right) = \tilde{N} + N. \tag{77}$$

The above equations are heuristically solved by the iterative manner:

$$\mu := \mu - \psi \log\left\{\frac{1}{\tilde{N} + N} \sum_{k=-\tilde{N}}^{N} 2\overline{\Psi}\left(\frac{t_k - \mu}{\psi}\right)\right\}, \tag{78}$$

$$\psi := \frac{\psi}{\tilde{N} + N} \sum_{k=-\tilde{N}}^{N} \frac{T_k - \mu}{\psi}\left(1 - 2\overline{\Psi}\left(\frac{T_k - \mu}{\psi}\right)\right). \tag{79}$$

Define

$$\overline{\Psi}^{(1)}(z) = \frac{1}{\psi}\int_{\psi z+\mu}^{\infty} 2\overline{\Psi}(u)d\Psi(u) = \frac{1}{(1+\exp(z))^2}, \tag{80}$$

$$\overline{\Psi}^{(2)}(z) = \frac{1}{\psi}\int_{\psi z+\mu}^{\infty} u\left(1-2\overline{\Psi}(u)\right)d\Psi(u) = \frac{1+(1+z)\exp(z)}{(1+\exp(z))^2}. \tag{81}$$

By using $\overline{\Psi}^{(1)}(z)$ and $\overline{\Psi}^{(2)}(z)$, the EM-step formulas for TLOGIS can be obtained as follows.

$$\tilde{\omega} := N^{(1)}, \tag{82}$$

$$\mu := \mu' - \psi' \log\left(\frac{T^{(1)}}{N^{(1)}}\right), \tag{83}$$

$$\psi := \psi' \frac{T^{(2)}}{N^{(1)}}, \tag{84}$$

where

$$N^{(1)} = K + \tilde{\omega}'\left(1 - \overline{\Psi}\left(-\frac{\mu'}{\psi'}\right) + \overline{\Psi}\left(\frac{T-\mu'}{\psi'}\right)\right), \tag{85}$$

$$T^{(1)} = \sum_{k=1}^{K} 2\overline{\Psi}\left(\frac{t_k-\mu'}{\psi'}\right) + \tilde{\omega}'\left(1 - \overline{\Psi}^{(1)}\left(-\frac{\mu'}{\psi'}\right) + \overline{\Psi}^{(1)}\left(\frac{T-\mu'}{\psi'}\right)\right), \tag{86}$$

$$T^{(2)} = \sum_{k=1}^{K} \frac{t_k-\mu'}{\psi'}\left(1 - 2\overline{\Psi}\left(\frac{t_k-\mu'}{\psi'}\right)\right) + \tilde{\omega}'\left(1 - \overline{\Psi}^{(2)}\left(-\frac{\mu'}{\psi'}\right) + \overline{\Psi}^{(2)}\left(\frac{T-\mu'}{\psi'}\right)\right). \tag{87}$$

After finishing the EM algorithm, $\omega := \tilde{\omega}\overline{\Psi}(-\mu'/\psi)$.

4.6 EM-Step Formulas for LLOGIS

Consider logarithm of complete samples: $\log T_1 < \cdots < \log T_N$, which obey a logistic distribution. When applying the fundamental EM formulas, we have the concrete EM-step formulas for LLOGIS:

$$\omega := N^{(1)}, \tag{88}$$

$$\mu := \mu' - \psi' \log\left(\frac{T^{(1)}}{N^{(1)}}\right), \tag{89}$$

$$\psi := \psi' \frac{T^{(2)}}{N^{(1)}}, \tag{90}$$

where

$$N^{(1)} = K + \omega' \overline{\Psi}\left(\frac{\log T - \mu'}{\psi'}\right), \tag{91}$$

$$T^{(1)} = \sum_{k=1}^{K} 2\overline{\Psi}\left(\frac{\log t_k - \mu'}{\psi'}\right) + \omega' \overline{\Psi}^{(1)}\left(\frac{\log T - \mu'}{\psi'}\right), \tag{92}$$

$$T^{(2)} = \sum_{k=1}^{K} \frac{\log t_k - \mu'}{\psi'}\left(1 - 2\overline{\Psi}\left(\frac{\log t_k - \mu'}{\psi'}\right)\right) + \omega' \overline{\Psi}^{(2)}\left(\frac{\log T - \mu'}{\psi'}\right). \tag{93}$$

In the above formulas, we also use the heuristic parameter update formulas, Eqs. (78) and (79).

4.7 EM-Step Formulas for TEVMAX

Suppose that failure times are truncated at both sides $t = 0$ and $t = T$. From similar arguments to TNORM and TLOGIS, we get the likelihood equation for maximum extreme value distribution parameters under the complete samples $T_{-\tilde{N}} < \cdots < T_{-1} < T_1 < \cdots < T_N$:

$$\hat{\theta} = \frac{1}{\tilde{N} + N} \sum_{k=-\tilde{N}}^{N} T_k - \frac{\sum_{k=-\tilde{N}}^{N} T_k \exp(-T_k/\hat{\theta})}{\sum_{k=-\tilde{N}}^{N} \exp(-T_k/\hat{\theta})}, \tag{94}$$

$$\hat{\mu} = -\hat{\theta} \log\left(\frac{1}{\tilde{N} + N} \sum_{k=-\tilde{N}}^{N} \exp(-T_k/\hat{\theta})\right). \tag{95}$$

Since, the above likelihood equation cannot be explicitly solved, we employ the following heuristic method to compute the MLEs:

$$\mu := \mu - \theta \log\left(\frac{1}{\tilde{N} + N} \sum_{k=-\tilde{N}}^{N} \exp\left(-\frac{T_k - \mu}{\theta}\right)\right), \tag{96}$$

$$\theta := \frac{\theta}{\tilde{N}+N} \sum_{k=-\tilde{N}}^{N} \left(\frac{T_k-\mu}{\theta}\right)\left(1-\exp\left(-\frac{T_k-\mu}{\theta}\right)\right). \tag{97}$$

Define the following functions calculated by integral operation on the standard extreme value distribution at maximum:

$$\begin{aligned}\overline{\Theta}^{(1)}(z) &= \frac{1}{\theta}\int_{\theta z+\mu}^{\infty} \exp(-u)d\Theta(u) \\ &= 1-(1+\exp(-z))\exp(-\exp(-z)), \qquad (98)\\ \overline{\Theta}^{(2)}(z) &= \frac{1}{\theta}\int_{\theta z+\mu}^{\infty} u(1-\exp(-u))d\Theta(u) \\ &= 1-\exp(-\exp(-z))(1-z\exp(-z)). \qquad (99)\end{aligned}$$

By using $\overline{\Theta}^{(1)}(z)$ and $\overline{\Theta}^{(2)}(z)$, the EM-step formulas for TEVMAX can be obtained as follows.

$$\tilde{\omega} := N^{(1)}, \tag{100}$$

$$\mu := \mu' - \theta' \log\left(\frac{T^{(1)}}{N^{(1)}}\right), \tag{101}$$

$$\theta := \theta' \frac{T^{(2)}}{N^{(1)}}, \tag{102}$$

where

$$N^{(1)} = K + \tilde{\omega}'\left(1-\overline{\Theta}\left(-\frac{\mu'}{\theta'}\right)+\overline{\Theta}\left(\frac{T-\mu'}{\theta'}\right)\right), \tag{103}$$

$$T^{(1)} = \sum_{k=1}^{K} \exp\left(-\frac{t_k-\mu'}{\theta'}\right) + \tilde{\omega}'\left(1-\overline{\Theta}^{(1)}\left(-\frac{\mu'}{\theta'}\right)+\overline{\Theta}^{(1)}\left(\frac{T-\mu'}{\theta'}\right)\right), \tag{104}$$

$$\begin{aligned}T^{(2)} &= \sum_{k=1}^{K} \frac{t_k-\mu'}{\theta'}\left(1-\exp\left(-\frac{t_k-\mu'}{\theta'}\right)\right) \\ &\quad + \tilde{\omega}'\left(1-\overline{\Theta}^{(2)}\left(-\frac{\mu'}{\theta'}\right)+\overline{\Theta}^{(2)}\left(\frac{T-\mu'}{\theta'}\right)\right). \qquad (105)\end{aligned}$$

After the EM algorithm converges to MLEs, we take $\omega := \tilde{\omega}\overline{\Theta}(-\mu/\theta)$.

4.8 EM-Step Formulas for LEVMAX

By taking logarithm of complete samples, we obtain the EM-step formulas for LEVMAX:

$$\omega := N^{(1)}, \tag{106}$$

$$\mu := \mu' - \theta' \log\left(\frac{T^{(1)}}{N^{(1)}}\right), \tag{107}$$

$$\theta := \theta' \frac{T^{(2)}}{N^{(1)}}, \tag{108}$$

where

$$N^{(1)} = K + \omega' \overline{\Theta}\left(\frac{\log T - \mu'}{\theta'}\right), \tag{109}$$

$$T^{(1)} = \sum_{k=1}^{K} \exp\left(-\frac{\log t_k - \mu'}{\theta'}\right) + \omega' \overline{\Theta}^{(1)}\left(\frac{\log T - \mu'}{\theta'}\right), \tag{110}$$

$$T^{(2)} = \sum_{k=1}^{K} \frac{\log t_k - \mu'}{\theta'}\left(1 - \exp\left(-\frac{\log t_k - \mu'}{\theta'}\right)\right)$$
$$+ \omega' \overline{\Theta}^{(2)}\left(\frac{\log T - \mu'}{\theta'}\right). \tag{111}$$

4.9 EM-Step Formulas for TEVMIN

In general, if a random variable X obeys a minimum extreme value distribution, the random variable $-X$ becomes a maximum extreme value random variable. That is, the parameter estimation for TEVMIN can be developed by considering the sign reversed failure time data.

$$\tilde{\omega} := N^{(1)}, \tag{112}$$

$$\mu := \mu' - \theta' \log\left(\frac{T^{(1)}}{N^{(1)}}\right), \tag{113}$$

$$\theta := \theta' \frac{T^{(2)}}{N^{(1)}}, \tag{114}$$

where

$$N^{(1)} = K + \tilde{\omega}'\left(1 - \overline{\Theta}\left(-\frac{T+\mu'}{\theta'}\right) + \overline{\Theta}\left(-\frac{\mu'}{\theta'}\right)\right), \tag{115}$$

$$T^{(1)} = \sum_{k=1}^{K} \exp\left(\frac{t_k+\mu'}{\theta'}\right) + \tilde{\omega}'\left(1 - \overline{\Theta}^{(1)}\left(-\frac{T+\mu'}{\theta'}\right) + \overline{\Theta}^{(1)}\left(-\frac{\mu'}{\theta'}\right)\right), \tag{116}$$

$$T^{(2)} = -\sum_{k=1}^{K} \frac{t_k+\mu'}{\theta'}\left(1 - \exp\left(\frac{t_k+\mu'}{\theta'}\right)\right) + \tilde{\omega}'\left(1 - \overline{\Theta}^{(2)}\left(-\frac{T+\mu'}{\theta'}\right) + \overline{\Theta}^{(2)}\left(-\frac{\mu'}{\theta'}\right)\right). \tag{117}$$

After the EM algorithm converges to MLEs, we take $\omega := \tilde{\omega}\Theta(-\mu/\theta)$.

4.10 EM-Step Formulas for LEVMIN

Consider sign inversion of logarithm of original samples. Then we have the EM-step formulas for LEVMIN.

$$\tilde{\omega} := N^{(1)}, \tag{118}$$

$$\mu := \mu' - \theta' \log\left(\frac{T^{(1)}}{N^{(1)}}\right), \tag{119}$$

$$\theta := \theta' \frac{T^{(2)}}{N^{(1)}}, \tag{120}$$

where

$$N^{(1)} = K + \omega'\Theta\left(-\frac{\log T+\mu'}{\theta'}\right), \tag{121}$$

$$T^{(1)} = \sum_{k=1}^{K} \exp\left(\frac{\log t_k+\mu'}{\theta'}\right) + \omega'\left(1 - \overline{\Theta}^{(1)}\left(-\frac{\log T+\mu'}{\theta'}\right)\right), \tag{122}$$

$$T^{(2)} = -\sum_{k=1}^{K} \frac{\log t_k+\mu'}{\theta'}\left(1 - \exp\left(\frac{\log t_k+\mu'}{\theta'}\right)\right) + \omega'\left(1 - \overline{\Theta}^{(2)}\left(-\frac{\log T+\mu'}{\theta'}\right)\right). \tag{123}$$

5 Remarks on EM Algorithm for NHPP-Based SRMs

5.1 Termination Condition

In general, it is proved that the estimates by EM algorithms gradually approach to the MLEs as the number of steps increases. However, the convergence speed of EM algorithms is relatively slower than those of Newton's and quasi-Newton's methods. Hence, it is important to decide the timing when the algorithm stops, i.e., termination condition.

Intuitively, reasonable conditions for termination condition are based on the differences of likelihood and estimates. Let $\boldsymbol{\lambda}'$ and $\boldsymbol{\lambda}$ denote parameter vectors before and after one EM-step, respectively. Then, the conditions based on the differences of likelihood and estimates are given by

$$\left| \frac{\mathrm{LLF}(\boldsymbol{\lambda}) - \mathrm{LLF}(\boldsymbol{\lambda}')}{\mathrm{LLF}(\boldsymbol{\lambda}')} \right| < \varepsilon_l \quad \text{and} \quad \frac{||\boldsymbol{\lambda} - \boldsymbol{\lambda}'||}{||\boldsymbol{\lambda}'||} < \varepsilon_p, \tag{124}$$

respectively, where $||\cdot||$ is a norm, ε_l and ε_p are error tolerances. The condition based on estimates is empirically more effective than the difference of likelihood.

5.2 Initial Guesses

Unlike the classical estimation procedures such as Newton's and quasi-Newton's methods, the proposed EM algorithms are not sensitive to the starting parameters, i.e., initial guesses of the algorithms, because of the global convergence property of EM algorithm. However, even if we apply the EM algorithm, initial guesses affect the performance of estimation. The simple adjustment is to use the data information. Table 2 presents typical initial guesses for NHPP-based SRMs. Although several scale parameters should depend on the maximum failure time t_K, the other parameters are given by suitable constants independent from data statistics. This is the advantage of using the EM algorithm.

5.3 Model Selection

The model selection is one of the most practical problems in utilizing NHPP-based SRMs. In general, information criteria are available to select the NHPP-based SRMs after computing the MLEs. The information criteria consist of the maximum log-likelihood and the penalty term concerning the number of free parameters, and different penalty terms are different information criteria. The well-known information criteria are AIC (Akaike's Information Criterion) and BIC (Bayesian Information Criterion), which are defined as follows.

Table 2 Typical initial guesses for model parameters

Model	Initial guesses
EXP	$\omega = K, \beta = 1/t_K$
GAMMA	$\omega = K, \alpha = 1, \beta = 1/t_K$
TNORM	$\omega = K, \mu = 0, \sigma = t_K$
LNORM	$\omega = K, \mu = 0, \sigma = \log(t_K)$
TLOGIS	$\omega = K, \mu = 0, \psi = t_K$
LLOGIS	$\omega = K, \mu = 0, \psi = \log(t_K)$
TEVMAX	$\omega = K, \mu = 0, \theta = t_K$
LEVMAX	$\omega = K, \mu = 0, \theta = \log(t_K)$
TEVMIN	$\omega = K, \mu = 0, \theta = t_K$
LEVMIN	$\omega = K, \mu = 0, \theta = \log(t_K)$

$$\text{AIC}(n) = -2(\text{maximum log-likelihood}) + 2p, \tag{125}$$

and

$$\text{BIC}(n) = -2(\text{maximum log-likelihood}) + p \log n, \tag{126}$$

where p and n denote the number of free parameters and the number of data records used in estimating the parameters. In most application, the model which has the least information criterion is selected as the best model fitted to the observation. However, since ML estimation of NHPP-based SRMs is not always regular condition [24], AIC and BIC do not work well in the case where there are few samples of failure time. The model selection still has several theoretical problems and is the future work in the software reliability assessment.

5.4 Local Maximum Problem

In general, the local maximum problem arises in the EM algorithm as well as Newton's method. It is not guaranteed that the proposed EM algorithm converges to the global maxima. However, the LLF of NHPP-based SRMs presented in the chapter becomes an unimodal function if we have the data consisting of the number of failures experienced is sufficient to provide ML estimates. For example, the uniqueness of MLEs of EXP is proved if the parameters satisfy the following condition [11]:

$$t_K > \frac{2}{K} \sum_{k=1}^{K} t_k. \tag{127}$$

Similar to EXP, it is expected that the MLEs are also unique for other NHPP-based SRMs, if there exist. Thus in the practical situation, we do not need to pay attention to the local maximum problem in the context of EM algorithms.

5.5 *Maximum a Posterior*

We extend the EM procedure for ML estimation to the procedure for maximum a posterior (MAP) estimation. As shown in 5.4, there are the cases where MLEs do not exist in finite domain. Even in such a case, MAP estimation gives finite estimates of model parameters by applying appropriate prior distributions. In the case of parameter estimation of NHPP-based SRMs, the parameter ω is essentially estimated from only one sample. Therefore, it is effective to apply the prior distribution for the parameter ω in order to derive finite estimates even in the case where there are few failure times.

Let $p(\omega)$ be a gamma prior density for the parameter ω with hyper parameters a and b, i.e.,

$$p(\omega) = \frac{b^{a+1}\omega^{a}e^{-b\omega}}{\Gamma(a)}, \quad \omega \geq 0. \tag{128}$$

Then the problem is to find the parameter maximizing;

$$\log p(D_T; \omega, \boldsymbol{\lambda})p(\omega). \tag{129}$$

Based on EM principle, M-step formulas of Eqs. (31) and (32) can be rewritten in the form:

$$\omega := \frac{\mathrm{E}[N|D_T; \omega', \boldsymbol{\lambda}'] + a}{b+1} \tag{130}$$

$$\boldsymbol{\lambda} := \operatorname*{argmax}_{\boldsymbol{\lambda}} \mathrm{E}\left[\sum_{i=1}^{N} \log f(T_i; \boldsymbol{\lambda}) \middle| D_T; \omega', \boldsymbol{\lambda}'\right]. \tag{131}$$

On the other hand, since the prior density $p(\omega)$ does not include unobserved values, the expected values $\mathrm{E}[N|D_T; \omega', \boldsymbol{\lambda}']$ and $\mathrm{E}\left[\sum_{i=1}^{N} \log f(T_i; \boldsymbol{\lambda})|D_T; \omega', \boldsymbol{\lambda}'\right]$ can be computed by the same formulas presented in Sects. 3 and 4. That is, by modifying only the M-step formulas, we have the EM algorithms to obtain MAP estimates for NHPP-based SRMs.

6 Numerical Example

We investigate the numerical characteristics of the EM algorithms in practical situation, and first present the difference with the classical Newton's method in the viewpoint of updating estimates.

Figure 1 shows the locus of estimates for EXP by the EM algorithm and Newton's method from the same initial points. In the figure, the contour plot indicates the LLF of EXP with the failure data collected from the existing software project [19]. The failure data consist of 136 software failure times. The color in the figure becomes

Fig. 1 Behavior of estimates updated in EXP; initial parameters are $(\omega, \beta) = (100, 6.0e^{-5})$

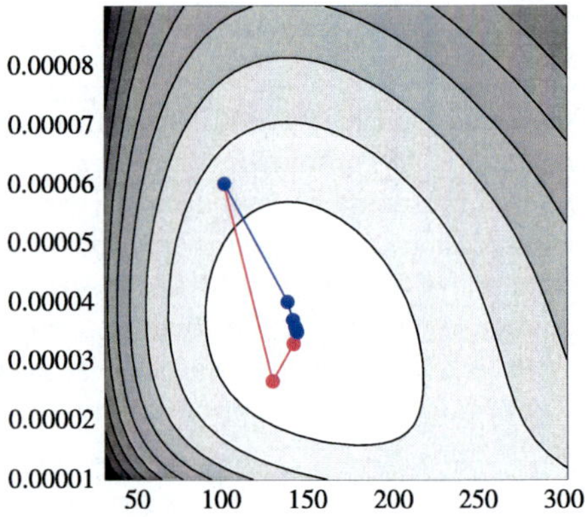

Fig. 2 Behavior of estimates updated in EXP; initial parameters are $(\omega, \beta) = (100, 8.0e^{-5})$

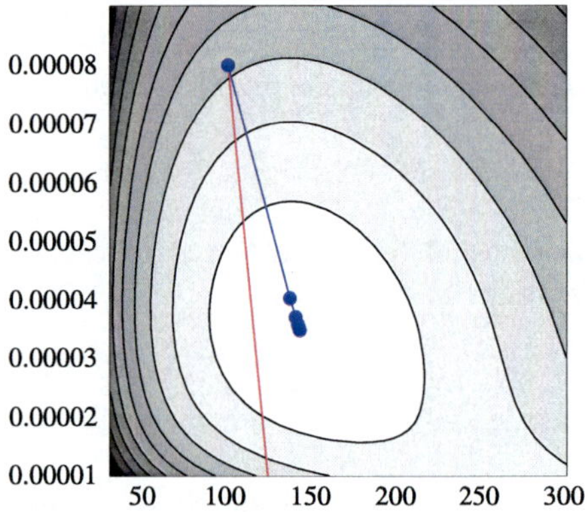

bright as the log-likelihood increases. In Fig. 1, the initial parameters are set as $(\omega, \beta) = (100, 6.0e^{-5})$. In this case, both estimates converge to the MLE, $(\hat{\omega}, \hat{\beta}) = (141.93, 3.48e^{-5})$. From the figure, we find that Newton's method needs less updates of parameters until the estimates converge to the MLE than the EM algorithm does. That is, the convergence speed of Newton's method is faster than the EM algorithm. Figure 2 shows the similar loci of both methods. In this case, the initial parameters are set as $(\omega, \beta) = (100, 8.0e^{-5})$. Although the EM algorithm converges to the MLE, Newton's method fails and oversteps the implicit parameter constraints, $\omega > 0$ and $\beta > 0$ at the first step. In such case, the numerical exception occurs at the first step.

These figures present the difference between the local convergence property of Newton's method and the global convergence property of the EM algorithm. In this example, Newton's method fails to estimating the parameters by changing $\beta = 6.0e^{-5}$ to $8.0e^{-5}$. Therefore, in practical application, it is quite difficult to determine the initial parameters of Newton's method. On the other hand, the EM algorithm converges to the MLEs without numerical exception in both cases.

7 Conclusions

This chapter has considered the problem on ML estimation for NHPP-based SRMs and has introduced an iterative scheme to estimate MLEs. In particular, we have developed the estimation procedures based on EM algorithm, and have presented concrete EM-step formulas for 10 typical NHPP-based SRMs. In the numerical experiment, we have compared the EM algorithm and Newton's method from the viewpoint of convergence properties. As a result, the proposed EM algorithm is effective to reduce the computation effort including selection of initial guesses.

In future, we develop an integrated EM procedure for NHPP-based SRMs even for grouped data. In addition, we will develop a software reliability assessment tool on spreadsheet application, which involves the developed EM algorithm.

Acknowledgments This research was supported by the Ministry of Education, Science, Sports and Culture, Grant-in-Aid for Scientific Research (C), Grant No. 23500047 (2011–2013) and Grant No. 23510171 (2011–2013). The authors are grateful to Prof. Shunji Osaki and his teaching. Everything he teaches was valuable for our lives as well as research activities.

Appendix

Derivation of Eq. (40)

For notational simplification, $\mathrm{E}[\cdot|D_T;\omega,\boldsymbol{\lambda}]$ is written by $\mathrm{E}[\cdot|D_T]$. From the left-hand side of Eq. (40), we have

$$\begin{aligned}\mathrm{E}\left[\sum_{k=1}^{N} h(T_k)\middle|D_T\right] &= \mathrm{E}\left[\sum_{k=1}^{K} h(T_k)\middle|D_T\right] + \mathrm{E}\left[\sum_{k=K+1}^{N} h(T_k)\middle|D_T\right] \\ &= \sum_{k=1}^{K} \mathrm{E}[h(T_k)|D_T] + \mathrm{E}\left[\sum_{k=K+1}^{N} h(T_k)\middle|D_T\right] \\ &= \sum_{k=1}^{K} h(t_k) + \mathrm{E}\left[\sum_{k=K+1}^{N} h(T_k)\middle|D_T\right]. \end{aligned} \tag{132}$$

Since $T_1,\ldots,T_N$ are IID samples, the first term of right-hand side of the above equation can be easily obtained. Then we focus on the derivation of the second term of right-hand side of Eq. (40). From the posterior distribution of N, we obtain

$$\mathrm{E}\Bigg[\sum_{k=K+1}^{N} h(T_k)\Bigg| D_T\Bigg]$$
$$= \left(\sum_{r=0}^{\infty} e^{-\omega}\omega^{K+r} \prod_{k=1}^{K} f(t_k) \int_T^{\infty}\int_{u_1}^{\infty}\cdots\int_{u_{r-1}}^{\infty} \sum_{j=1}^{r} h(u_j) \prod_{j=1}^{r} f(u_j) du_r \cdots du_1\right)$$
$$\Bigg/\left(\sum_{r=0}^{\infty} e^{-\omega}\omega^{K+r} \prod_{k=1}^{K} f(t_k) \int_T^{\infty}\int_{u_1}^{\infty}\cdots\int_{u_{r-1}}^{\infty} \prod_{j=1}^{r} f(u_j) du_r \cdots du_1\right), \tag{133}$$

where r corresponds to the number of failures experienced after time T. The integrals in Eq. (133) can be changed to

$$\int_{u_1}^{\infty}\cdots\int_{u_{r-1}}^{\infty} \sum_{j=1}^{r} h(u_j) \prod_{j=1}^{r} f(u_j) du_r \cdots du_1$$
$$= \frac{r}{r!}\int_T^{\infty} h(u) f(u) du \left(\int_T^{\infty} f(u) du\right)^{r-1}, \tag{134}$$

and

$$\int_{u_1}^{\infty}\cdots\int_{u_{r-1}}^{\infty} \prod_{j=1}^{r} f(u_j) du_r \cdots du_1 = \frac{1}{r!}\left(\int_T^{\infty} f(u) du\right)^{r}. \tag{135}$$

By canceling the constants of numerator and denominator, Eq. (133) is reduced to

$$\mathrm{E}\Bigg[\sum_{k=K+1}^{N} h(T_k)\Bigg| D_T\Bigg]$$
$$= \left(\sum_{r=1}^{\infty} \omega^r \frac{1}{(r-1)!}\int_T^{\infty} h(u) f(u) du \overline{F}(T)^{r-1}\right)\Bigg/\left(\sum_{r=0}^{\infty} \omega^r \frac{1}{r!}\overline{F}(T)^r\right)$$
$$= \omega \int_T^{\infty} h(u) f(u) du. \tag{136}$$

Derivation of Eq. (56)

From the left-hand side of Eq. (56), we have

$$
\begin{aligned}
\mathrm{E}\Big[\sum_{k=-\tilde{N}}^{N} h(T_k)\Big|D_T\Big] &= \mathrm{E}\Big[\sum_{k=1}^{K} h(T_k)\Big|D_T\Big] \\
&\quad + \mathrm{E}\Big[\sum_{k=1}^{\tilde{N}} h(T_{-k})\Big|D_T\Big] + \mathrm{E}\Big[\sum_{k=K+1}^{N} h(T_k)\Big|D_T\Big] \\
&= \sum_{k=1}^{K} h(t_k) + \mathrm{E}\Big[\sum_{k=1}^{\tilde{N}} h(T_{-k})\Big|D_T\Big] + \mathrm{E}\Big[\sum_{k=K+1}^{N} h(T_k)\Big|D_T\Big].
\end{aligned} \tag{137}
$$

Similar to the previous section, we consider the second and third terms of right-hand side of Eq. (137). Then we have

$$
\begin{aligned}
&\mathrm{E}\Big[\sum_{k=1}^{\tilde{N}} h(T_{-k})\Big|D_T\Big] \\
&= \Big(\sum_{l=0}^{\infty}\sum_{r=0}^{\infty} e^{-\omega}\omega^{K+l+r} \int_{-\infty}^{0}\int_{-\infty}^{u_1}\cdots\int_{-\infty}^{u_{l-1}} \sum_{j=1}^{l} h(u_l)\prod_{j=1}^{l} f(u_j) \\
&\quad \times \prod_{k=1}^{K} f(t_k)\int_{T}^{\infty}\int_{u_1}^{\infty}\cdots\int_{u_{r-1}}^{\infty}\prod_{j=1}^{r} f(u_j)du_r\cdots du_1\Big) \\
&\quad \Big/\Big(\sum_{l=0}^{\infty}\sum_{r=0}^{\infty} e^{-\omega}\omega^{K+l+r} \int_{-\infty}^{0}\int_{-\infty}^{u_1}\cdots\int_{-\infty}^{u_{l-1}} \prod_{j=1}^{l} f(u_j) \\
&\quad \times \prod_{k=1}^{K} f(t_k)\int_{T}^{\infty}\int_{u_1}^{\infty}\cdots\int_{u_{r-1}}^{\infty}\prod_{j=1}^{r} f(u_j)du_r\cdots du_1\Big) \\
&= \Big(\sum_{l=1}^{\infty}\omega^{l}\frac{1}{(l-1)!}\int_{-\infty}^{0} h(u)f(u)du F(0)^{l-1}\Big)\Big/\Big(\sum_{l=0}^{\infty}\omega^{l}\frac{1}{l!}F(0)^{l}\Big) \\
&= \omega\int_{-\infty}^{0} h(u)f(u)du
\end{aligned} \tag{138}
$$

and

$$
\begin{aligned}
&\mathrm{E}\Big[\sum_{k=K+1}^{N} h(T_k)\Big|D_T\Big] \\
&= \Big(\sum_{l=0}^{\infty}\sum_{r=0}^{\infty} e^{-\omega}\omega^{K+l+r} \int_{-\infty}^{0}\int_{-\infty}^{u_1}\cdots\int_{-\infty}^{u_{l-1}} \prod_{j=1}^{l} f(u_j) \\
&\quad \times \prod_{k=1}^{K} f(t_k)\int_{T}^{\infty}\int_{u_1}^{\infty}\cdots\int_{u_{r-1}}^{\infty}\sum_{j=1}^{l} h(u_r)\prod_{j=1}^{r} f(u_j)du_r\cdots du_1\Big)
\end{aligned}
$$

$$
\Big/ \Big(\sum_{l=0}^{\infty} \sum_{r=0}^{\infty} e^{-\omega} \omega^{K+l+r} \int_{-\infty}^{0} \int_{-\infty}^{u_1} \cdots \int_{-\infty}^{u_{l-1}} \prod_{j=1}^{l} f(u_j)
$$

$$
\times \prod_{k=1}^{K} f(t_k) \int_{T}^{\infty} \int_{u_1}^{\infty} \cdots \int_{u_{r-1}}^{\infty} \prod_{j=1}^{r} f(u_j) du_r \cdots du_1 \Big)
$$

$$
= \Big(\sum_{r=1}^{\infty} \omega^r \frac{1}{(r-1)!} \int_{T}^{\infty} h(u) f(u) du \overline{F}(T)^{r-1} \Big) \Big/ \Big(\sum_{r=0}^{\infty} \omega^r \frac{1}{r!} \overline{F}(T)^r \Big)
$$

$$
= \omega \int_{T}^{\infty} h(u) f(u) du. \tag{139}
$$

References

1. Achcar JA, Dey DK, Niverthi M (1998) A Bayesian approach using nonhomogeneous poisson processes for software reliability models. In: Basu AP, Basu KS, Mukhopadhyay S (eds) Frontiers in reliability. World Scientific, Singapore, pp 1–18
2. Ando T, Okamura H, Dohi T (2006) Estimating Markov modulated software reliability models via EM algorithm. In: Proceedings of 2nd IEEE international symposium on dependable, autonomic and secure computing (DASC'06), pp 111–118. IEEE Computer Society Press, Los Alamitos
3. Chen Y, Singpurwalla ND (1997) Unification of software reliability models by self-exciting point processes. Adv Appl Probab 29:337–352
4. Cheung RC (1980) A user-oriented software reliability model. IEEE Trans Softw Eng SE–6(2):118–125
5. Dempster AP, Laird NM, Rubin DB (1977) Maximum likelihood from incomplete data via the EM algorithm. J Roy Stat Soc B B–39:1–38
6. Goel AL (1985) Software reliability models: assumptions, limitations and applicability. IEEE Transactions on Software Engineering SE–11:1411–1423
7. Goel AL, Okumoto K (1978) An imperfect debugging model for reliability and other quantitative measures of software systems. Technical report 78–1, Department of IE& OR, Syracuse University
8. Goel AL, Okumoto K (1979) Time-dependent error-detection rate model for software reliability and other performance measures. IEEE Trans Reliab R–28:206–211
9. Gokhale SS, Philip T, Marinos PN, Trivedi KS (1996) Unification of finite failure non-homogeneous Poisson process models through test coverage. In: Proceedings of 7th international symposium on software, reliability engineering, pp 299–307
10. Gokhale SS, Trivedi KS (1998) Log-logistic software reliability growth model. In: Proceedings of 3rd IEEE international symposium on high-assurance systems engineering (HASE-1998), pp 34–41. IEEE CS Press
11. Hossain SA, Dahiya RC (1993) Estimating the parameters of a non-homogeneous Poisson-process model for software reliability. IEEE Trans Reliab 42(4):604–612
12. Jelinski Z, Moranda PB (1972) Software reliability research. In: Freiberger W (ed) Statistical computer performance evaluation. Academic, New York, pp 465–484
13. Jeske DR, Pham H (2001) On the maximum likelihood estimates for the Goel-Okumoto software reliability model. Am Stat 55(3):219–222

14. Joe H (1989) Statistical inference for general-order-statistics and nonhomogeneous-Poisson-process software reliability models. IEEE Trans Softw Eng 15(11):1485–1490
15. Kimura M, Yamada S (2003) Software reliability management: techniques and applications. In: Pham H (ed) Handbook of reliability engineering. Springer, London, pp 265–284
16. Knafl G, Morgan J (1996) Solving ML equations for 2-parameter Poisson-process model for ungrouped software failure data. IEEE Trans Reliab 45:42–53
17. Kuo L, Yang TY (1996) Bayesian computation for nonhomogeneous Poisson processes in software reliability. J Am Stat Assoc 91:763–773
18. Langberg N, Singpurwalla ND (1985) Unification of some software reliability models. SIAM J Sci Comput 6(3):781–790
19. Lyu MR (ed) (1996) Handbook of Software Reliability Engineering. McGraw-Hill, New York
20. Miller DR (1986) Exponential order statistic models of software reliability growth. IEEE Trans Softw Eng SE–12:12–24
21. Musa JD (1999) Software reliability engineering. McGraw-Hill, New York
22. Musa JD, Iannino A, Okumoto K (1987) Software reliability, measurement, prediction, application. McGraw-Hill, New York
23. Musa JD, Okumoto K (1984) A logarithmic Poisson execution time model for software reliability measurement. In: Proceedings of 7th international conferece on software engineering (ICSE-1084), pp 230–238. IEEE CS Press/ACM (1984)
24. Nayak TK, Bose S, Kundu S (2008) On inconsistency of estimators of parameters of non-homogeneous Poisson process models for software reliability. Stat Prob Lett 78:2217–2221
25. Ohba M (1984) Inflection S-shaped software reliability growth model. In: Osaki S, Hatoyama Y (eds) Stochastic models in reliability theory. Springer, Berlin, pp 144–165
26. Ohba M (1984) Software reliability analysis. IBM J Res Dev 28:428–443
27. Ohishi K, Okamura H, Dohi T (2009) Gompertz software reliability model: estimation algorithm and empirical validation. J Sys Softw 82(3):535–543
28. Okamura H, Dohi T, Osaki S (2004) EM algorithms for logistic software reliability models. In: Proceedings of 22nd IASTED international conference on software engineering, pp 263–268. ACTA Press
29. Okamura H, Murayama A, Dohi T (2004) EM algorithm for discrete software reliability models: a unified parameter estimation method. In: Proceedings of 8th IEEE international symposium on high assurance systems engineering, pp 219–228
30. Okamura H, Watanabe Y, Dohi T (2003) An iterative scheme for maximum likelihood estimation in software reliability modeling. In: Proceedings of 14th international symposium on software reliability engineering, pp 246–256
31. Pham H (2000) Software reliability. Springer, Singapore
32. Pham H, Zhang X (1997) An NHPP software reliability models and its comparison. Int J Reliab Qual Safe Eng 4:269–282
33. Ramani S, Gokhale SS, Trivedi KS (2000) Srept: software reliability estimation and prediction tool. Perf Eval 39:37–60
34. Shanthikumar JG (1981) A general software reliability model for performance prediction. Microelect Reliab 21:671–682
35. Wu CFJ (1983) On the convergence properties of the EM algorithm. Ann Stat 11:95–103
36. Xie M (1991) Software reliability modelling. World Scientific, Singapore
37. Yamada S (1992) A stochastic software reliability growth model with Gompertz curve. Trans Inf Proces Soc Japan (in Japanese) 33(7):964–969
38. Yamada S, Ohba M, Osaki S (1983) S-shaped reliability growth modeling for software error detection. IEEE Trans Reliab R–32:475–478
39. Zhao M, Xie M (1996) On maximum likelihood estimation for a general non-homogeneous Poisson process. Scand J Stat 23:597–607

Closed-Form Approach for Epistemic Uncertainty Propagation in Analytic Models

Kesari Mishra and Kishor S. Trivedi

1 Introduction

System dependability or performance is often studied using stochastic models. These models capture the natural uncertainty in the system being studied, known as aleatory uncertainty. Randomness in events of interest like times to failure/recovery of components, ability to detect failures, ability to perform recovery action, inter-arrival time, service time, etc., are taken into account in the models, by means of their distributions. The models are usually solved at fixed parameter values. However, the model input parameter values have uncertainty associated with them as they are derived either from a finite number of observations (from lifetime determining experiments or field data) or are based upon expert guesses. This uncertainty in model input parameter values, known as epistemic uncertainty, is not normally taken into account by the stochastic aleatory model.

The uncertainty in model output, due to epistemic uncertainty in model input parameters should be distinguished from the modeling error in the aleatory model. Modeling error in the aleatory model causes the aleatory model to not be a faithful representation of the behavior of the real system. It can be due to incorrect understanding of system behavior, omissions in components/ states, incorrect assumptions of distributions of various events in the aleatory model, incorrect use of constructs of modeling paradigm, or simply incorrect implementation of the model; while epistemic uncertainty is the uncertainty in the parameters of the aleatory model due to incomplete information. Assuming the aleatory model of the system behavior and the aleatory distributional assumptions to be correct, this chapter discusses propagation

K. Mishra (✉) · K. S. Trivedi
Department of ECE, Duke University, Durham NC 27708, USA
e-mail: km@ee.duke.edu

K. S. Trivedi
e-mail: kst@ee.duke.edu

T. Dohi and T. Nakagawa (eds.), *Stochastic Reliability and Maintenance Modeling*, Springer Series in Reliability Engineering 9, DOI: 10.1007/978-1-4471-4971-2_14,

of epistemic uncertainty through stochastic aleatory model, to obtain the uncertainty in the model output metric.

The epistemic uncertainty in the model parameters may be expressed in the form of distribution of parameter values themselves (epistemic distribution) or in the form of bounds or confidence interval of parameter values, obtained from manufacturer datasheets. The model output value computed using fixed values of the model parameters can be considered to be conditional upon the parameter values used. To propagate the epistemic uncertainty of model input parameters to the model output, it needs to be unconditioned. Applying the theorem of total probability, unconditioning of the model output can be performed by means of multi-dimensional integration. Various analytic and numerical techniques can be employed to solve this integration.

Depending on the nature of the stochastic models (analytic or simulation) and their complexity, different techniques may be applied to perform the unconditioning integration. Simple analytic aleatory models may be solved analytically to get the model output as closed-form expressions of input parameters. For more complex analytic models, only analytic-numerical solutions using tools like SHARPE [16] or SPNP [4] may be possible. Alternatively, the aleatory model may be a simulation model. Different methods of propagating the epistemic uncertainty need to be employed for each of these cases. For analytic models which have simple closed-form model outputs, direct analytic integration may be possible. For more complex analytic models without closed-form solutions, numerical integration maybe needed. Sampling-based uncertainty propagation [3, 15] can be applied to complex analytic models (can also be used when the model can be solved either analytically or analytic-numerically), as well as simulation models. Figure 1 summarizes the applicability of various methods of epistemic uncertainty propagation for different types of aleatory models.

In this chapter, epistemic uncertainty propagation by direct analytical integration of closed-form expressions of system reliability is discussed. This method is applied to compute uncertainty in reliability of some nonrepairable systems. The results are then analyzed to gain insight into the uncertainty in system reliability, due to the epistemic uncertainty in model input parameters. The limiting behavior of metrics of uncertainty in model output is also studied. As the expression for model output

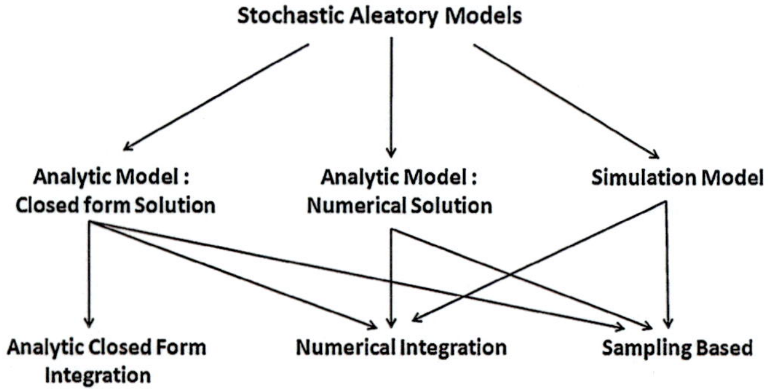

Fig. 1 Epistemic uncertainty propagation for different aleatory model types

becomes more complex, analytic closed-form integration of the model output may not be possible. For such an example, the expectation and variance of model output, due to epistemic uncertainties in the model input parameters, are obtained using numerical integration to perform the unconditioning integrations.

2 Uncertainty Propagation

Due to the epistemic uncertainties, the input parameters of a reliability model can be considered a random vector. Therefore, the reliability obtained by solving the model can be considered a random variable that is a function of these input random variables. If random variables $\{\Lambda_i, i = 1, 2, \ldots, l\}$ are the set of l input parameters of the model, the reliability $R(t)$, at time t, can be viewed as a random variable (function) g of the l input parameters as $R(t) = g(\Lambda_1, \Lambda_2, \ldots, \Lambda_l)$. Due to the uncertainty associated with the model parameters, computing the reliability at specific parameter values can be seen as computing the conditional reliability $R(t|\Lambda_1 = \lambda_1, \Lambda_2 = \lambda_2, \ldots, \Lambda_l = \lambda_l)$ (denoted by $R(t|.)$ in Eq. 1). Applying the theorem of total probability [14], this can be unconditioned to compute the distribution of reliability via the joint epistemic density $f_{\Lambda_1,\Lambda_2,\ldots,\Lambda_l}(\lambda_1, \lambda_2, \ldots, \lambda_l)$ of the input parameters (denoted by $f(.)$ in Eq. 1).

$$F_{R(t)}(p) = \int \ldots \int I(R(t|.) \leq p) f(.) d\lambda_1 \ldots d\lambda_l \tag{1}$$

where $I(Event)$ is the indicator variable of the event $Event$. The unconditional expected reliability at time t can be computed as:

$$E[R(t)] = \int \ldots \int R(t|.) f(.) d\lambda_1 \ldots d\lambda_l \tag{2}$$

Similarly, the second moment of reliability, $E[R(t)^2]$ can be computed, as:

$$E[(R(t))^2] = \int \ldots \int (R(t|.))^2 f(.) d\lambda_1 \ldots d\lambda_l \tag{3}$$

With the second moment and the expected value, the variance of reliability at time t, $Var[R(t)]$ can be computed.

If the reliability model is simple, it can be solved analytically to obtain a closed-form expression of reliability, in terms of the model input parameters. In such cases, the above integrations can theoretically, be directly performed on the expression for reliability, to propagate the epistemic uncertainty. For simpler expressions, the integration can be performed analytically, while for more complex expressions of reliability, numerical integration [10] may be performed. However, the task of evaluating these integrals quickly becomes intractable for complex expressions of system reliability or for larger numbers of model input parameters. Apart from the computational problem, the joint epistemic density of all the model parameters also needs to be specified. The problem becomes somewhat simpler if the epistemic random variables can be assumed to be independent, as the joint probability density functions can then be factored into the product of marginals.

While in this chapter we assume the model input parameters as random variables to be independent, in real life, there may be dependencies between them, originating from the fact that they may be obtained from a common data or information source. Ignoring epistemic dependencies among parameters can lead to errors and biases in the output metrics, depending on the degree of correlation and the parameters which are correlated [12]. Sampling-based epistemic uncertainty propagation methods can take the epistemic dependencies into account, relatively easily, via methods like the one proposed by Iman and Conover [5], which can introduce rank correlation between the parameter values sampled from the marginal distributions. It should be noted that assuming epistemic independence between the model input parameters as random variables, does not rule out considering dependency of any kind between events in the aleatory model (e.g., dependency between failure or repair events of components or dependency between failure modes of components). Dependence can always be allowed in the aleatory model via Markov chains, stochastic Petri nets, or other state-space models [1, 7, 14], even when independence is assumed among the epistemic variables. Table 1 summarizes the differences between dependence in the aleatory model and the epistemic dependence between model input parameters as random variables.

In this chapter, propagation of parametric epistemic uncertainty through analytic integration of closed-form expression of system reliability is considered. In the examples discussed here, the model parameters as random variables are considered to be independent (epistemic independence assumed).

Clearly, for this approach, the epistemic distributions of each of the parameters as random variables need to be determined or known first. Determination of epistemic distribution of parameters as random variables, from observed values of times to failure, is discussed next.

Table 1 Difference between epistemic dependence and dependence in aleatory model

	Epistemic Dependence	Dependence in Aleatory model
Dependence Between	Model input parameters as random variables	Events in the aleatory model of the system
Reason for Dependence	Common source of data used to compute parameter values; Parameter values guessed at by the same expert	Failure or repair events of components dependent on each other or on the state of the system; Failure modes of components dependent on each other or on state of the system
Accounting for Dependence	Use joint pdf of all model input parameters; Employ methods like Iman-Conover method [5] to introduce correlation between sampled values from marginals	Use state-space models like Markov chains, stochastic Petri nets/stochastic reward nets. Hierarchical models may also be used [1, 7, 14]

2.1 Determining Epistemic Distribution

Assume that the time to failure, X, of a component, follows a distribution (aleatory) with parameter Λ. If $X_1, X_2, \ldots, X_r$ are the independent and identically distributed (iid) random variables corresponding to the r observed values of X, then the probability of these observations, given $\Lambda = \lambda$ (likelihood), is given by $f_{X_1,X_2,\ldots,X_r|\Lambda}(x_1, x_2, \ldots, x_r|\lambda)$. Applying the continuous form of Bayes' theorem, the probability density function for Λ (or the epistemic density function), given the set of observed values, can be obtained by:

$$f_{\Lambda|X_1,X_2,\ldots,X_r}(\lambda|x_1, x_2, \ldots, x_r) = \frac{f_\Lambda(\lambda) f_{X_1,X_2,\ldots,X_r|\Lambda}(x_1, x_2, \ldots, x_r|\lambda)}{\int f_\Lambda(\lambda) f_{X_1,X_2,\ldots,X_r|\Lambda}(x_1, x_2, \ldots, x_r|\lambda) d\lambda} \tag{4}$$

where, $f_{\Lambda|X_1,X_2,\ldots,X_r}(\lambda|x_1, x_2, \ldots, x_r)$ is the likelihood function, based on the aleatory distribution and $f_\Lambda(\lambda)$ is the prior density function. Clearly the epistemic posterior density function, determined in Eq. (4), will be different for different prior density functions. Assuming that we do not have much information about the epistemic distribution of the parameter beforehand, we choose a non-informative (or objective) prior based on Jeffreys' rule [6, 11]. In case of bounded parameter space (e.g., coverage probability), the prior is assumed to be uniform over the entire parameter space (every value equally likely). For parameter λ of exponential distribution, it is chosen to be $\propto 1/\lambda$.

Epistemic Distribution for Rate Parameter of Exponential Distribution

If the time to failure of a component, X, is exponentially distributed with rate parameter λ, then the random variable S, such that $S = \sum_{i=1}^{r} X_i$, will have an $r - stage$ Erlang distribution with parameter λ. Therefore, probability density function (pdf) of S, given $\Lambda = \lambda$, can be shown as:

$$f_{S|\Lambda}(s|\lambda) = \frac{\lambda^r s^{r-1} e^{-\lambda s}}{(r-1)!} \tag{5}$$

Then, applying Bayes' theorem as in Eq. (4), the pdf of Λ, given $S = s$, will be given by:

$$f_{\Lambda|S}(\lambda|s) = \frac{f_\Lambda(\lambda) \frac{\lambda^r s^{r-1} e^{-\lambda s}}{(r-1)!}}{\int_0^\infty f_\Lambda(\lambda) \frac{\lambda^r s^{r-1} e^{-\lambda s}}{(r-1)!} d\lambda} \tag{6}$$

Using Jeffreys' prior for Λ, as $f_\Lambda(\lambda) = s/\lambda$ [11], upon evaluating the integral and performing simple algebraic manipulations, Eq. (6) reduces to pdf of $r - stage$ Erlang distribution with rate parameter s, as:

$$f_{\Lambda|S}(\lambda|s) = \frac{\lambda^{r-1} s^r e^{-\lambda s}}{(r-1)!} = Erlang\ pdf(r; s) \tag{7}$$

This provides the epistemic probability density function for the rate parameter λ, when the time to failure is exponentially distributed.

Using the above equations, the rest of the chapter discusses closed-form method of uncertainty propagation through different nonrepairable systems.

3 Reliability of a Single Component System

Reliability of a single component system at time t, when the time to failure of the component follows the exponential distribution with parameter λ, is given by $R(t) = e^{-\lambda t}$ [14]. In this section, we discuss obtaining the distribution, expectation, and variance of reliability of such a single component system.

3.1 Distribution of Reliability

Due to the epistemic uncertainty in parameter λ, the reliability $R(t)$, computed at a fixed value of λ, can be seen to be conditioned on the value of λ used. If the point estimate $\hat{\lambda}$ of parameter λ were computed from r observations of times to failure, then applying the theorem of total probability and using Eqs. (1) and (7), the cumulative distribution function (CDF) of reliability at time t can be computed as shown below:

$$F_{R(t)}(p) = \int_0^\infty I(R(t) \leq p)\frac{\lambda^{r-1}s^r e^{-\lambda s}}{(r-1)!}d\lambda \tag{8}$$

As $I(.)$ is the indicator function, the above integral is nonzero only for values of λ for which $R(t) \leq p$. Since $R(t) = e^{-\lambda t}$, it can be shown that the integral will be nonzero for $\lambda \geq \lambda_a$, such that $\lambda_a = -\ln p/t$. Using the expression for CDF of an Erlang distributed random variable [14] and knowing that $s = r/\hat{\lambda}$, the above equation reduces to:

$$\begin{aligned} F_{R(t)}(p) &= \int_{\lambda_a}^\infty \frac{\lambda^{r-1}s^r e^{-\lambda s}}{(r-1)!}d\lambda = 1 - \underbrace{\int_0^{\lambda_a} \frac{\lambda^{r-1}s^r e^{-\lambda s}}{(r-1)!}d\lambda}_{Erlang\ CDF} \\ &= \sum_{i=0}^{r-1} e^{-s\lambda_a}\frac{(s\lambda_a)^i}{i!} \\ &= \sum_{i=0}^{r-1} e^{\frac{r\ln p}{\hat{\lambda}t}}\frac{(\frac{-r\ln p}{\hat{\lambda}t})^i}{i!} \end{aligned} \tag{9}$$

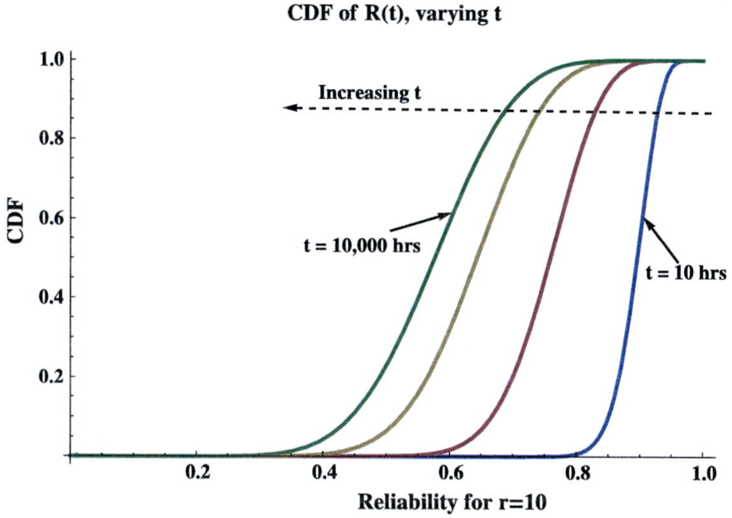

Fig. 2 CDF of reliability of a single-component system, for r = 10

At any t the reliability $R(t)$ of the system will have a distribution given by Eq. (9) and hence will have an expected value, a variance, and a confidence interval. Figure 2, shows the distribution of reliability, $F_{R(t)}(p)$, at different values of t. As t increases, the CDF shifts to the left (i.e., the reliability tends to 0, as t increases, as expected). In this figure, the value of $\hat{\lambda}$ is chosen to be 5.7078×10^{-5} $hrs.$, corresponding to an MTTF of $17,520$ $hrs.$, used for failure of software in [13] and the number of observations, r, is chosen to be 10.

Next we discuss the limiting behavior of this distribution as r is varied. As provided by the Central Limit Theorem (CLT) [14], for very large values of r, the Erlang distribution for Λ as derived in Eq. (7) will tend to normal distribution with mean $\mu_{Normal} = n/s = \hat{\lambda}$ and standard deviation $\sigma = \sqrt{r}/s = \hat{\lambda}/\sqrt{r}$. As $r \to \infty, \sigma \to 0$ and the normal pdf tends to Dirac-delta function [9]. Hence the CDF of Λ tends to Heaviside step function (or unit step function) $H[\lambda - \hat{\lambda}]$ [9], such that, for $\lambda \geq \hat{\lambda}$, the CDF evaluates to 1 and for all other values of λ, it evaluates to 0. In other words, the only possible value of λ as $r \to \infty$ is $\hat{\lambda}$, which being an unbiased estimate, is the true value of λ. Therefore, the only value of $R(t)$ possible as $r \to \infty$ is $e^{-\hat{\lambda}t}$ and the CDF of reliability, $F_{R(t)}(p)$ tends to Heaviside step function $H[p - e^{-\hat{\lambda}t}]$. Figure 3 shows the CDF of reliability, $R(t)$, at time $t = 5000$ $hours$, as the number of observations r, is varied from 10 to 1000. It can be seen clearly that as r increases, the CDF tends to the step function. The value of $\hat{\lambda}$ is chosen as earlier to be 5.7078×10^{-5} $hours$.

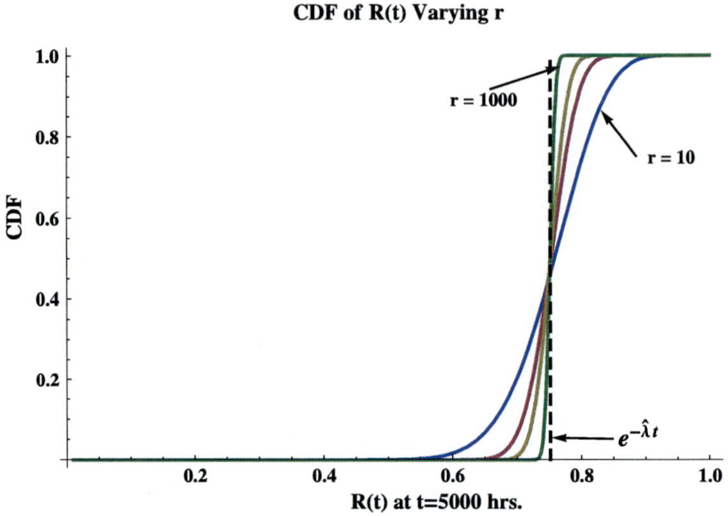

Fig. 3 CDF of reliability of a single-component system at t = 5000 hours

3.2 Expected Reliability

If the point estimate $\hat{\lambda}$ of parameter λ, were computed from r observations of times to failure, then using Eqs. (2) and (7), the unconditional expected reliability at time t can be computed as shown in Eq. (10).

$$E[R(t)] = \int_0^\infty e^{-\lambda t} . \underbrace{\frac{\lambda^{r-1} s^r e^{-\lambda s}}{(r-1)!}}_{Erlang\ pdf} d\lambda$$
$$= \left(\frac{s}{s+t}\right)^r = \left(\frac{1}{1+\hat{\lambda}t/r}\right)^r \tag{10}$$

The above equation makes use of the expression $\hat{\lambda} = r/s$, used to calculate the Maximum Likelihood Estimate (MLE) of λ. Using the identity $\lim_{h\to\infty}(1+1/h)^h = e$ [14], the limiting value of unconditional expectation of reliability at time t, can be shown in Eq. (11) to be $e^{-\hat{\lambda}t}$.

$$\lim_{r\to\infty} E[R(t)] = \frac{1}{\lim_{r\to\infty}\left(1+\frac{\hat{\lambda}t}{r}\right)^r} = e^{-\hat{\lambda}t} \tag{11}$$

Since $\hat{\lambda}$ is an unbiased estimate, it tends to the true value of λ, as $r \to \infty$. Figure 4 plots the expected value of $R(t)$, at $t = 1000$, as a function of r. The value of $\hat{\lambda}$ is

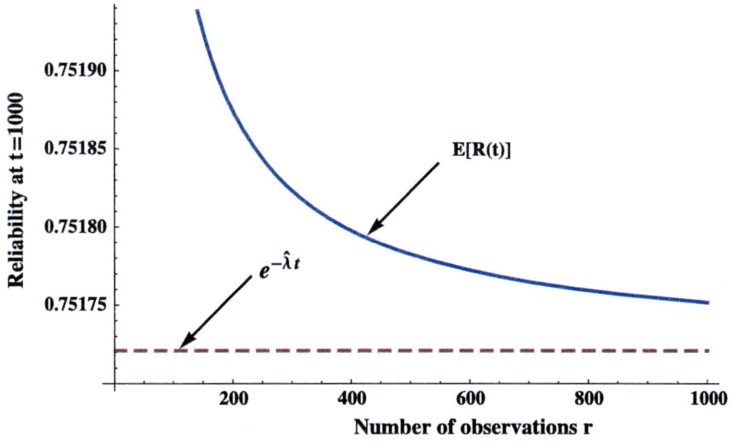

Fig. 4 Expected reliability of a single-component system at t=1000 hours

chosen to be 5.7078×10^{-5} *hrs.*, corresponding to an MTTF of 17, 520 *hrs.*, used for failure of software in [13]. It can be seen that $E[R(t)]$ tends to $e^{-\hat{\lambda}t}$ as r increases.

3.3 Variance of Reliability

The relation $Var[Y] = E[Y^2] - (E[Y])^2$, where Y is a random variable, can be used to derive the variance of reliability at time t, $Var[R(t)]$. $E[(R(t))^2]$ can be derived in a similar fashion as $E[R(t)]$ has been derived in Eq. 10. Variance of reliability of a singe-component system, at time t, is shown in Eq. (12).

$$Var[R(t)] = \underbrace{\left(\frac{s}{s+2t}\right)^r}_{E[(R(t))^2]} - \underbrace{\left(\frac{s}{s+t}\right)^{2r}}_{E[R(t)]^2}$$
$$= \left(\frac{1}{1+2\hat{\lambda}t/r}\right)^r - \left(\frac{1}{1+\hat{\lambda}t/r}\right)^{2r} \tag{12}$$

It follows from Eq. (12) that $Var[R(t)] \to 0$ as $r \to \infty$. Using the same value of $\hat{\lambda}$ as in Sect. 3.2, Figure 5 shows the variance of reliability at time $t = 1000$ *hours*. It is clear from the figure that $Var[R(t)]$ tends to 0 as r increases.

Extending the distribution function and the expressions for expected reliability and variance of a single-component system, to obtain the same for an n-component series system is trivial (assuming each component having independent and identically distributed times to failure following the exponential distribution with parameter λ).

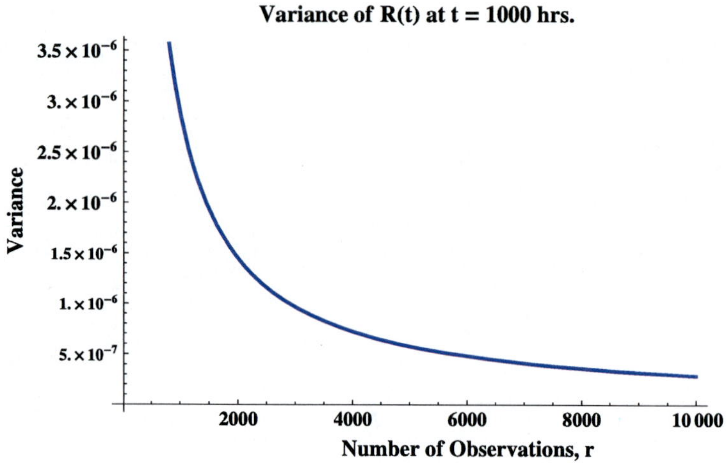

Fig. 5 Variance of reliability of a single-component system at t=1000 hours

For ease of reference in the later parts of the chapter, the expressions for expected reliability of an n-component series system with identical components, $E[R_{nseries}(t)]$ and variance of its reliability, $Var[R_{nseries}(t)]$, are provided below in Eqs. (13) and (14), respectively.

$$E[R_{nseries}(t)] = \left(\frac{s}{s+nt}\right)^r = \left(\frac{1}{1+n\hat{\lambda}t/r}\right)^r \tag{13}$$

$$\begin{aligned} Var[R_{nseries}(t)] &= \left(\frac{s}{s+2nt}\right)^r - \left(\frac{s}{s+nt}\right)^{2r} \\ &= \left(\frac{1}{1+2n\hat{\lambda}t/r}\right)^r - \left(\frac{1}{1+n\hat{\lambda}t/r}\right)^{2r} \end{aligned} \tag{14}$$

4 Reliability of a k-out-of-n System

A $k-out-of-n$ system is considered to be operational as long as at least k of the total n components in the system are operational [14]. Consider a $k-out-of-n$ system, where each of the n components has an independent and identically distributed time to failure, following the exponential distribution with parameter λ. The reliability $R(t)$ of such a system is given by:

$$R(t) = \sum_{i=k}^{n} \binom{n}{i} \binom{i-1}{k-1} (-1)^{i-k} e^{-i\lambda t} \tag{15}$$

Since $e^{-i\lambda t}$ is the same expression as the reliability of an i-component series system, using the linearity property of expectation [14], the expected reliability of a $k-out-of-n$ system can be derived based on Eq. (13). As in the earlier sections, assume that the number of observations of times to failure of a component, used to compute the point estimate $\hat{\lambda}$, is r and that the value of random variable S, denoting the sum of observed times to failure, is s. Eq. (16) provides the expression for the expected reliability, $E[R(t)]$, of a $k-out-of-n$ system with all components having independent and identically distributed times to failure.

$$\begin{aligned} E[R(t)] &= \sum_{i=k}^{n} \binom{n}{i} \binom{i-1}{k-1} (-1)^{i-k} \left(\frac{s}{s+it}\right)^r \\ &= \sum_{i=k}^{n} \binom{n}{i} \binom{i-1}{k-1} (-1)^{i-k} \left(\frac{1}{1+i\hat{\lambda}t/r}\right)^r \end{aligned} \tag{16}$$

The limiting value of expected reliability of a $k-out-of-n$ system, as $r \to +\infty$ can easily be shown to be tending to the value of $R(t)$, Eq. (15), computed at point estimate $\hat{\lambda}$, which, being an unbiased estimate, is also equal to the true value, as $r \to \infty$.

Since the components have independent failures, from Eq. (15) it follows that the variance of reliability of a $k-out-of-n$ system can be derived as:

$$\begin{aligned} Var[R(t)] &= \sum_{i=k}^{n} \left(\binom{n}{i} \binom{i-1}{k-1} (-1)^{i-k}\right)^2 Var[e^{-i\lambda t}] \\ &= \sum_{i=k}^{n} \left(\binom{n}{i} \binom{i-1}{k-1}\right)^2 \\ &\left(\left(\frac{1}{1+2i\hat{\lambda}t/r}\right)^r - \left(\frac{1}{1+i\hat{\lambda}t/r}\right)^{2r}\right) \end{aligned} \tag{17}$$

The variance of the $k-out-of-n$ system can be shown to tend to 0, as $r \to \infty$, as each of the terms in the summation in Eq. (17) tend to 0, when $r \to \infty$.

The distribution of reliability of a $k-out-of-n$ system, due to the epistemic uncertainty in parameter λ can be derived using Eqs. (1) and (7), as shown in Sect. 3. The CDF of reliability of a $k-out-of-n$ system is given by:

$$F_{R_{kofn}(t)}(p) = \int_0^\infty I(R_{kofn}(t) \le p) \frac{\lambda^{r-1} s^r e^{-\lambda s}}{(r-1)!} d\lambda$$

$$= \int_{\lambda_{akofn}}^{\infty} \frac{\lambda^{r-1} s^r e^{-\lambda s}}{(r-1)!} d\lambda$$

$$= 1 - \underbrace{\int_{0}^{\lambda_{akofn}} \frac{\lambda^{r-1} s^r e^{-\lambda s}}{(r-1)!} d\lambda}_{Erlang\ CDF}$$

$$= \sum_{i=0}^{r-1} e^{-s\lambda_{akofn}} \frac{(s\lambda_{akofn})^i}{i!} \tag{18}$$

where λ_{akofn} is the value of λ for which $R_{kofn}(t) - p \leq 0$.

Using different values for k and n, the above expressions can be used to compute the expected reliability and variance of reliability, due to epistemic uncertainty in failure rate parameter, for different series, parallel or k-out-of-n (also called N-modular redundant (NMR) [8]) systems.

5 Duplex System with a Spare

Consider a duplex system with a spare, where 2 components are initially in operation (active) and a third component is a warm spare (de-energized standby). Each active component has an exponentially distributed time to failure with parameter λ (independent and identically distributed), while the standby component has an exponentially distributed time to failure with parameter μ (where, μ is usually expected to be less than λ). Upon failure of an active component, the standby component is brought into active operation and then follows the exponential failure law with parameter λ. The system is considered to be operational as long as 2 of the components are in active operation [14]. The reliability $R(t)$ of such a system is:

$$R(t) = \left(2\frac{\lambda}{\mu} + 1\right) e^{-2\lambda t} - 2\frac{\lambda}{\mu} e^{(-2\lambda+\mu)t} \tag{19}$$

While the epistemic random variables (Λ and M) are assumed to be independent for uncertainty propagation purposes, the aleatory model takes into account different failure rates of components in different states (active or standby).

5.1 Expected Reliability : Duplex System with a Spare

The expected reliability of a duplex system with a spare, taking into account the epistemic uncertainties in parameters λ and μ, can be derived based on Eq. (2).

Assume that the point estimate $\hat{\lambda}$ is computed using r observations of times to failure and that the value of random variable representing sum of observed times to failure S, is s, while the point estimate $\hat{\mu}$ (of random variable M corresponding to failure rate in standby mode) is derived using u observations of times to failure of component in standby mode, with the value of sum of observed times to failure random variable B, being b. The expected reliability derived, is shown below in 20.

$$\begin{aligned} E[R(t)] &= \int_0^\infty \int_0^\infty \left(\left(2\frac{\lambda}{\mu} + 1 \right) e^{-2\lambda t} - 2\frac{\lambda}{\mu} e^{(-2\lambda+\mu)t} \right) \\ &\quad \underbrace{\frac{\lambda^{r-1} s^r e^{-\lambda s}}{(r-1)!}}_{pdf\ of\ \Lambda} \underbrace{\frac{\mu^{u-1} b^u e^{-\mu b}}{(u-1)!}}_{pdf\ of\ M} d\lambda d\mu \\ &= \left[\left(\frac{s}{s+2t} \right)^r \frac{1}{s+2t} \left(\frac{b}{b+t} \right)^{u-1} \frac{(-2)b \times r}{u-1} \right] \\ &\quad + \left[\left(\frac{s}{s+2t} \right)^r \frac{1}{s+2t} \frac{2b \times r}{u-1} \right] + \left(\frac{s}{s+2t} \right)^r \end{aligned} \tag{20}$$

Using the number of observations r and the point estimate $\hat{\lambda}$, the value s, of sum of observed times to failure random variable S can be expressed as, $s = r/\hat{\lambda}$. Similarly b can be expressed as $b = u/\hat{\mu}$. Equation. (20) can then be rewritten as:

$$\begin{aligned} E[R(t)] &= \underbrace{\left[\left(\frac{1}{1+2\hat{\lambda}t/r} \right)^r \frac{1}{r/\hat{\lambda}+2t} \left(\frac{1}{1+\hat{\mu}t/u} \right)^{u-1} \frac{(-2)u \times r}{\hat{\mu}(u-1)} \right]}_{I} \\ &\quad + \underbrace{\left[\left(\frac{1}{1+2\hat{\lambda}t/r} \right)^r \frac{1}{r/\hat{\lambda}+2t} \frac{2u \times r}{\hat{\mu}(u-1)} \right]}_{II} + \underbrace{\left(\frac{1}{1+2\hat{\lambda}t/r} \right)^r}_{III} \end{aligned} \tag{21}$$

It can be shown, similar to Eq. (11), that as $r \to \infty$ and $u \to \infty$, the term in Eq. (21) marked by I, tends to $-2e^{-(2\hat{\lambda}+\hat{\mu})t}\hat{\lambda}/\hat{\mu}$, term II tends to $2e^{-2\hat{\lambda}t}\hat{\lambda}/\hat{\mu}$, and term marked by III tends to $e^{-2\hat{\lambda}t}$. Thus in the limiting case, the expected reliability of a duplex system with a spare, tends to the reliability evaluated at $\hat{\lambda}$ and $\hat{\mu}$ (which being unbiased estimates, are also the true values of λ and μ as r and u tend to ∞, respectively).

Figure 6 plots the expected reliability, $E[R(t)]$ of a duplex system with one spare, at time $t = 1000$ *hours*. The value for $\hat{\lambda}$ is chosen to be 5.7078×10^{-5} *hours*, corresponding to an MTTF of 17,520 *hours* and $\hat{\mu}$ is chosen to be half of $\hat{\lambda}$. While Eq. (21) allows different values for r and u, in Fig. 6, the number of observations r and u are kept equal at each point, for ease of illustration. It can be seen that as r and u increase, the expected reliability $E[R(t)]$ tends to the value of reliability computed

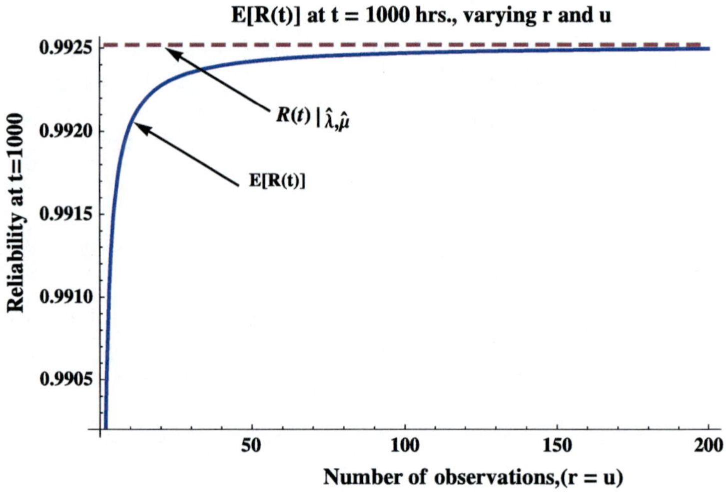

Fig. 6 Expected reliability of a duplex system with one spare, at t=1000 hours

at point estimates $\hat{\lambda}$ and $\hat{\mu}$ (which are the true values for λ and μ, as r and u tend to ∞).

5.2 *Variance of Reliability : Duplex System with a Spare*

For a duplex system with one spare, the variance of reliability, due to epistemic uncertainty in the rate parameters of times to failure distribution for components in active and standby mode, can be computed using the expected reliability $E[R(t)]$ and the second moment of reliability, $E[(R(t))^2]$. While $E[R(t)]$ has already been derived in Eq. (21), $E[(R(t))^2]$ is derived below in Eq. (22).

$$E[(R(t))^2] = \left(\frac{s}{s+4t}\right)^r \left(\frac{1}{s+4t}\right) \left[\frac{4b^2 r(r+1)}{(u-1)(u-2)} \left(\frac{1}{s+4t}\right) - \frac{8r(r+1)b^2}{(u-1)(u-2)} \left(\frac{b}{b+t}\right)^{u-2} \left(\frac{1}{s+4t}\right) + \frac{4r(r+1)b^2}{(u-1)(u-2)} \left(\frac{b}{b+2t}\right)^{u-2} \left(\frac{1}{s+4t}\right) + \frac{4b \times r}{u-1} - \frac{4r \times b}{u-1} \left(\frac{b}{b+t}\right)^{u-1}\right] + \left(\frac{s}{s+4t}\right)^r$$

$$= \left(\frac{1}{1+4\hat{\lambda}t/r}\right)^r \left(\frac{1}{r/\hat{\lambda}+4t}\right) \left[\frac{4u^2 r(r+1)}{\hat{\mu}^2(u-2)}\left(\frac{1}{r/\hat{\lambda}+4t}\right) - \frac{8r(r+1)u^2}{(u-1)(u-2)\hat{\mu}^2}\left(\frac{1}{1+\hat{\mu}t/u}\right)^{u-2}\left(\frac{1}{r/\hat{\lambda}+4t}\right) + \frac{4r(r+1)u^2}{\hat{\mu}^2(u-1)(u-2)}\left(\frac{1}{1+2\hat{\mu}t/u}\right)^{u-2}\left(\frac{1}{r/\hat{\lambda}+4t}\right) + \frac{4r \times u}{\hat{\mu}(u-1)} - \frac{4r \times u}{\hat{\mu}(u-1)}\left(\frac{1}{1+\hat{\mu}t/u}\right)^{u-1}\right] + \left(\frac{1}{1+4\hat{\lambda}t/r}\right)^r \tag{22}$$

The variance $Var[R(t)]$ can be derived next from Eqs. (21) and (22), simply using the relation $Var[R(t)] = E[(R(t))^2] - (E[R(t)])^2$. It can also be shown that as the number of observations used to compute the estimates of parameters λ and μ tend to ∞, the variance of reliability tends to 0, as expected. Figure 7 shows the variance of reliability of this system, as r and u are varied. It is clear that the variance tends to 0 as r and u increase.

6 Analytic-Numeric Epistemic Uncertainty Propagation

Quite often, the expression for model output is complex and difficult to integrate analytically. For complex expressions of model output, which cannot be analytically integrated easily, numerical integration can be used to perform the unconditioning integrals explained in Eqs. (1), (2), and (3).

We use the $M/M/1$ queuing system with server breakdown and repair, explained in [2], as an example to illustrate analytic-numeric epistemic uncertainty propagation. The system is modeled using a Markov chain (with approximations). An approximate expression for expected number of customers in the system is derived in [2] as:

$$\bar{N} = \frac{\rho}{1-\rho} + \frac{\lambda\gamma}{\tau(\gamma+\tau)(1-\rho)} \tag{23}$$

where, $\rho = \lambda/\mu$, λ, and μ are the customer arrival and service rates, respectively, while γ and τ are the server failure and repair rates, respectively.

Since the parameters in Eq. (23), are all rate parameters of exponential aleatory distribution, the epistemic density function of each parameter can be derived as shown in Eq. (7). The expected value and variance due to the epistemic uncertainties in the model input parameters, for the expected number of customers in the system are obtained by computing the integrations shown in Eqs. (2) and (3), numerically using Global Adaptive method of numerical integration (a built-in method, supported by NIntegrate function in Mathematica [17]). We exploit various properties of expectation to simplify the computations (linearity property and expectation of product of

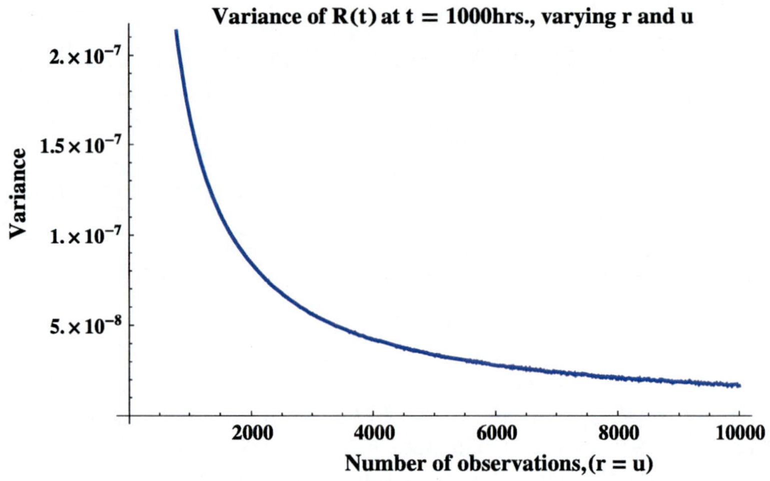

Fig. 7 Variance of reliability of a duplex system with one spare, at t=1000 hours

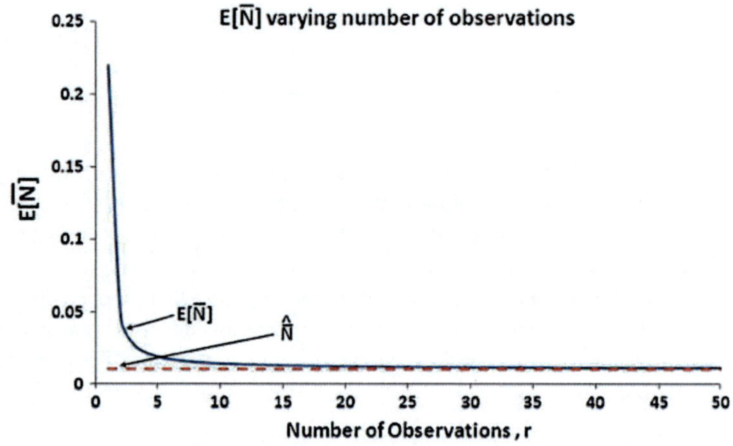

Fig. 8 Expected value of expected number of customers in the system

independent random variables). Figure 8 shows the expectation of $\bar{N}$, as the number of observations used to compute the point estimates of the parameters is increased. As in the other examples in this chapter, when the number of observations, r, used to compute the point estimates of the parameters increases, the expectation tends to expected number of customers $\bar{N}$, computed at point estimates of the parameters (which tends to the true value as the number of observations is increased). While the uncertainty propagation method allows different number of observations for different parameters, the same value of r is used for all the parameters at each point in Fig. 8 for ease of illustration. Similarly, based on Eqs. (2) and (3), the variance of expected number of customers in the system, due to the epistemic uncertainty in model input parameters, can be obtained. The variance thus computed tends to 0 as the number of observations used to compute the point estimate of $\bar{N}$ tends to ∞ (indicating the point estimate of $\bar{N}$ approaching the true value).

7 Summary

In this chapter, an approach for propagating parametric epistemic uncertainty through analytic stochastic models is presented. This approach can be applied when the model output is a closed-form expression of input parameters. The method for deriving closed-form expressions of CDF, expected value, and variance of reliability due to epistemic uncertainty in input parameter values is discussed. Closed-form expressions for the distribution function, expected value, and variance of reliability are derived for some nonrepairable systems. Limiting behavior of the CDF, expected value and variance of reliability, is also studied. As the number of observations used to determine the time to failure distribution parameter tends to ∞, the variance of reliability, due to the epistemic uncertainty in the input parameter, tends to zero. The expected value of reliability tends to the reliability evaluated at point estimates of the parameters, as the number of observations tend to ∞ (the point estimates of parameters being unbiased estimates, tend to the true value as number of observations tend to ∞). The CDF of reliability tends to a Heaviside step function in the limiting case. While closed-form expressions for distribution function, expected value, and variance of reliability can be obtained for simpler expressions of system reliability, the task becomes difficult for more complex cases. Numerical integration or sampling-based methods may need to be applied for epistemic uncertainty propagation in such cases. Numerical integration is used to perform uncertainty propagation for one such case.

References

1. Ajmone-Marsan M, Balbo G, Conte G, Franceschinis SG (1995) Modeling with generalized stochastic petri nets. Wiley, Donatelli
2. Bobbio A, Trivedi KS (1986) An aggregation technique for the transient analysis of stiff markov chains. IEEE Trans Comp C-35(9):803–814
3. Devaraj A, Mishra K, Trivedi K (2010) Uncertainty propagation in analytic availability models. In 29th IEEE international symposium on reliable distributed systems, SRDS, pp 121–130
4. Hirel C, Tuffin B, Trivedi KS (2000) Spnp: Stochastic petri nets. version 6.0. In computer performance evaluation. Modelling techniques and tools, vol 1786. Springer Berlin/Heidelberg, pp 354–357
5. Iman RL, Conover WJ (1982) A distribution-free approach to inducing rank correlation among input variables. Comm Stat Simul Comput 11(3):311–334
6. Kass-Robert E, Wasserman L (1996) The selection of prior distributions by formal rules. J Am Stat Assoc 91(435):1343–1370

7. Muppala J, Fricks R, Trivedi KS (2000) Techniques for system dependability evaluation. In: Grassman W (ed) Computational probability. Kluwer Academic Publishers, New York, pp 445–480
8. Ng YW, Avizienis AA (1980) A unified reliability model for fault-tolerant computers. IEEE Trans Comp C–29(211):1002–1011
9. Oppenheim AV, Schafer RW, Buck JR (1999) Discrete-time signal processing. Prentice Hall, New York
10. Rabinowitz P, Davis PJ (2007) Methods of numerical integration. Dover Publications, New York
11. Singpurwalla ND (2006) Reliability and risk: a Bayesian perspective, 1st edn. Wiley, New York
12. Smith AE, Ryan PB, Evans JS (1992) The effect of neglection correlations when propagating uncertainty and estimating the population distribution risk. Risk Anal 12(4):467–474
13. Smith W, Trivedi K, Tomek L, Ackaret J (2008) Availability analysis of blade server systems. IBM Syst J 47(4):155
14. Trivedi K (2001) Probability and statistics with reliability, queuing and computer science applications. Wiley, New York
15. Trivedi K, Mishra K (2010) A non-obtrusive method for uncertainty propagation in analytic dependability models. In: 4th Asia-Pacific international symposium on advanced reliability and maintenance modeling (APARM 2010)
16. Trivedi KS, Sahner R (2009) Sharpe at the age of twenty two. SIGMETRICS Perf Eval Rev 36(4):52–57
17. Wolfram Research, Inc Wolfram mathematica 6. http://www.wolfram.com/products/mathematica/index.html

Generational Garbage Collection Policies

Xufeng Zhao, Syouji Nakamura and Cunhua Qian

1 Introduction

In the computer science community, the technique of *garbage collection* [5] is an automatic process of memory recycling, which refers to those objects in the memory no longer referenced by programs are called *garbage* and should be thrown away. A *garbage collector* determines which objects are garbage and makes the heap space occupied by such garbage available again for the subsequent new objects. Garbage collection plays an important role in Java's security strategy, however, it adds a large overhead that can deteriorate the program performances. From related studies which are summarized in [5], a garbage collector spends between 25 and 40 percent of execution time of programs for its work in general, and delays caused by such a garbage collection are obtrusive.

In recent years, *generational garbage collection* [1, 17, 19, 20] has been popular with programmers as it can be made more efficiently. Compared with classical tracing collectors, e.g., reference counting collector, mark-sweep collector, mark-compact collector, and copying collector, a *generational garbage collector* is effective in computer programs with the characteristic that it is unnecessary to mark or copy all active data of the whole heap for every collection, i.e., the collector concentrates effort on

X. Zhao (✉)
Department of Business Administration, Aichi Institute of Technology,
1247 Yachigusa, Yakusa-cho, Toyota 470-0392, Japan
e-mail: g09184gg@aitech.ac.jp

S. Nakamura
Department of Human Life and Information, Kinjo Gakuin University,
1723 Omori 2-chome, Moriyama-ku, Nagoya 463-8521, Japan
e-mail: snakam@kinjo-u.ac.jp

C. Qian
School of Economics and Management, Nanjing University of Technology,
30 Puzhu Road, 211816 Nanjing, China
e-mail: qch64317@njut.edu.cn

T. Dohi and T. Nakagawa (eds.), *Stochastic Reliability and Maintenance Modeling*, Springer Series in Reliability Engineering 9, DOI: 10.1007/978-1-4471-4971-2_15,

those objects that are most likely to be garbage. Based on the weak generational hypothesis [17] which asserts that most objects are short-lived after their allocation, a generational garbage collector segregates objects by age into two or more regions called *generations* or *multiple generations*. The survival rates of younger generations are always much lower than those of older ones, which means that younger generations are more likely to be garbage and can be collected more frequently than older ones. Although such generational collections cost much shorter time than that of a full collection, the problems of pointers from older generations to younger ones and the size of root sets for younger generations become more complicated. For these reasons, many generational collectors are limited to just two or three generations [5]. This generational technique is now in widespread use for memory management. For instance, the garbage collector, which is used in Sun's HotSpot Java Virtual Machine (JVM), manages heap space for both young and old generations [19]: New objects space *Eden*, two equal survivor spaces *SS♯1* and *SS♯2* for surviving objects, and tenured objects space *Old (Tenured)*, where Eden, SS♯1 and SS♯2 are for younger generations, and Old (Tenured) is for older ones.

A generational garbage collector uses *minor collection* and *tenuring collection* [1] for younger generations and *major collection* for multi-generations [5]. Most generational garbage collectors are copying collectors, although it is possible to use mark-sweep collectors [2]. In this chapter, we concentrate on a generational garbage collector using copying collection. However, for every garbage collection, the manner of *stop and copy* pauses all application threads to collect the garbage. The duration of time for which the collector has worked is called *pause time* [5], which is an important parameter for interactive systems, and depends largely upon the volume of surviving objects and the type of collections. That is, pause time suffered for minor collection increases with the number of collections and is less than that of tenuring collection; major collection pause time is the longest among the three.

With regard to garbage collection modelings, there have been few research papers that studied analytical expressions of optimal policies for a generational garbage collector. Most problems were concerned with several ways to introduce garbage collection methods in techniques and how to tune the garbage collector by simulations, which is more complex and time-consuming due to the random accesses of programs in the memory in practice [4, 6, 7, 16, 18]. We propose that garbage collection is a *stochastic decision making process* and should be analyzed by the theory of stochastic processes from the viewpoints of management. As some applications of damage models, a garbage collection model for a database in the computer system [14] was studied, but the theoretical point of garbage collection was not considered essentially, and optimal policies for a generational garbage collector with tenuring threshold and major collection times according to practical working schemes [21, 22] were studied recently.

[1] Tenuring collection is also a kind of minor collection [5]. We define tenuring collection as distinct from minor collection because there may be some surviving objects tenured from survivor space into Old.

This chapter considers a pause time goal which is called time cost or cost for simplicity, and our problem is to obtain optimal collection times which minimize the expected cost rates. Using the techniques of cumulative processes and reliability theory [8–10], optimal tenuring collection times and major collection times are discussed. Furthermore, increase in objects might be unclear at discrete times for the high frequency of computer processes. According to [1, 19], it would be more practical to assume that surviving objects that should be copied increase with time continuously and roughly according to some mathematical laws. Applying the techniques of degradation processes [11, 15] and continuous wear processes [9], optimal tenuring collection times are discussed analytically and numerically.

2 Working Schemes

In general, the frequency of garbage collections depends on whether the computer processes are busy or not. Hence, it is practical to assume that garbage collections occur at a nonhomogeneous Poisson process with an intensity function $\lambda(t)$ and a mean-value function $R(t) \equiv \int_0^t \lambda(u)\mathrm{d}u$. Then, the probability that collections occur exactly j times in $(s, t]$ is [12]

$$H_j(s,t) \equiv \frac{[R(t)-R(s)]^j}{j!}\mathrm{e}^{-[R(t)-R(s)]} \quad (j = 0, 1, 2, \ldots).$$

Letting $F_j(s,t)$ $(j = 1, 2, \ldots)$ denote the probability that collections occur at least j times in the time interval $(s, t]$,

$$F_j(s,t) = \int_s^t H_{j-1}(s,u)\lambda(u)\mathrm{d}u = \sum_{i=j}^{\infty} H_i(s,t), \tag{1}$$

where $F_0(s,t) \equiv 1$ and

$$H_j(t) \equiv H_j(0,t) = \frac{[R(t)]^j}{j!}\mathrm{e}^{-R(t)},$$

$$F_j(t) \equiv F_j(0,t) = \sum_{i=j}^{\infty} H_i(t).$$

Further, the volume X_i of new objects in Eden at the ith collection has an identical distribution $G(x) \equiv \Pr\{X_i \le x\}$ $(i = 1, 2, \ldots)$, and survivor rate α_i $(0 \le \alpha_i < 1; i = 1, 2, \ldots)$, where $1 > \alpha_1 > \alpha_2 > \cdots > \alpha_i > \cdots \ge 0$, means that new objects will survive $100\alpha_i$ percent at the ith minor collection. That is, detailed working schemes of a generational garbage collector that have been introduced in [5, 19, 21, 22] are given as the following steps (Fig. 1):

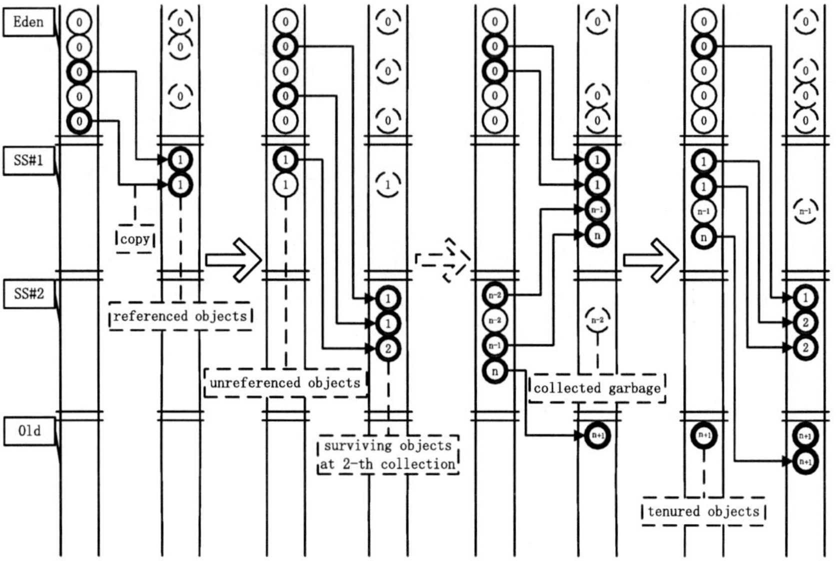

Fig. 1 Working schemes of a generational garbage collector

1. New objects X_1 are allocated in Eden.
2. When the first minor collection occurs, surviving objects $\alpha_1 X_1$ from Eden are copied into SS♮1.
3. When the second minor collection occurs, surviving objects $\alpha_1 X_2$ from Eden and $\alpha_2 X_1$ from SS♮1 are copied into SS♮2.
4. In the fashions of 1–3, minor collections copy surviving objects between SS♮1 and SS♮2 until they become tenured, *i.e.*, tenuring collection occurs when some parameter meets the tenuring threshold, and then, the older or the oldest objects are copied into Old.
5. When Old fills up, major collection of the whole heap occurs, and surviving objects from Old are kept in Old, while objects from Eden and survivor space are kept in survivor space.

In practice, tenuring threshold mentioned in step 4 above is adaptive, which is called *adaptive tenuring* [5] and can be modified at any time. In this chapter, we propose two cases of working schemes according to the properties of adaptive tenuring:

Based on [17], new objects can be tenured only if they survive at least one minor collection, because objects that survive two minor collections are much less than those that survive just one. In other words, surviving objects are likely to reduce slightly with the number of minor collections beyond two. That is, for step 4:

4a. When tenuring collection occurs, surviving objects from Eden and survivor space are copied into the other survivor space and Old, respectively. That is, if tenuring collection is made at the jth ($j = 1, 2, \ldots$) collection, surviving objects $\alpha_1 X_j$

and $\alpha_2 X_{j-1} + \alpha_3 X_{j-2} + \cdots + \alpha_j X_1$ are copied into survivor space and Old, respectively.

4b. After tenuring collection, the same collection cycle begins with step 1. The collector works $1 \to 2 \to 3 \to 4a \to 4b \to 1 \to \cdots$. In this case, tenuring collections can be consider as renewal points of the collection processes, because Old will be filled with tenured objects slowly and major collection occurs rarely, especially when the tenuring threshold is high and the survivor rates are low. Modelings and optimizations of tenuring collection times are discussed in Sects. 3 and 5.

From [19], the oldest objects can be tenured from survivor space into Old at every collection time when tenuring collection begins. That is, for step 4:

4c. When tenuring collection occurs, the oldest objects from survivor space are copied into Old, and the other surviving objects from Eden and survivor space are copied into the other survivor space. That is, if tenuring collection is made at the jth ($j = 1, 2, \ldots$) collection, surviving objects $\alpha_1 X_j + \alpha_2 X_{j-1} + \cdots + \alpha_{j-1} X_2$ and $\alpha_j X_1$ are copied into survivor space and Old, respectively.

4d. When the next collection occurs, the collector works as the same rule as 4c. That is, when the second tenuring collection occurs, surviving objects $\alpha_1 X_{j+1} + \alpha_2 X_j + \cdots + \alpha_{j-1} X_3$ and $\alpha_j X_2$ are copied into the other survivor space and Old, respectively. The collector works $1 \to 2 \to 3 \to 4c \to 4d \to 5 \to 1 \to \cdots$. In this case, major collections can be consider as renewal points of the collection processes, because there are always some surviving objects tenured from survivor space into Old at every collection time when tenuring collection begins, especially when the tenuring threshold is low and the survivor rates are high. Related optimization problems of major collection times are discussed in Sect. 4.

From the above discussions, if tenuring collection is made at the jth ($j = 1, 2, \ldots$) collection, surviving objects that should be copied at the ith ($i = 0, 1, 2, \ldots, j-1$) minor collection, copied objects and tenured objects at the kth ($k = 1, 2, \ldots$) tenuring collection are, respectively,

$$\sum_{n=0}^{i-1} \alpha_{n+1} X_{i-n} < K, \qquad \sum_{n=1}^{j} \alpha_n X_{j+k-n} \geq K \qquad \text{and} \quad \alpha_j X_k, \tag{2}$$

where $\sum_{n=0}^{-1} \equiv 0$, and K is tenuring threshold in step 4, which means that the total volume of surviving objects has exceeded level K. It could be easily seen that copied objects increase with the number of minor collections and are relatively stable with the number of tenuring collections. We define that the distribution of the total surviving objects at the ith minor collection is

$$G_i(x) \equiv \Pr\left\{ \sum_{n=0}^{i-1} \alpha_{n+1} X_{i-n} \leq x \right\} \quad (i = 0, 1, 2, \ldots), \tag{3}$$

where $G_i(x)$ decreases with i, and $G_0(x) \equiv 1$ means that there are no objects in the heap space at time 0. The probability that the total surviving objects exceed exactly a threshold level K at the $(i+1)$th $(i = 0, 1, 2, \ldots)$ minor collection is

$$p_i(K) \equiv \int_0^K \overline{G}(K-x)\mathrm{d}G_i(x) = G_i(K) - G_{i+1}(K), \tag{4}$$

where $\overline{V}(x) \equiv 1 - V(x)$ for any distribution $V(x)$.

Letting $c_S + c_M(x)$ be the cost suffered for every minor collection, where c_S is the constant cost of scanning surviving objects and x is the surviving objects that should be copied, $c_M(x)$ increases with x and $c_M(0) \equiv 0$. Then, the expected cost of the ith minor collection is

$$C(i, K) \equiv \frac{1}{G_i(K)} \int_0^K [c_S + c_M(x)]\,\mathrm{d}G_i(x) \qquad (i = 0, 1, 2, \ldots), \tag{5}$$

where $C(0, K) \equiv 0$ and $C(i, K)$ increases with i.

3 Tenuring Collection Times

Suppose that minor collections are made when the garbage collector begins to work, tenuring collection is made at a planned time T $(0 < T \le \infty)$ or at the first collection time when surviving objects have exceeded a threshold level K $(0 < K \le \infty)$, whichever occurs first. Then, the probability that tenuring collection is made at time T is

$$P_T = \sum_{j=0}^{\infty} H_j(T) G_j(K), \tag{6}$$

and the probability that tenuring collection is made at level K is

$$P_K = \sum_{j=0}^{\infty} F_{j+1}(T) p_j(K), \tag{7}$$

where note that $P_T + P_K \equiv 1$. The mean time to tenuring collection is

$$\begin{aligned} E_1(L) &= T \sum_{j=0}^{\infty} H_j(T) G_j(K) + \sum_{j=0}^{\infty} p_j(K) \int_0^T t\mathrm{d}F_{j+1}(t) \\ &= \sum_{j=0}^{\infty} G_j(K) \int_0^T H_j(t)\mathrm{d}t. \end{aligned} \tag{8}$$

The expected cost suffered for minor collections until tenuring collection is

$$\begin{aligned} C_M &= \sum_{j=1}^{\infty}\sum_{i=1}^{j} C(i,K)H_j(T)G_j(K) + \sum_{j=1}^{\infty}\sum_{i=1}^{j} C(i,K)F_{j+1}(T)p_j(K) \\ &= \sum_{j=1}^{\infty} C(j,K)F_j(T)G_j(K). \end{aligned} \tag{9}$$

Then, the expected cost until tenuring collection is

$$E_1(C) = c_K - (c_K - c_T)\sum_{j=0}^{\infty} H_j(T)G_j(K) + \sum_{j=1}^{\infty} C(j,K)F_j(T)G_j(K), \tag{10}$$

where c_T and c_K $(c_T,\ c_K > c_S + c_M(K))$ are the costs suffered for tenuring collections at time T and when surviving objects have exceeded K, respectively. Therefore, from (8) to (10), by using the theory of renewal reward process [13], the expected cost rate is

$$C_1(T,K) = \frac{\begin{array}{c} c_K - (c_K - c_T)\sum_{j=0}^{\infty} H_j(T)G_j(K) \\ +\sum_{j=1}^{\infty} C(j,K)F_j(T)G_j(K) \end{array}}{\sum_{j=0}^{\infty} G_j(K)\int_0^T H_j(t)\mathrm{d}t}. \tag{11}$$

3.1 Optimal Policies

1. Optimal T_1^*: When tenuring collection is made only at time T,

$$C_1(T) \equiv \lim_{K\to\infty} C_1(T,K) = \frac{1}{T}\left\{\sum_{j=1}^{\infty} F_j(T)\int_0^{\infty}[c_S + c_M(x)]\,\mathrm{d}G_j(x) + c_T\right\}. \tag{12}$$

Letting $f_j(t)$ be a density function of $F_j(t)$, i.e., $f_j(t) \equiv \mathrm{d}F_j(t)/\mathrm{d}t$. Then, differentiating $C_1(T)$ with respect to T and setting it equal to zero,

$$\sum_{j=1}^{\infty}[Tf_j(T) - F_j(T)]\int_0^{\infty}[c_S + c_M(x)]\,\mathrm{d}G_j(x) = c_T. \tag{13}$$

Letting $L_1(T)$ be the left-hand side of (13),

$$L_1(0) \equiv \lim_{T\to 0} L(T) = 0,$$

$$L_1'(T) = \lambda'(T)T\sum_{j=0}^{\infty}H_j(T)\int_0^{\infty}[c_S + c_M(x)]\,\mathrm{d}G_{j+1}(x) + \lambda(T)^2T\sum_{j=0}^{\infty}H_j(T)\int_0^{\infty}p_{j+1}(x)\mathrm{d}c_M(x).$$

Thus, if $\lambda(t)$ increases with t and $L_1(\infty) > c_T$, there exists a finite and unique T_1^* $(0 < T_1^* < \infty)$ which satisfies (13), and the resulting cost rate is

$$C_1(T_1^*) = \lambda(T_1^*)\sum_{j=0}^{\infty}F_j(T_1^*)\int_0^{\infty}p_j(x)\mathrm{d}c_M(x).$$

In particular, when $H_j(t) = [(\lambda t)^j/j!]\mathrm{e}^{-\lambda t}$ $(j = 0, 1, 2, \ldots)$, i.e., garbage collections occur at a Poisson process with rate λ, (13) becomes

$$\sum_{j=1}^{\infty}jF_{j+1}(T)\int_0^{\infty}p_j(x)\mathrm{d}c_M(x) = c_T. \tag{14}$$

Differentiating the left-hand side of (14) with respect to T,

$$\lambda\sum_{j=1}^{\infty}jH_j(T)\int_0^{\infty}p_j(x)\mathrm{d}c_M(x) > 0.$$

Thus, if the left-hand side of (14) is greater than c_T, then there exists a finite and unique T_1^* $(0 < T_1^* < \infty)$ which satisfies (14).

2. Optimal K_1^*: When tenuring collection is made only at level K,

$$C_1(K) \equiv \lim_{T\to\infty}C_1(T, K) = \frac{\sum_{j=1}^{\infty}\int_0^K[c_S + c_M(x)]\mathrm{d}G_j(x) + c_K}{\sum_{j=0}^{\infty}G_j(K)\int_0^{\infty}H_j(t)\mathrm{d}t}. \tag{15}$$

Letting $g_i(x)$ be a density function of $G_i(x)$ in (3), i.e., $g_i(x) \equiv \mathrm{d}G_i(x)/\mathrm{d}x$. Differentiating $C_1(K)$ with respect to K and setting it equal to zero,

$$Q_1(K)\sum_{j=0}^{\infty}G_j(K)\int_0^{\infty}H_j(t)\mathrm{d}t - \sum_{j=1}^{\infty}\int_0^K[c_S + c_M(x)]\,\mathrm{d}G_j(x) = c_K, \tag{16}$$

where

$$Q_1(K) \equiv \frac{[c_S + c_M(K)]\sum_{j=1}^{\infty}g_j(K)}{\sum_{j=1}^{\infty}g_j(K)\int_0^{\infty}H_j(t)\mathrm{d}t}.$$

Letting $L_1(K)$ be the left-hand side of (16),

$$L_1(0) \equiv \lim_{K\to 0} L_1(K) = Q_1(0)\int_0^\infty H_0(t)\mathrm{d}t,$$

$$L_1'(K) = Q_1'(K)\sum_{j=0}^{\infty} G_j(K)\int_0^\infty H_j(t)\mathrm{d}t.$$

Thus, if $Q_1(K)$ increases with K and $L_1(0) < c_K < L_1(\infty)$, then there exists a finite and unique K_1^* $(0 < K_1^* < \infty)$ which satisfies (16), and the resulting cost rate is

$$C_1(K_1^*) = Q_1(K_1^*).$$

In particular, when $H_j(t) = [(\lambda t)^j/j!]\mathrm{e}^{-\lambda t}$, (16) becomes

$$c_M(K) + \int_0^K [c_M(K) - c_M(x)]\,\mathrm{d}M(x) = c_K - c_S, \tag{17}$$

whose left-hand side increases with K from 0 to ∞, where $M(x) \equiv \sum_{j=1}^{\infty} G_j(x)$. Thus, there exists a finite and unique K_1^* $(0 < K_1^* < \infty)$ which satisfies (17).

3.2 Numerical Examples

When $\lambda(t) = \lambda$, X_i $(i = 1, 2, \dots)$ has a normal distribution $N(\mu, \sigma^2)$, $\alpha_i = \alpha/i$ $(0 \le \alpha < 1; i = 1, 2, \dots)$ and $c_M(x) = c_M x$. Then

$$F_j(t) = 1 - \sum_{i=0}^{j-1} \frac{(\lambda t)^i}{i!}\mathrm{e}^{-\lambda t}, \qquad G_j(x) = \Phi\left(\frac{x - \alpha\mu\nu_j}{\alpha\sigma\sqrt{\omega_j}}\right), \tag{18}$$

where $\Phi(x)$ is the standard normal distribution with mean 0 and variance 1, i.e., $\Phi(x) \equiv (1/\sqrt{2\pi})\int_{-\infty}^{x} \mathrm{e}^{-u^2/2}\mathrm{d}u$, and

$$\nu_j \equiv \sum_{n=1}^{j} \frac{1}{n}, \qquad \omega_j \equiv \sum_{n=1}^{j} \frac{1}{n^2}.$$

Tables 1 and 2 present λT_1^*, $C_1(T_1^*)/\lambda$, K_1^* and $C_1(K_1^*)/\lambda$ for $c_T = c_K = 20$, 30, 40, $\mu = 8$, 10 and $\alpha = 0.40, 0.45, 0.50, 0.55, 0.60$ when $c_S = 10$, $c_M = 1$ and $\sigma = 1$. These show that optimal tenuring collection times λT_1^* increase with cost c_T and decrease with both the volume of new objects in Eden at collection time μ and the survivor rate α, optimal tenuring collection times K_1^* increase with all of c_K, μ and α, and $C_1(T_1^*)/\lambda$ and $C_1(K_1^*)/\lambda$ increase with all of c_T or c_K, μ and α. We can explain all the results and obtain some interesting conclusions as follows:

Table 1 Optimal λT_1^* and $C_1(T_1^*)/\lambda$ when $c_S = 10$, $c_M = 1$ and $\sigma = 1$

μ	α	$c_T = 20$		$c_T = 30$		$c_T = 40$	
		λT_1^*	$C_1(T_1^*)/\lambda$	λT_1^*	$C_1(T_1^*)/\lambda$	λT_1^*	$C_1(T_1^*)/\lambda$
	0.40	8.99	19.24	12.48	20.18	15.84	20.89
	0.45	8.24	20.11	11.34	21.14	14.35	21.92
8	0.50	7.61	20.95	10.42	22.07	13.15	22.92
	0.55	7.08	21.77	9.66	22.98	12.17	23.90
	0.60	6.64	22.58	9.03	23.86	11.34	24.85
	0.40	7.61	20.95	10.42	22.07	13.14	22.91
	0.45	6.96	21.98	9.49	23.20	11.95	24.14
10	0.50	6.44	22.97	8.75	24.30	10.97	25.31
	0.55	6.01	23.95	8.13	25.37	10.17	26.47
	0.60	5.64	24.91	7.61	26.43	9.49	27.60

Table 2 Optimal K_1^* and $C_1(K_1^*)/\lambda$ when $c_S = 10$, $c_M = 1$ and $\sigma = 1$

μ	α	$c_K = 20$		$c_K = 30$		$c_K = 40$	
		K_1^*	$C_1(K_1^*)/\lambda$	K_1^*	$C_1(K_1^*)/\lambda$	K_1^*	$C_1(K_1^*)/\lambda$
	0.40	8.76	18.76	9.61	19.61	10.71	20.71
	0.45	9.25	19.25	11.04	21.04	11.57	21.57
8	0.50	9.71	19.71	12.03	22.03	12.39	22.39
	0.55	10.12	20.12	12.69	22.69	13.18	23.18
	0.60	10.49	20.49	13.32	23.32	13.94	23.94
	0.40	10.72	20.72	12.05	22.05	12.41	22.41
	0.45	11.24	21.24	12.87	22.87	13.39	23.39
10	0.50	11.64	21.64	13.64	23.64	14.33	24.33
	0.55	12.08	22.08	14.37	24.37	15.23	25.23
	0.60	12.44	22.44	15.07	25.07	16.09	26.09

- When tenuring collection cost c_T or c_K increases, it is not economical to make tenuring collections frequently, then T_1^* or K_1^* should be postponed.
- When μ or α increases, cost suffered for minor collections will increase in a shorter time, because of faster increase in copied objects. If cost c_T or c_K is constant in this case, T_1^* should be advanced. For K_1^*, it costs much shorter time to increase copied objects until level K, then K_1^* would increase suitably to decrease both the frequency of tenuring collections and the total minor collection cost.
- The resulting cost rates $C_1(T_1^*)$ or $C_1(K_1^*)$ increase with all μ, α and c_T or c_K, because the total expected cost of one cycle increases but the expected time decreases.
- It is interesting that $C_1(K_1^*)$ are always less than $C_1(T_1^*)$ for the same parameters, i.e., tenuring collections at level K are better than those at time T. In fact, from Tables 1 and 2, we know that the expected number of minor collections until tenuring collection for two models are almost the same. That is, from the assumption

of $\alpha_i = \alpha/i$, we can derive

$$1+\frac{1}{2}+\frac{1}{3}+\cdots+\frac{1}{[\lambda T_1^*]}<\frac{K_1^*}{\alpha\mu}<1+\frac{1}{2}+\frac{1}{3}+\cdots+\frac{1}{[\lambda T_1^*]+1}, \tag{19}$$

where $[x]$ denotes the greatest integer contained in x. For example, when $c_T = c_K = 20$, $\mu = 8$ and $\alpha = 0.4$, $\lambda T_1^* = 8.99$ and $K_1^* = 8.76$, and hence

$$1+\frac{1}{2}+\cdots+\frac{1}{8}=2.55<\frac{8.76}{0.4\times 8}=2.74<1+\frac{1}{2}+\cdots+\frac{1}{9}=2.83.$$

We can estimate approximate values K_1^* from T_1^* using the relationship of the two policies in (19), and vice versa.

4 Major Collection Times

4.1 Model 1

Suppose that minor collections are made before surviving objects exceed a threshold level K $(0 < K < \infty)$, and when they have exceeded K, tenuring collections are always made. Further, major collection is made at time T $(0 < T \leq \infty)$ or at the Nth $(N = 1, 2, \ldots)$ collection including minor and tenuring collections, whichever occurs first. Furthermore, Letting c_{kT} $(k = 1, 2, \ldots)$ be the cost suffered for the kth tenuring collection, where $c_S + c_M(K) < c_{1T} < c_{2T} < \cdots$, and c_F $(c_F > c_{kT})$ be the cost suffered for major collection. Then, the probability that major collection is made at time T is

$$P_T = \sum_{j=0}^{N-1} H_j(T)G^{(j)}(K) + \sum_{j=1}^{N-1}\sum_{i=1}^{j-1} H_j(T)p_i(K) = 1 - F_N(T), \tag{20}$$

and the probability that major collection is made at collection N is

$$P_N = F_N(T)G^{(N)}(K) + \sum_{j=0}^{N-1} F_N(T)p_j(K) = F_N(T), \tag{21}$$

where note that $P_T + P_N \equiv 1$. The mean time to major collection is

$$E_2(L) = \int_0^T t\,\mathrm{d}F_N(t) + T\sum_{j=0}^{N-1} H_j(T) = \int_0^T [1 - F_N(t)]\,\mathrm{d}t. \tag{22}$$

The expected costs suffered for minor collections and tenuring collections when major collection is made at time T are, respectively,

$$C_{TM} = \sum_{j=1}^{N-1} H_j(T)\left[\sum_{i=1}^{j} C(i,K)G^{(j)}(K) + \sum_{i=1}^{j-1}\sum_{k=1}^{i} C(k,K)p_i(K)\right]$$

$$= \sum_{j=1}^{N-1} H_j(T)\sum_{i=1}^{j} C(i,K)G^{(i)}(K), \tag{23}$$

$$C_{TT} = \sum_{j=1}^{N-1} H_j(T)\sum_{i=0}^{j-1}\sum_{k=1}^{j-i} c_{kT}\, p_i(K)$$

$$= \sum_{j=1}^{N-1} H_j(T)\sum_{i=0}^{j-1}\left[c_{(i+1)T} - c_{(j-i)T}G^{(i+1)}(K)\right], \tag{24}$$

and the expected costs suffered for minor collections and tenuring collections when major collection is made at collection N are, respectively,

$$C_{NM} = F_N(T)\left[\sum_{j=1}^{N} C(j,K)G^{(N)}(K) + \sum_{j=1}^{N-1}\sum_{i=1}^{j} C(i,K)p_j(K)\right]$$

$$= F_N(T)\sum_{j=1}^{N} C(j,K)G^{(j)}(K), \tag{25}$$

$$C_{NT} = F_N(T)\sum_{j=0}^{N-1}\sum_{i=1}^{N-j} c_{iT}\, p_j(K)$$

$$= F_N(T)\sum_{j=0}^{N-1}\left[c_{(j+1)T} - c_{(N-j)T}G^{(j+1)}(K)\right]. \tag{26}$$

Thus, the total expected cost until major collection is, summing up from (23) to (26) and adding the cost c_F of major collection,

$$E_2(C) = c_F + \sum_{j=1}^{N} C(j,K)F_j(T)G^{(j)}(K)$$
$$+ \sum_{j=1}^{N} F_j(T)\left[c_{jT} - \sum_{i=0}^{j-1} G^{(j-i)}(K)(c_{(i+1)T} - c_{iT})\right]. \tag{27}$$

Therefore, the expected cost rate is, from (22) and (27),

$$C_2(T,N) = \frac{c_F + \sum_{j=1}^{N} F_j(T)A_j}{\int_0^T [1 - F_N(t)]\mathrm{d}t}, \tag{28}$$

where

$$A_j \equiv c_{jT} + \int_0^K [c_S + c_M(x)]\,\mathrm{d}G^{(j)}(x) - \sum_{i=0}^{j-1} G^{(j-i)}(K)(c_{(i+1)T} - c_{iT}).$$

It can be easily proved that A_j increases with j because

$$A_{j+1} - A_j = (c_{1T} - c_S - c_M(K))p_j(K) + \int_0^K p_j(x)\mathrm{d}c_M(x) + \sum_{i=1}^{j} p_{j-i}(K)(c_{(i+1)T} - c_{iT}) > 0.$$

1. Optimal T_2^*: When major collection is made only at time T,

$$C_2(T) \equiv \lim_{N\to\infty} C_2(T,N) = \frac{1}{T}\left[\sum_{j=1}^{\infty} F_j(T)A_j + c_F\right]. \tag{29}$$

Differentiating $C_2(T)$ in (29) with respect to T and setting it equal to zero,

$$\sum_{j=1}^{\infty} A_j\left[T\lambda(T)H_{j-1}(T) - F_j(T)\right] = c_F,$$

that is,

$$\sum_{j=1}^{\infty} A_j \int_0^T t\mathrm{d}\left[\lambda(t)H_{j-1}(t)\right] = c_F. \tag{30}$$

Letting $L_2(T)$ be the left-hand side of (30),

$$L_2'(T) = \sum_{j=0}^{\infty} A_{j+1}\int_0^T t\lambda'(t)H_j(t)\mathrm{d}t + \sum_{j=0}^{\infty}(A_{j+2} - A_{j+1})\int_0^T t[\lambda(t)]^2 H_j(t)\mathrm{d}t,$$

$$L_2(\infty) = \sum_{j=1}^{\infty} A_j \int_0^{\infty} t\mathrm{d}[\lambda(t)H_{j-1}(t)].$$

Thus, if $\lambda(t)$ increases with t and $L_2(\infty) > c_F$, then there exists a finite and unique T_2^* $(0 < T_2^* < \infty)$ which satisfies (30).

In particular, when $\lambda(t) = \lambda$,

$$L_2'(T) = \sum_{j=0}^{\infty}(j+1)F_{j+2}(T)(A_{j+2} - A_{j+1}),$$

$$L_2(\infty) = \sum_{j=1}^{\infty}(A_\infty - A_j).$$

Therefore, if $\sum_{j=1}^{\infty}(A_\infty - A_j) > c_F$, then there exists a finite and unique T_2^* $(0 < T_2^* < \infty)$, and the resulting cost rate is

$$\frac{C_2(T_2^*)}{\lambda} = \sum_{j=0}^{\infty} H_j(T_2^*)A_{j+1}.$$

2. Optimal N_2^*: When major collection is made only at collection N,

$$C_2(N) \equiv \lim_{T\to\infty} C_2(T,N) = \frac{\sum_{j=1}^{N} A_j + c_F}{\int_0^\infty [1 - F_N(t)]\mathrm{d}t} \quad (N = 1, 2, \ldots). \tag{31}$$

From the inequality $C_2(N+1) - C_2(N) \geq 0$,

$$\sum_{j=0}^{N-1}\left[\frac{A_{N+1}}{\int_0^\infty H_N(t)\mathrm{d}t}\int_0^\infty H_j(t)\mathrm{d}t - A_{j+1}\right] \geq c_F. \tag{32}$$

Letting $L_2(N)$ be the left-hand side of (32),

$$L_2(N+1) - L_2(N) = \left[\frac{A_{N+2}}{\int_0^\infty H_{N+1}(t)\mathrm{d}t} - \frac{A_{N+1}}{\int_0^\infty H_N(t)\mathrm{d}t}\right]\int_0^\infty \left[1 - F_{N+1}(t)\right]\mathrm{d}t. \tag{33}$$

Thus, if $A_{N+1}/\int_0^\infty H_N(t)\mathrm{d}t$ increases with N and $L_2(\infty) > c_F$, then there exists a finite and unique minimum N_2^* $(1 \leq N_2^* < \infty)$ which satisfies (32).

In particular, when $\lambda(t) = \lambda$,

$$L_2(N) = \sum_{j=1}^{N}(A_{N+1} - A_j),$$

$$L_2(N+1) - L_2(N) = (N+1)(A_{N+2} - A_{N+1}) > 0.$$

It is assumed that $A_\infty \equiv \lim_{j\to\infty} A_j < \infty$. Then,

$$L_2(\infty) = \sum_{j=1}^{\infty}(A_\infty - A_j).$$

Further, because $\sum_{j=1}^{N}(A_{N+1} - A_j) \geq A_{N+1} - A_1$ $(N = 1, 2, \ldots)$, if $A_\infty = \infty$, then $L_2(\infty) = \infty$. Therefore, if $\sum_{j=1}^{\infty}(A_\infty - A_j) > c_F$, then there exists a finite and unique minimum N_2^* $(1 \leq N_2^* < \infty)$, and the resulting cost rate is

$$A_{N_2^*} \leq \frac{C_2(N_2^*)}{\lambda} < A_{N_2^*+1}.$$

It is of interest that when collections occur at a Poisson process with rate λ, if $\sum_{j=1}^{\infty}(A_\infty - A_j) > c_F$, then both finite and unique T_2^* and N_2^* exist.

4.2 Model 2

Suppose that minor collections are made before surviving objects exceed a threshold level K, and after they have exceeded K, tenuring collections are always made. Further, major collection is made at time T $(0 < T \leq \infty)$ or at collection N $(N = 1, 2, \ldots)$ including tenuring collections, whichever occurs first. Then, the probability that major collection is made at time T is

$$P_T = \sum_{j=0}^{\infty}\sum_{i=0}^{N-2} p_j(K) \int_0^\infty H_i(u, u+T)\mathrm{d}F_{j+1}(u), \tag{34}$$

and the probability that major collection is made at collection N is

$$P_N = \sum_{j=0}^{\infty}\sum_{i=N-1}^{\infty} p_j(K) \int_0^\infty H_i(u, u+T)\mathrm{d}F_{j+1}(u). \tag{35}$$

The mean time to major collection is

$$E_3(L) = \sum_{j=0}^{\infty} p_j(K) \int_0^\infty \left[\int_0^T (u+t)\mathrm{d}F_{N-1}(u, u+t)\right]\mathrm{d}F_{j+1}(u)$$
$$+ \sum_{j=0}^{\infty}\sum_{i=0}^{N-2} p_j(K) \int_0^\infty (u+T)H_i(u, u+T)\mathrm{d}F_{j+1}(u)$$

$$
\begin{aligned}
&=\sum_{j=0}^{\infty} p_j(K) \int_0^{\infty} u \mathrm{d} F_{j+1}(u) \\
&\quad+\sum_{j=0}^{\infty} p_j(K) \int_0^{\infty}\left\{\int_0^T\left[1-F_{N-1}(u, u+t)\right] \mathrm{d} t\right\} \mathrm{d} F_{j+1}(u) .
\end{aligned}
\tag{36}
$$

The expected costs suffered for minor collections and tenuring collections when major collection is made at time T are, respectively,

$$
C_{TM}=\sum_{j=0}^{\infty} \sum_{i=1}^{j} c_{iM} p_j(K) \int_0^{\infty}\left[1-F_{N-1}(u, u+T)\right] \mathrm{d} F_{j+1}(u),
\tag{37}
$$

$$
C_{TT}=\sum_{j=0}^{\infty} \sum_{i=0}^{N-2} \sum_{k=1}^{i+1} c_{kT} p_j(K) \int_0^{\infty} H_i(u, u+T) \mathrm{d} F_{j+1}(u),
\tag{38}
$$

and the expected costs suffered for minor collections and tenuring collections when major collection is made at collection N are, respectively,

$$
C_{NM}=\sum_{j=0}^{\infty} \sum_{i=1}^{j} c_{iM} p_j(K) \int_0^{\infty} F_{N-1}(u, u+T) \mathrm{d} F_{j+1}(u),
\tag{39}
$$

$$
C_{NT}=\sum_{j=0}^{\infty} \sum_{i=1}^{N} c_{iT} p_j(K) \int_0^{\infty} F_{N-1}(u, u+T) \mathrm{d} F_{j+1}(u).
\tag{40}
$$

Thus, the total expected cost until major collection is, summing up from (37) to (40) and adding the cost c_F of major collection,

$$
\begin{aligned}
E_3(C)= & c_F+\sum_{j=1}^{\infty} \sum_{i=1}^{j} c_{iM} p_j(K) \\
& +\sum_{j=0}^{\infty} \sum_{i=1}^{N} c_{iT} p_j(K) \int_0^{\infty} F_{i-1}(u, u+T) \mathrm{d} F_{j+1}(u) .
\end{aligned}
\tag{41}
$$

Therefore, from (36) to (41), the expected cost rate is

$$
C_3(T, N)=\frac{E_3(C)}{E_3(L)} .
\tag{42}
$$

1. Optimal T_3^*: When major collection is made only at time T,

$$C_3(T) \equiv \lim_{N\to\infty} C_3(T,N) = \frac{\begin{array}{l} c_F + \sum_{j=1}^{\infty}\sum_{i=1}^{j} c_{iM}p_j(K) \\ +\sum_{j=0}^{\infty}\sum_{i=1}^{\infty} c_{iT}p_j(K) \\ \times \int_0^{\infty} F_{i-1}(u,u+T)\mathrm{d}F_{j+1}(u) \end{array}}{\sum_{j=0}^{\infty} p_j(K)\int_0^{\infty} u\mathrm{d}F_{j+1}(u)+T}. \tag{43}$$

Differentiating $C_3(T)$ with respect to T and setting it equal to zero,

$$\sum_{j=0}^{\infty} p_{j+1}(K)\int_0^{\infty} Q_3(u,T)\mathrm{d}F_{j+1}(u) = c_F + \sum_{j=1}^{\infty} c_{jM}G^{(j)}(K), \tag{44}$$

where

$$\begin{aligned} Q_3(u,T) &\equiv \sum_{i=1}^{\infty} c_{iT}\int_0^{\infty} (l+x)\,\mathrm{d}\big[\lambda(u+x)H_{i-2}(u,u+x)\big] \\ &= \sum_{i=1}^{\infty} c_{iT}\int_0^{\infty} (l+x)\,\lambda'(u+x)H_{i-2}(u,u+x)\mathrm{d}x \\ &\quad + \sum_{i=1}^{\infty}(c_{(i+3)T}-c_{(i+2)T})\int_0^{\infty} (l+x)\,[\lambda(u+x)]^2 H_i(u,u+x)\mathrm{d}x, \end{aligned}$$

and

$$l \equiv \sum_{j=1}^{\infty} p_j(K)\int_0^{\infty} t\mathrm{d}F_j(t),$$

which represents the mean time until surviving objects have exceeded K. Letting $L_3(T)$ be the left-hand side of (44). Thus, if $\lambda(t)$ increases with t, $L_3(T)$ increases with T. Therefore, if $L_3(\infty) > c_F + \sum_{j=1}^{\infty} c_{jM}G^{(j)}(K)$, then there exists a finite and unique T_3^* $(0 < T_3^* < \infty)$ which satisfies (44).

In particular, when $\lambda(t) = \lambda$, then $l = [1+M(K)]/\lambda$, and

$$\begin{aligned} Q_3(u,T) &= [1+M(K)]\sum_{j=1}^{\infty} F_j(T)(c_{(j+2)T}-c_{(j+1)T}) \\ &\quad + \sum_{j=1}^{\infty} jF_{j+1}(T)(c_{(j+2)T}-c_{(j+1)T}), \\ L_3(\infty) &= \sum_{j=1}^{\infty}(c_{\infty T}-c_{(j+1)T}) + [1+M(K)](c_{\infty T}-c_{2T}). \end{aligned}$$

Therefore, if

$$\sum_{j=1}^{\infty}(c_{\infty T}-c_{(j+1)T})+[1+M(K)](c_{\infty T}-c_{2T})>c_F+\sum_{j=1}^{\infty}c_{jM}G^{(j)}(K),$$

then there exists a finite and unique T_3^* ($0 < T_3^* < \infty$), and the resulting cost rate is

$$\frac{C_3(T_3^*)}{\lambda}=\sum_{j=0}^{\infty}H_j(T_3^*)c_{(j+2)T}.$$

2. Optimal N_3^*: When major collection is made only at collection N,

$$C_3(N)\equiv\lim_{T\to\infty}C_3(T,N)=\frac{c_F+\sum_{j=1}^{\infty}\sum_{i=1}^{j}c_{iM}p_j(K)+\sum_{j=1}^{N}c_{jT}}{\sum_{j=0}^{\infty}p_j(K)\int_0^{\infty}[1-F_{j+N}(t)]\mathrm{d}t}$$
$$(N=1,2,\ldots). \tag{45}$$

From the inequality $C_3(N+1)-C_3(N)\geq 0$,

$$Q_3(N)c_{(N+1)T}-\sum_{j=1}^{N}c_{jT}\geq c_F+\sum_{j=1}^{\infty}c_{jM}G^{(j)}(K), \tag{46}$$

where

$$Q_3(N)\equiv\frac{\sum_{j=0}^{\infty}p_j(K)\int_0^{\infty}[1-F_{j+N}(t)]\mathrm{d}t}{\sum_{j=0}^{\infty}p_j(K)\int_0^{\infty}H_{j+N}(t)\mathrm{d}t.}$$

Letting $L_3(N)$ be the left-hand side of (46),

$$L_3(N+1)-L_3(N)=\left[\widetilde{Q}_3(N+1)-\widetilde{Q}_3(N)\right]\sum_{j=0}^{\infty}p_j(K)\int_0^{\infty}\left[1-F_{j+N+1}(t)\right]\mathrm{d}t,$$

where

$$\widetilde{Q}_3(i)\equiv\frac{c_{(i+1)T}}{\sum_{j=0}^{\infty}p_j(K)\int_0^{\infty}H_{j+i}(t)\mathrm{d}t}.$$

Thus, if $\widetilde{Q}_3(i)$ increases with i, $L_3(N)$ increases with N. Therefore, if $L_3(\infty) > c_F+\sum_{j=1}^{\infty}c_{jM}G^{(j)}(K)$, then there exists a finite and unique minimum N_3^* ($1 \leq N_3^* < \infty$) which satisfies (46).

In particular, when $\lambda(t)=\lambda$, then $Q_3(N)=M(K)+N$, where $M(x)\equiv\sum_{j=1}^{\infty}G^{(j)}(x)$ is the expected number of minor collections before surviving objects exceed x, and

$$L_3(N) = \sum_{j=1}^{N}(c_{(N+1)T} - c_{jT}) + M(K)c_{(N+1)T},$$

$$L_3(N+1) - L_3(N) = [M(K) + N + 1]\,(c_{(N+2)T} - c_{(N+1)T}) > 0.$$

It is assumed that $c_{\infty T} \equiv \lim_{j\to\infty} c_{jT} < \infty$. Then,

$$L_3(\infty) = \sum_{j=1}^{\infty}(c_{\infty T} - c_{jT}) + M(K)c_{\infty T}.$$

Clearly, if $c_{\infty T} = \infty$, then $L_2(\infty) = \infty$. Therefore, if

$$\sum_{j=1}^{\infty}(c_{\infty T} - c_{jT}) + M(K)c_{\infty T} > c_F + \sum_{j=1}^{\infty} c_{jM}G^{(j)}(K),$$

then there exists a finite and unique minimum N_3^* ($1 \leq N_3^* < \infty$) which satisfies (46), and the resulting cost rate is

$$c_{N_3^*T} \leq \frac{C_3(N_3^*)}{\lambda} < c_{(N_3^*+1)T}.$$

4.3 Numerical Examples

It is assumed that $c_{kT} = c_T + k\beta$ ($\beta > 0$; $k = 1, 2, \ldots$), and other assumptions are the same as in Sect. 3.2. We give numerical examples of each model as follows:

Tables 3–6 present optimal λT_i^* and $C_i(T_i^*)/\lambda$ ($i = 2, 3$), N_i^* and $C_i(N_i^*)/\lambda$ ($i = 2, 3$), when $c_F = 100$, $c_T = c_N = 20$, $c_S = 10$, $c_M = 1$, $\mu = 10$ and $\sigma = 1$ for different α and β. These show that both λT_2^* and N_2^* decrease with α or β, both λT_3^* and N_3^* increase with α and decrease with β, all $C_i(T_i^*)/\lambda$ ($i = 2, 3$) and $C_i(N_i^*)/\lambda$ ($i = 2, 3$) increase with α or β.

It can be explained as follows:

- When α or β increases, it means that the total cost suffered for minor collections or tenuring collections increases, then optimal major collection times should be advanced, but even then the expected cost rates increase.
- The differences between Tables 3 and 5, Tables 4 and 6, are that when α increases, $M(K)$ decreases, then optimal major collection times should be postponed, because it is not economic to make major collection frequently.
- Compared Tables 3 with 4, Tables 5 with 6, these show that $C_2(T_2^*) > C_2(N_2^*)$ and $C_3(T_3^*) > C_3(N_3^*)$ for the same parameters, that is, major collections made at N_2 or N_3 are better than those at T_2 or T_3. It is interesting that $C_2(N_2^*) \approx C_3(N_3^*)$ and

Table 3 Optimal λT_2^* and $C_2(T_2^*)/\lambda$ when $c_F = 100$, $c_T = 20$, $c_S = 10$, $c_M = 1$, $\mu = 10$ and $\sigma = 1$

α	$\beta = 1$		$\beta = 2$		$\beta = 5$	
	λT_2^*	$C_2(T_2^*)/\lambda$	λT_2^*	$C_2(T_2^*)/\lambda$	λT_2^*	$C_2(T_2^*)/\lambda$
0.3	17.98	24.4699	15.51	25.1134	13.14	26.0229
0.4	14.09	28.1939	10.66	31.5677	7.77	35.1867
0.5	13.80	31.6197	9.95	35.5256	6.60	42.2212
0.6	12.86	32.8499	9.86	37.6922	6.33	46.6393
0.7	12.86	33.6762	9.86	39.2143	6.27	50.0259

Table 4 Optimal N_2^* and $C_2(N_2^*)/\lambda$ when $c_F = 100$, $c_N = 20$, $c_S = 10$, $c_M = 1$, $\mu = 10$ and $\sigma = 1$

α	$\beta = 1$		$\beta = 2$		$\beta = 5$	
	N_2^*	$C_2(N_2^*)/\lambda$	N_2^*	$C_2(N_2^*)/\lambda$	N_2^*	$C_2(N_2^*)/\lambda$
0.3	17	24.1785	16	24.3642	15	24.7538
0.4	14	28.6941	11	30.5818	8	32.8485
0.5	14	31.1212	10	34.5257	7	39.7804
0.6	14	32.3506	10	36.6942	6	44.1853
0.7	14	33.1763	10	38.2160	6	47.5548

Table 5 Optimal λT_3^* and $C_3(T_3^*)/\lambda$ when $c_F = 100$, $c_T = 20$, $c_S = 10$, $c_M = 1$, $\mu = 10$ and $\sigma = 1$

α	$\beta = 1$		$\beta = 2$		$\beta = 5$	
	λT_3^*	$C_3(T_3^*)/\lambda$	λT_3^*	$C_3(T_3^*)/\lambda$	λT_3^*	$C_3(T_3^*)/\lambda$
0.3	1.95	23.9629	0.06	24.1471	0.01	24.3401
0.4	6.92	28.9179	3.42	30.8389	0.57	32.8532
0.5	9.45	31.4559	5.54	35.0679	2.10	40.4869
0.6	10.75	32.7357	6.69	37.3694	3.06	45.3683
0.7	11.60	33.5878	7.48	38.9639	3.80	49.0298

Table 6 Optimal N_3^* and $C_3(N_3^*)/\lambda$ when $c_F = 100$, $c_N = 20$, $c_S = 10$, $c_M = 1$, $\mu = 10$ and $\sigma = 1$

α	$\beta = 1$		$\beta = 2$		$\beta = 5$	
	N_3^*	$C_3(N_3^*)/\lambda$	N_3^*	$C_3(N_3^*)/\lambda$	N_3^*	$C_3(N_3^*)/\lambda$
0.3	3	23.9071	2	24.1387	1	24.3365
0.4	8	28.6662	5	30.5026	2	32.5947
0.5	10	31.1223	7	34.4997	3	39.6811
0.6	12	32.3455	8	36.6799	4	44.1223
0.7	13	33.1731	9	38.2105	5	47.4478

$C_2(T_2^*) \approx C_3(T_3^*)$, that is, although the two policies are different, the resulting expected cost rates are almost the same.

- We can derive the relationship of the two polices, that is,

$$\lambda T_2^* \approx 1 + M(K) + \lambda T_3^*,$$
$$N_2^* \approx M(K) + N_3^*.$$

For example, when $\alpha = 0.3$ and $\beta = 1$, $M(K) = 14.8$, then

$$\lambda T_2^* = 17.98, \qquad 1 + M(K) + \lambda T_3^* = 1 + 14.8 + 1.95 = 17.75,$$
$$N_2^* = 17, \qquad M(K) + N_3^* = 14.8 + 3 = 17.8.$$

Therefore, the concrete performances of the two kinds of policies would depend on the program engineers and software system structures at the beginning, and so on.

5 Continuous Models

From the related studies in Sect. 2, we know that the volume of surviving objects that should be copied increases with the number of minor collections and is relatively stable with the number of tenuring collections. However, it may be difficult to inspect the survivor rates exactly at collection times. Hence, in this section, we assume that the total volume of surviving objects in Eden and survivor space at time t is $Z(t) = A(t)t + \sigma B(t)$ with distribution $\Pr\{Z(t) \le x\} = W(t, x)$, where both $A(t)$ and $B(t)$ are random variables of time t. Then, the expected cost of minor collection at time t is

$$C(t, K) = \frac{1}{W(t, K)} \int_0^K [c_S + c_M(x)]\, \mathrm{d}W(t, x), \tag{47}$$

where $C(0, K) \equiv 0$. Letting $r(t, x)$ be the failure rate of $W(t, x)$, i.e., $r(t, x) \equiv -[\mathrm{d}W(t, x)/\mathrm{d}t]/W(t, x)$ [3]. It is clear that if $r(t, x)$ increases with t for any $x \ge 0$, $C(t, K)$ increases with t for any $K \ge 0$.

Suppose that garbage collections occur at a nonhomogeneous Poisson process in Sect. 2, minor collections are made when the garbage collector begins to work, tenuring collection is made at a planned time T $(0 < T \le \infty)$, or when surviving objects have exceeded a threshold level K $(0 < K \le \infty)$, whichever occurs first. Then, the mean time to tenuring collection is

$$E_4(L) = TW(T, K) + \int_0^T t\mathrm{d}\overline{W}(t, K) = \int_0^T W(t, K)\mathrm{d}t, \tag{48}$$

where $\overline{V}(t, x) \equiv 1 - V(t, x)$ for any distribution $V(t, x)$.

The expected cost suffered for minor collections until tenuring collection is

$$
\begin{aligned}
C_M = & W(T,K)\sum_{j=1}^{\infty}\int_0^T C(t,K)\mathrm{d}F_j(t) \\
& + \int_0^T \left[\sum_{j=1}^{\infty}\int_0^t C(u,K)\mathrm{d}F_j(u)\right]\mathrm{d}\overline{W}(t,K) \\
= & \int_0^T \lambda(t)C(t,K)W(t,K)\mathrm{d}t.
\end{aligned} \tag{49}
$$

Then, the expected cost until tenuring collection is

$$
E_4(C) = c_K - (c_K - c_T)W(T,K) + \int_0^T \lambda(t)C(t,K)W(t,K)\mathrm{d}t. \tag{50}
$$

Therefore, from (48) to (50), the expected cost rate is

$$
C_4(T,K) = \frac{\begin{array}{c} c_K - (c_K - c_T)W(T,K) \\ + \int_0^T \lambda(t)C(t,K)W(t,K)\mathrm{d}t \end{array}}{\int_0^T W(t,K)\mathrm{d}t}. \tag{51}
$$

5.1 Optimal Policies

It can be seen that $C_4(T,K)$ includes the following collection polices:

- Tenuring collection is made at time T for a given K, the reason why making such a policy is $c_T < c_K$.
- Tenuring collection is made at level K for a given T. In this case, $c_K < c_T$.
- Tenuring collection is made only at time T or only at level K. In these two cases, $c_K = c_T$.

1. Optimal T_4^*: When $c_T < c_K$, we find an optimal T_4^* which minimizes $C_4(T,K)$ in (51) for a given K. Differentiating $C_4(T,K)$ with respect to T and setting it equal to zero,

$$
\begin{aligned}
& (c_K - c_T)\left[r(T,K)\int_0^T W(t,K)\mathrm{d}t - \overline{W}(T,K)\right] \\
& + \int_0^T [\lambda(T)C(T,K) - \lambda(t)C(t,K)]\, W(t,K)\mathrm{d}t = c_T.
\end{aligned} \tag{52}
$$

Letting $L_4(T)$ be the left-hand side of (52),

$$
L_4(0) \equiv \lim_{T\to 0} L_4(T) = 0,
$$

$$L_4'(T) = (c_K - c_T)r'(T, K)\int_0^T W(t, K)\mathrm{d}t + \left[\lambda'(T)C(T, K) + \lambda(T)C'(T, K)\right]\int_0^T W(t, K)\mathrm{d}t.$$

Thus, if both $r(t, K)$ and $\lambda(t)$ increase with t, then the left-hand side of (52) increases with t from 0. Therefore, there exists a unique optimal T_4^* $(0 < T_4^* \leq \infty)$ which satisfies (52), and the resulting cost rate is

$$C_4(T_4^*, K) = (c_K - c_T)r(T_4^*, K) + \lambda(T_4^*)C(T_4^*, K).$$

2. Optimal K_4^*: When $c_K < c_T$, we find an optimal K_4^* which minimizes $C_4(T, K)$ in (51) for a given T. Letting $w(t, x)$ be a density function of $W(t, x)$, i.e., $w(t, x) \equiv \mathrm{d}W(t, x)/\mathrm{d}x$. Then, differentiating $C_4(T, K)$ with respect to K and setting it equal to zero,

$$(c_T - c_K)\left[Q_4(T, K)\int_0^T W(t, K)\mathrm{d}t - W(T, K)\right] + \int_0^T \left[\widetilde{Q}_4(T, K) - \lambda(t)C(t, K)\right]W(t, K)\mathrm{d}t = c_K, \tag{53}$$

where

$$Q_4(T, K) \equiv \frac{w(T, K)}{\int_0^T w(t, K)\mathrm{d}t}, \qquad \widetilde{Q}_4(T, K) \equiv \frac{[c_S + c_M(K)]\int_0^T \lambda(t)w(t, K)\mathrm{d}t}{\int_0^T w(t, K)\mathrm{d}t}.$$

Letting $L_4(K)$ be the left-hand side of (53),

$$L_4(0) \equiv \lim_{K\to 0} L_4(K) = 0,$$

$$L_4'(K) = (c_T - c_K)Q_4'(T, K)\int_0^T W(t, K)\mathrm{d}t + \widetilde{Q}_4'(T, K)\int_0^T W(t, K)\mathrm{d}t.$$

Thus, if both $Q_4(T, K)$ and $\widetilde{Q}_4(T, K)$ increase with K, then the left-hand side of (53) increases with K from 0. Therefore, there exists a unique optimal K_4^* $(0 < K_4^* \leq \infty)$ which satisfies (53), and the resulting cost rate is

$$C_4(T, K_4^*) = (c_T - c_K)Q_4(T, K_4^*) + \widetilde{Q}_4(T, K_4^*).$$

3. Optimal $\widetilde{T}_4^*$: When $c_K = c_T$, putting that $K = \infty$ in (51), the expected cost rate is

$$\widetilde{C}_4(T) \equiv \lim_{K\to\infty} C_4(T, K) = \frac{1}{T}\left[\int_0^T \lambda(t)C(t, \infty)\mathrm{d}t + c_T\right], \tag{54}$$

where

$$C(t,\infty) \equiv \int_0^{\infty} [c_S + c_M(x)]\,\mathrm{d}W(t,x) = c_S + \int_0^{\infty} \overline{W}(t,x)\mathrm{d}c_M(x).$$

From (52), if $\lambda(t)$ increases with t, then an optimal tenuring collection time $\widetilde{T}_1^*$ which minimizes (54) is given by a unique solution of the equation

$$\int_0^T [\lambda(T)C(T,\infty) - \lambda(t)C(t,\infty)]\,\mathrm{d}t = c_T, \tag{55}$$

and the resulting cost rate is

$$\widetilde{C}_4(\widetilde{T}_4^*) = \lambda(\widetilde{T}_4^*)C(\widetilde{T}_4^*,\infty).$$

In particular, when $\lambda(t) = \lambda$, (55) becomes

$$\int_0^{\infty} \left\{\int_0^T [W(t,x) - W(T,x)]\mathrm{d}t\right\} \mathrm{d}c_M(x) = \frac{c_T}{\lambda}, \tag{56}$$

which increases with T, and the resulting cost rate is

$$\frac{\widetilde{C}_4(\widetilde{T}_4^*)}{\lambda} = c_S + \int_0^{\infty} \overline{W}(\widetilde{T}_4^*,x)\mathrm{d}c_M(x).$$

4. Optimal $\widetilde{K}_4^*$: When $c_K = c_T$, putting that $T = \infty$ in (51), the expected cost rate is

$$\widetilde{C}_4(K) = \lim_{T\to\infty} C_4(T,K) = \frac{\int_0^{\infty} \lambda(t)C(t,K)W(t,K)\mathrm{d}t + c_K}{\int_0^{\infty} W(t,K)\mathrm{d}t}. \tag{57}$$

From (53), if $\widetilde{Q}_4(\infty,K)$ increases with K, then an optimal tenuring collection time $\widetilde{K}_4^*$ which minimizes (57) is given by a unique solution of the equation

$$\int_0^{\infty} \left[\widetilde{Q}_4(\infty,K) - \lambda(t)C(t,K)\right] W(t,K)\mathrm{d}t = c_K, \tag{58}$$

and the resulting cost rate is

$$\widetilde{C}_4(\widetilde{K}_4^*) = \widetilde{Q}_4(\infty,\widetilde{K}_4^*).$$

In particular, when $\lambda(t) = \lambda$, (58) becomes

$$\int_0^{\infty} \left[\int_0^K W(t,x)\mathrm{d}c_M(x)\right] \mathrm{d}t = \frac{c_K}{\lambda}, \tag{59}$$

Table 7 Optimal T_4^* and $C_4(T_4^*, K)$ when $c_T = 10$ and $c_S = \lambda = \mu = \sigma = 1$

K	c_K	$c_M = 0.1$		$c_M = 0.5$		$c_M = 1.0$	
		T_4^*	$C_4(T_4^*, K)$	T_4^*	$C_4(T_4^*, K)$	T_4^*	$C_4(T_4^*, K)$
	20	4.73	0.4723	4.09	0.5462	3.57	0.6324
	30	3.25	0.5473	3.06	0.6087	2.87	0.6822
5	40	2.80	0.5891	2.69	0.6453	2.56	0.7133
	50	2.57	0.6191	2.49	0.6718	2.40	0.7361
	20	7.43	0.2990	5.77	0.4285	4.50	0.5586
	30	6.37	0.3160	5.40	0.4339	4.42	0.5596
10	40	5.91	0.3252	5.18	0.4378	4.36	0.5604
	50	5.63	0.3320	5.03	0.4408	4.31	0.5612

which increases with K and the resulting cost rate is

$$\frac{\widetilde{C}_4(\widetilde{K}_4^*)}{\lambda} = c_S + c_M(\widetilde{K}_4^*).$$

5.2 Numerical Examples

We compute numerical examples of the models discussed above for $Z(t) = \mu t + \sigma B(t)$ when $B(t)$ is normally distributed with mean 0 and variance t or for $Z(t) = A(t)t$ when $A(t)$ is normally distributed with mean μ and variance σ^2/t, that is,

$$W(t, x) = \Phi\left(\frac{x - \mu t}{\sigma\sqrt{t}}\right), \tag{60}$$

where $\Phi(x)$ is the standard normal distribution with mean 0 and variance 1, *i.e.*, $\Phi(x) \equiv (1/\sqrt{2\pi}) \int_{-\infty}^{x} e^{-u^2/2} du$.

From Tables 7–9, we can obtain the following results:

- Optimal tenuring collection times increase with the initial parameters and decrease with minor or tenuring collection cost, however, the resulting cost rates have the opposite tendencies, that is, they decrease with the initial parameters and increase with minor or tenuring collection cost. Take T_4^* and $C_4(T_4^*, K)$ in Table 7 for an example: T_4^* increase with K and decrease with c_K or c_M. Increasing in K, c_K or c_M means that tenuring collection time made at a given level K is postponed, tenuring or minor collection cost is increased, respectively, so that tenuring collection times should be postponed for K or be advanced for c_K or c_M to decrease the frequency of tenuring collections or to decrease the total minor collection cost. $C_4(T_4^*, K)$ decrease with K and increase with c_K or c_M for the reason that the

Table 8 Optimal K_4^* and $C_4(T, K_4^*)$ when $c_K = 10$ and $c_S = \lambda = \mu = \sigma = 1$

T	c_T	$c_M = 0.1$		$c_M = 0.5$		$c_M = 1.0$	
		K_4^*	$C_4(T, K_4^*)$	K_4^*	$C_4(T, K_4^*)$	K_4^*	$C_4(T, K_4^*)$
	20	4.35	0.4677	3.77	0.5342	3.23	0.6063
	30	3.45	0.5500	3.14	0.6048	2.80	0.6661
5	40	3.01	0.6153	2.79	0.6633	2.54	0.7183
	50	2.67	0.6719	2.55	0.7155	2.36	0.7659
	20	7.43	0.2973	5.30	0.4209	3.92	0.5328
	30	6.37	0.3202	4.93	0.4221	3.80	0.5390
10	40	5.80	0.3372	4.67	0.4318	3.71	0.5448
	50	5.42	0.3512	4.48	0.4405	3.62	0.5503

Table 9 Optimal $\widetilde{T}_4^*$, $\widetilde{C}_4(\widetilde{T}_4^*)$, $\widetilde{K}_4^*$ and $\widetilde{C}_4(\widetilde{K}_4^*)$ when $c_T = c_K = 10$ and $c_S = \lambda = \mu = \sigma = 1$

c_M	$\widetilde{T}_4^*$	$\widetilde{C}_4(\widetilde{T}_4^*)$	$\widetilde{K}_4^*$	$\widetilde{C}_4(\widetilde{K}_4^*)$
0.1	14.24	0.2424	14.15	0.2413
0.2	10.11	0.3021	9.99	0.2997
0.3	8.28	0.3486	8.15	0.3444
0.4	7.20	0.3882	7.05	0.3820
0.5	6.46	0.4233	6.30	0.4151
0.6	5.91	0.4554	5.75	0.4449
0.7	5.49	0.4851	5.32	0.4722
0.8	5.15	0.5130	4.97	0.4976
0.9	4.86	0.5394	4.68	0.5114
1.0	4.62	0.5546	4.44	0.5239

frequency of tenuring collections is decreased and tenuring or minor collection cost is increased.

- Compared with Tables 7 and 8, we can derive that $T_4^* \approx K_4^*$, in fact, this means that $\mu T_4^* \approx K_4^*$, which corresponds to the assumption of $Z(t)$. $C_4(T_4^*, K) \approx C_4(T, K_4^*)$, however, $C_4(T_4^*, K)$ are sometimes greater than and sometimes less than $C_4(T, K_4^*)$. That is, we can not compare them exactly.
- $\widetilde{C}_4(T)$ and $\widetilde{C}_4(K)$ are the particular cases of $C_4(T, K)$. Take T_4^* and $\widetilde{T}_4^*$ in Tables 7 and 9 for an example, when $c_M = 0.1, 0.5, 1.0$, $\widetilde{T}_4^*$ should be greater than T_4^* and $C_4(T_4^*, K)$ should be less than $\widetilde{C}_4(\widetilde{T}_4^*)$ when $K = 10$ and $c_K = 20$.

6 Conclusions

This chapter has discussed the problems of when to make tenuring and major collections for a generational garbage collector to meet the pause time goal. According to the properties of adaptive tenuring, two cases of working schemes have been introduced first, where tenuring and major collections have been considered as renewal points of the collection processes, respectively. Second, analyses of the costs suffered for collections, including minor, tenuring and major collections, have been given. Third, using the techniques of cumulative processes and degradation processes or continuous wear processes, expected cost rates for the two cases have been derived, and optimal tenuring collection times and major collection times are discussed analytically. Fourth, numerical examples have been given and some comparisons of the policies have been made. Such theoretical analyses would be applied to actual garbage collections by suitable modifications and extensions.

Acknowledgments This work is partially supported by the Grant-in-Aid for Scientific Research (C) of Japan Society for the Promotion of Science (22500897, 21530318), National Natural Science Foundation of China (70471017, 70801036) and Humanities and Social Science Research Foundation of China (05JA630027).

References

1. Appel AW (1989) Simple generational garbage collection and fast allocation. Softw Pract Exper 19:171–183
2. Armstrong J, Virding R (1995) One-pass real-time generational mark-sweep garbage collection. In: Proceedings of international workshop on memory managementx (Lecture notes in computer science), vol 986. Springer, Berlin, pp 313–322
3. Barlow RE, Proschan F (1965) Mathematical theory of reliability. Wiley, New York
4. Clinger WD, Rojas FV (2006) Linear combinations of radioactive decay models for generational garbage collection. Sci Comput Program 62:184–203
5. Jones R, Lins R (1996) Garbage collection: algorithms for automatic dynamic memory management. Wiley, Chichester
6. Kaldewaij A, Vries L (2001) Optimal real-time garbage collection for acyclic pointer structures. Inf Process Lett 77:151–157
7. Lee WH, Chang JM (2004) A garbage collection policy based on empirical behavior. Inf Sci 167:129–146
8. Nakagawa T (2005) Maintenance theory of reliability. Springer, London
9. Nakagawa T (2007) Shock and damage models in reliability theory. Springer, London
10. Nakamura S, Nakagawa T (2010) Stochastic reliability modeling, optimization and applications. World Scientific, Singapore
11. Nikulin MS, Balakrishnan N (2010) Advances in degradation modeling: applications to reliability, survival analysis, and finanace. Birkhöuser, Boston
12. Osaki S (1992) Applied stochastic system modeling. Springer, Berlin
13. Ross SM (1983) Stochastic processes. Wiley, New York
14. Satow T, Yasui K, Nakagawa T (1996) Optimal garbage collection policies for a database in a computer system. RAIRO Oper Res 30:359–372
15. Sato K (2001) Basic results on Lévy processes. In: Bandorff-Nielsen O, Mikosch T, Resnick S (eds) Lévy processes, theory and applications. Birkhöuser, Boston
16. Soman S, Krintz C (2007) Application-specific garbage collection. J Syst Softw 80:1037–1056
17. Ungar D (1984) Generation scavenging: A non-disruptive high performance storage reclamation algorithm. ACM Sigplan Not 19:157–167
18. Ungar D, Jackson F (1992) An adaptive tenuring policy for generation scavengers. ACM Trans Program Lang Syst 14:1–27

19. Vengerov D (2009) Modeling, analysis and throughput optimization of a generational garbage collector. Technical Report, Sun Labs, TR-2009-179
20. Wilson PR (1992) Uniprocessor garbage collection techniques. In: International workshop on memory management, (Lecture notes in computer science), vol 637. Springer, London, pp 1–42
21. Zhao XF, Nakamura S, Nakagawa T (2010) Optimal policies for random and periodic garbage collections with tenuring threshold. In: Tomar GS, Chang RS, Gervasi O, Kim T, Bandyopadhyay SK (eds) vol 74 Communications in computers and information science. Springer, Berlin, pp 125–135
22. Zhao XF, Nakamura S, Nakagawa T (2011) Two generational garbage collection models with major collection time. IEICE transactions on fundamentals of electronics communications and computer sciences, E94-A:1558–1566